Antimicrobial Peptides Aka Host Defense Peptides—From Basic Research to Therapy

Antimicrobial Peptides Aka Host Defense Peptides—From Basic Research to Therapy

Editor

Jitka Petrlova

MDPI • Basel • Beijing • Wuhan • Barcelona • Belgrade • Manchester • Tokyo • Cluj • Tianjin

Editor
Jitka Petrlova
Lunds Universitet
Sweden

Editorial Office
MDPI
St. Alban-Anlage 66
4052 Basel, Switzerland

This is a reprint of articles from the Special Issue published online in the open access journal *Biomedicines* (ISSN 2227-9059) (available at: https://www.mdpi.com/journal/biomedicines/special_issues/Host_Defense_Peptides).

For citation purposes, cite each article independently as indicated on the article page online and as indicated below:

LastName, A.A.; LastName, B.B.; LastName, C.C. Article Title. *Journal Name* **Year**, *Volume Number*, Page Range.

ISBN 978-3-0365-5819-6 (Hbk)
ISBN 978-3-0365-5820-2 (PDF)

Cover image courtesy of Jitka Petrlova

Contents

About the Editor

Jitka Petrlova

After completing my PhD in Biochemistry in 2006 at Mendel University, Czech Republic, I have worked as a Postdoctoral Researcher at the School of Medicine, University of California Davis, USA. I joined the Medical Faculty, Lund University, Sweden, in the summer of 2010. Throughout my career, I have always been fascinated by the aggregation of proteins and peptides and the links to diverse physiological and pathological processes. My current research is focused on aggregation as a novel host-defense mechanism, which plays an important role in skin inflammation and wound healing. In particular, I study host-defense roles of blood proteins that belong to the apolipoprotein group, such as apolipoprotein E (abbreviated as ApoE). Our findings demonstrate a previously undisclosed role of ApoE that involves aggregation of endotoxins and bacteria, enabling clearance of bacterial toxins and microbial killing. The link between host defense and aggregation also suggests that persisting inflammation may lead to dysfunctional activation of such protein-based host defense pathways; therefore, our research also raises interesting perspectives on the relationship between antimicrobial and amyloidogenic peptides.

Preface to "Antimicrobial Peptides Aka Host Defense Peptides—From Basic Research to Therapy"

All wounds are at risk of becoming contaminated by pathogens, which could lead to infection. The ability to effectively overcome this danger is of evolutionary significance to our survival. It is, therefore, not surprising that multiple host-defense systems, such as host-defense peptides, have evolved. Antimicrobial peptides, such as host-defense peptides (HDPs), are produced by all living organisms, including bacteria, fungi, plants, invertebrates, and vertebrates. These peptides display activity against bacteria, fungi, viruses, and parasites. Today, we have clear evidence that HDPs exhibit not only broad-spectrum antimicrobial activity, but also have immunomodulatory effects, wound healing, LPS neutralization, chemotaxis, or anti-cancer effects. Moreover, the multifunctionality of HDPs has resulted in the increased interest of scholars in the development of new therapeutic strategies against infectious diseases.

Jitka Petrlova
Editor

Article

Membranolytic Mechanism of Amphiphilic Antimicrobial β-Stranded [KL]$_n$ Peptides

Fabian Schweigardt [1], **Erik Strandberg** [2,*], **Parvesh Wadhwani** [2], **Johannes Reichert** [2], **Jochen Bürck** [2], **Haroldo L. P. Cravo** [3], **Luisa Burger** [2] and **Anne S. Ulrich** [1,2,*]

[1]   Karlsruhe Institute of Technology (KIT), Institute of Organic Chemistry, Fritz-Haber-Weg 6, 76131 Karlsruhe, Germany
[2]   KIT, Institute of Biological Interfaces (IBG-2), P.O. Box 3640, 76021 Karlsruhe, Germany
[3]   Laboratório de Biofísica Molecular, Universidade de São Paulo, Ribeirão Preto 14040-901, SP, Brazil
*   Correspondence: erik.strandberg@kit.edu (E.S.); anne.ulrich@kit.edu (A.S.U.)

Abstract: Amphipathic peptides can act as antibiotics due to membrane permeabilization. KL peptides with the repetitive sequence [Lys-Leu]$_n$-NH$_2$ form amphipathic β-strands in the presence of lipid bilayers. As they are known to kill bacteria in a peculiar length-dependent manner, we suggest here several different functional models, all of which seem plausible, including a carpet mechanism, a β-barrel pore, a toroidal wormhole, and a β-helix. To resolve their genuine mechanism, the activity of KL peptides with lengths from 6–26 amino acids (plus some inverted LK analogues) was systematically tested against bacteria and erythrocytes. Vesicle leakage assays served to correlate bilayer thickness and peptide length and to examine the role of membrane curvature and putative pore diameter. KL peptides with 10–12 amino acids showed the best therapeutic potential, i.e., high antimicrobial activity and low hemolytic side effects. Mechanistically, this particular window of an optimum β-strand length around 4 nm (11 amino acids × 3.7 Å) would match the typical thickness of a lipid bilayer, implying the formation of a transmembrane pore. Solid-state ^{15}N- and ^{19}F-NMR structure analysis, however, showed that the KL backbone lies flat on the membrane surface under all conditions. We can thus refute any of the pore models and conclude that the KL peptides rather disrupt membranes by a carpet mechanism. The intriguing length-dependent optimum in activity can be fully explained by two counteracting effects, i.e., membrane binding versus amyloid formation. Very short KL peptides are inactive, because they are unable to bind to the lipid bilayer as flexible β-strands, whereas very long peptides are inactive due to vigorous pre-aggregation into β-sheets in solution.

Keywords: cationic antimicrobial peptides; length dependent activity; antimicrobial activity; hemolysis; vesicle leakage; solid-state ^{31}P-, ^{15}N- and ^{19}F-NMR; β-stranded peptides; β-sheets; structure and orientation of peptides in membranes

Citation: Schweigardt, F.; Strandberg, E.; Wadhwani, P.; Reichert, J.; Bürck, J.; Cravo, H.L.P.; Burger, L.; Ulrich, A.S. Membranolytic Mechanism of Amphiphilic Antimicrobial β-Stranded [KL]$_n$ Peptides. *Biomedicines* **2022**, *10*, 2071. https://doi.org/10.3390/biomedicines10092071

Academic Editor: Jitka Petrlova

Received: 26 July 2022
Accepted: 21 August 2022
Published: 24 August 2022

Publisher's Note: MDPI stays neutral with regard to jurisdictional claims in published maps and institutional affiliations.

1. Introduction

Membrane-permeabilizing amphiphilic peptides are a promising class of antimicrobial agents to combat multidrug-resistant bacteria that are responsible for an increasing number of nosocomial infections [1–3]. Antimicrobial peptides (AMPs) are found in almost all types of organisms and constitute a natural host defense system against microorganisms [4,5]. An important goal in biophysical studies of AMPs is to describe their molecular mechanism of action and to understand which factors are important for activity. It is, therefore, necessary to examine not only the biological action of such peptides, such as antimicrobial activity and hemolytic side effects but also their structural properties, especially when interacting with lipid bilayers. Macroscopically aligned membrane samples reveal a wealth of 3D information when studied by oriented circular dichroism (OCD) [6] and solid-state NMR spectroscopy (SSNMR) [7–11].

Many natural and man-made AMPs are cationic and fold into simple amphiphilic α-helices. Much fewer examples exist of another fundamental type of secondary structure, namely amphiphilic β-strands. A classic model sequence, [KIGAKI]$_3$-NH$_2$, had been designed from alternating polar and hydrophobic side chains [12–14]. Recently, we presented an even simpler group of so-called KL peptides, consisting of the alternating (lysine-leucine) dipeptide repeat: [KL]$_n$-NH$_2$. Recently, four KLx peptides of different lengths (x = number of residues, x = 2n) with 6–18 amino acids were compared in terms of their antimicrobial and hemolytic activities [15]. KL10 had a better antimicrobial effect than KL6, KL14 or KL18. Since KL14 and KL18 were highly hemolytic, KL10 was clearly the most promising antibiotic candidate. These KL peptides showed interesting aggregation characteristics, especially at high pH and/or in the presence of phosphate ions. Aggregation was faster not only at higher concentrations of peptides and phosphate but also for the longer peptides (even when compared at the same mass ratio), and the resulting aggregates were inactive against bacteria [15]. If we want to understand and optimize such peptides further, it is obviously important to take aggregation into account, both in solution and when membrane-bound. For example, we were able to demonstrate that the antimicrobial activity was strongly dependent on the detailed method of performing the minimum inhibitory concentration (MIC) assay because the order of mixing had a pronounced effect on aggregation [15].

In our previous study on four different KL peptides, the length had been varied in steps of four residues, i.e., in increments of about 15 Å (counting 3.7 Å per residue in an extended β-strand), which had given only a very rough estimate of the optimal peptide length. In the present study, we have thus compared 20 peptides to obtain a better length-dependent profile and explore further features (see Table 1). Peptides with lengths from 8 to 16 residues were prepared in increments of single residues, and we furthermore extended the overall range up to 26 amino acids. For odd-numbered peptides, the Lys-to-Leu ratio is no longer 1:1, which changes the balance between charged and hydrophobic residues. To investigate this subtle effect, we further designed some inverted LK peptides, starting with Leu. For example, KL9 ([KL]$_4$K) and LK9 (L[KL]$_4$) have different charges and hydrophobicity despite their same length.

Table 1. Synthesized peptides used in this study. The net charge and the theoretical length of peptides in an ideal β-strand conformation are shown.

Peptide	Sequence	Net Charge	Length [a]/Å
KL6	KLKLKL-NH$_2$	+4	22.2
KL8	KLKLKLKL-NH$_2$	+5	29.6
KL9	KLKLKLKLK-NH$_2$	+6	33.3
LK9	LKLKLKLKL-NH$_2$	+5	33.3
KL10	KLKLKLKLKL-NH$_2$	+6	37.0
LK10	LKLKLKLKLK-NH$_2$	+6	37.0
KL11	KLKLKLKLKLK-NH$_2$	+7	40.7
LK11	LKLKLKLKLKL-NH$_2$	+6	40.7
KL12	KLKLKLKLKLKL-NH$_2$	+7	44.4
KL13	KLKLKLKLKLKLK-NH$_2$	+8	48.1
LK13	LKLKLKLKLKLKL-NH$_2$	+7	48.1
KL14	KLKLKLKLKLKLKL-NH$_2$	+8	51.8
KL15	KLKLKLKLKLKLKLK-NH$_2$	+9	55.5
LK15	LKLKLKLKLKLKLKL-NH$_2$	+8	55.5
KL16	KLKLKLKLKLKLKLKL-NH$_2$	+9	59.2
KL18	KLKLKLKLKLKLKLKLKL-NH$_2$	+10	66.6
KL20	KLKLKLKLKLKLKLKLKLKL-NH$_2$	+11	74.0
KL22	KLKLKLKLKLKLKLKLKLKLKL-NH$_2$	+12	81.4
KL24	KLKLKLKLKLKLKLKLKLKLKLKL-NH$_2$	+13	88.8
KL26	KLKLKLKLKLKLKLKLKLKLKLKLKL-NH$_2$	+14	96.2
KL10-^{15}N	KLKLK-(^{15}N-Leu)-KLKL-NH$_2$	+6	37.0
KL14-^{15}N	KLKLK-(^{15}N-Leu)-KLKLKLKL-NH$_2$	+8	51.8
KL10-^{19}F	KLKLK-(CF$_3$-Bpg)-KLKL-NH$_2$	+6	37.0
KL14-^{19}F	KLKLK-(CF$_3$-Bpg)-KLKLKLKL-NH$_2$	+8	51.8

[a] Approximate length, assuming 3.7 Å per residue for an ideal β-sheet.

In addition to the standard functional assays used in our previous study (MIC, hemolysis, vesicle leakage, CD in solution), here we also analyzed the peptide structure and membrane alignment by means of OCD and SSNMR in oriented lipid bilayers. These complementary methods offer detailed information about the peptide conformation, the alignment of its backbone in the membrane, and its molecular mobility. This way, we should be able to discriminate between several conceivable models of membrane-bound structures, as illustrated in Figure 1, in order to find out which of these architectures is responsible for membrane permeabilization. It is reasonable to assume that, upon membrane contact, a monomer gets amphiphilically embedded in the bilayer surface and can diffuse laterally within the plane as a flexible peptide strand (Figure 1A). Such a "2-dimensionally disordered" state has already been demonstrated for the related KIGAKI peptides [14]. Given that the KL and KIGAKI peptides tend to precipitate from aqueous solution as H-bonded amyloid-like fibrils [13–15], they might do the same when embedded in a membrane. That is, once the amphiphiles are immersed and pre-ordered in the 2-dimensional plane, they could self-assemble into immobile amphiphilic β-sheets and eventually disrupt the two lipid monolayers due to the difference in lateral pressure in the inner and outer leaflet [16] (Figure 1B). Alternatively, it is conceivable that a number of monomers could induce local lipid clustering and/or trap some lipids in between them, thereby forming a condensed peptide-lipid assembly with largely restricted mobility (Figure 1C). Especially in the presence of anionic lipids, this would lead to lateral phase segregation into patches enriched in anionic lipids and cationic peptides, which could likely lead to membrane permeabilization [17]. In all of these scenarios, the peptide backbone would end up parallel to the bilayer plane, and the hydrophobic side chains would point along the membrane normal. For the NMR discussion below, the geometry of the two orientation-dependent NMR parameters is illustrated in the box at the top of Figure 1. The red vector depicts the alignment of the C_α-C_β bond, which is equivalent to the ^{19}F dipole-dipole interaction (DD) of the CF_3-labeled side chain used here; the black arrow represents the amide ^{15}N-^{1}H bond vector, which can be taken as the axis of the chemical shift anisotropy tensor (CSA). These two vectors are essentially orthogonal to one another, as is obviously the case for the N-H bonds and side chains in a β-strand and/or β-sheet. Note that the yellow surface of the KL peptide denotes the hydrophobic Leu side chains, while the blue surface stands for the cationic Lys residues. As an example, KL10 is shown in the insert.

If, on the other hand, the KL peptides permeabilize the membrane by assembling into discrete transmembrane pores, the β-strands would have to tilt or flip, and several different architectures are conceivable in that case. H-bonded transmembrane β-barrels, as are characteristic in outer membrane porins, typically consist of alternating polar and hydrophobic residues, just like the KL peptides. Such peptidic assemblies have been suggested for a membrane-bound form of the Alzheimer Aβ-peptide [18–20]. In the case of the KL peptides, one would expect the backbone to be strongly tilted or almost upright, depending on the oligomer number (Figure 1D). Alternatively, it is also conceivable that a toroidal wormhole might form, consisting of several monomeric cationic peptides that are separated by anionic lipid head groups (Figure 1E). This pore architecture has been described for the α-helical magainin-family and KIA peptides and was also proposed for the β-hairpin peptide protegrin-1 [21]. In both of these pore models—a β-barrel (model D) as well as a toroidal wormhole (model E)—the KL backbone should be strongly tilted or aligned more or less upright in the membrane, i.e., distinctly different from the surface-bound peptides in models A/B/C.

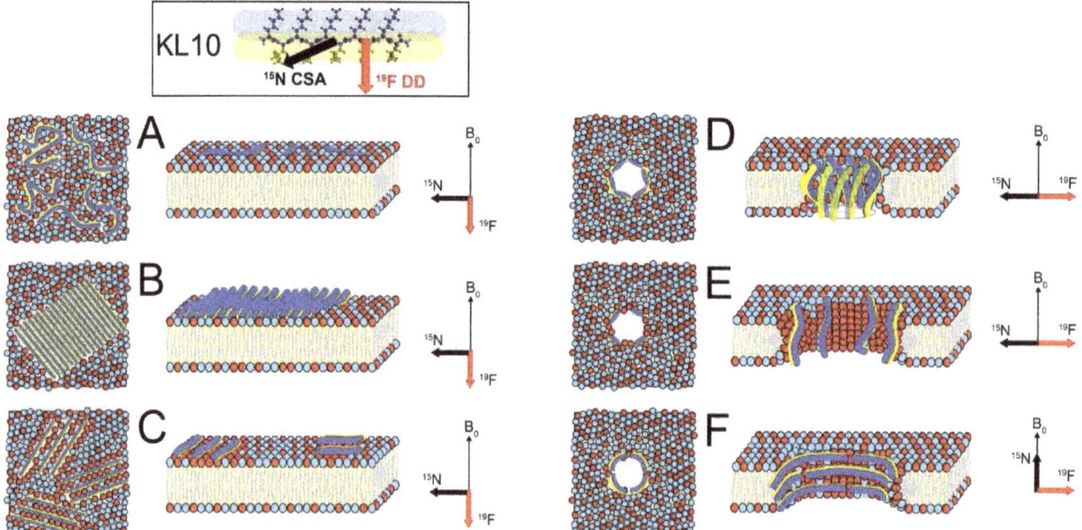

Figure 1. Different possible models of KL peptides in lipid bilayers, showing hydrophobic regions (lipid acyl chains, Leu side chains) in yellow, polar groups (uncharged lipid head groups) in light blue, cationic groups (Lys side chains) in dark blue, and anionic groups (charged lipid head groups) in red. Peptides are depicted in an extended β-strand conformation. The amphiphilic nature of the blue-yellow peptide structure of KL10 is seen in the box at the top. Superimposed on its molecular structure are the orientations of the relevant ^{15}N- and ^{19}F-NMR tensors as black and red arrows, respectively. The panels on the left show a top view of the membrane, and the central panels illustrate the respective side views. The panels on the right indicate the corresponding orientation of the NMR tensors in ^{15}N- and ^{19}F-labeled peptides when placed as an oriented membrane sample into the static magnetic field B_0 (which is aligned along the membrane normal). **Model A:** Monomeric peptides floating around on the membrane surface with high flexibility. **Model B:** Peptides on the membrane surface forming β-sheets that are stabilized by H-bonds between peptides. **Model C:** Peptides on the membrane surface that are indirectly assembled as an immobilized patch via clustering of anionic lipids between the strands. **Model D:** Peptides in a transmembrane orientation forming a β-barrel-type pore that is stabilized by H-bonds between peptides. **Model E:** Peptides forming a toroidal wormhole pore with peptides in a transmembrane orientation, with lipids in between the peptides, such that the pore interior is enriched in anionic lipids. **Model F:** Peptides lining the inside of a β-helix-type pore, with the long axis of the β-strand parallel to the membrane surface.

To complete the range of possibilities, one further conceivable type of backbone alignment could theoretically be constructed from stacks of β-strands (Figure 1F). In this architecture, the orientation of the peptide backbone would once more be completely orthogonal to both A/B/C as well as D/E. In such a hypothetical transmembrane pore, the backbone would be aligned horizontally or slightly tilted, reminiscent of the β-helix of Gramicidin A (though that sequence consists of alternating L- and D-amino acids) [22]. In this model F, the C_α-C_β vector of the side chains would also be aligned in the plane of the membrane, while the amide bond vector would be roughly parallel to the membrane normal. To differentiate between these distinctly different functional models (A/B/C vs. D/E vs. F), the backbone alignment (black amide bond vector) and the side-chain orientations (red arrow) need to be determined. Here, we have performed such analysis using SSNMR in oriented membrane samples by observing ^{15}N-labels in the backbone and ^{19}F-labels in the side-chains. NMR parameters reflecting the peptide mobility and any length-dependent change in alignment will allow further differentiation between the above-mentioned models A–F. We have thus compiled a battery of KL peptides with different lengths in order to elucidate the mechanism behind their intriguing length-dependent activity. To this aim, we compared their functional activities (antimicrobial, hemolytic, vesicle leakage) and determined their membrane-bound structures (oriented CD, ^{15}N-NMR, ^{19}F-NMR).

2. Materials and Methods

2.1. Materials

The lipids 1-palmitoyl-2-oleoyl-*sn*-glycero-3-phosphatidylglycerol (POPG), 1-palmitoyl-2-oleoyl-*sn*-glycero-3-phosphatidylcholine (POPC), 1,2-dimyristoyl-*sn*-glycero-3-phosphatidylglycerol (DMPG), 1,2-dimyristoyl-*sn*-glycero-3-phosphatidylcholine (DMPC), and 1,2-dierucoyl-*sn*-glycero-3-phosphatidylcholine (DErPC) were purchased from NOF Europe (Grobbendonk, Belgium). 1-myristoyl-2-hydroxy-*sn*-glycero-3-phosphatidylcholine (lyso-MPC), 1,2-dimyristoleoyl-*sn*-glycero-3-phosphatidylglycerol (DMoPG), 1,2-dimyristoleoyl-*sn*-glycero-3-phosphatidylcholine (DMoPC), and 1,2-dierucoyl-*sn*-glycero-3-phosphatidylglycerol (DErPG) were obtained from Avanti Polar Lipids (Alabaster, USA). Fmoc-protected amino acids and reagents for peptide synthesis were purchased from Merck Biosciences (Darmstadt, Germany) or Iris Biotech (Marktredwitz, Germany). ^{15}N-labelled amino acids for NMR experiments were purchased from Cambridge Isotope Laboratories (Andover, MA, USA) and were Fmoc-protected using Fmoc-Cl as described previously [23]. The ^{19}F-NMR label 3-(trifluoromethyl)-L-bicyclopent-[1.1.1]-1-ylglycine (CF$_3$-Bpg) was obtained from Enamine (Kiev, Ukraine). Solvents for peptide synthesis were obtained from Merck (Darmstadt, Germany) or Biosolve (Valkenswaard, The Netherlands), and solvents for HPLC purification were obtained from Fischer Scientific (Geel, Belgium). The fluorescent lipid 1,2-dioleoyl-*sn*-glycero-3-phosphoethanolamine-N-(lissamine rhodamine B sulfonyl) (Rhod-PE) was obtained from Avanti Polar Lipids. The fluorescent probes 8-aminonaphtalene-1,3,6-trisulfonic acid sodium salt (ANTS) and p-xylene-bis(pyridinium)bromide (DPX) were purchased from Invitrogen Molecular Probes (Karlsruhe, Germany).

2.2. Peptide Synthesis

KL peptides with lengths between 6 and 26 amino acids were synthesized with standard Fmoc solid-phase synthesis methods using a Syro II multiple peptide synthesizer (MultiSynTech, Witten, Germany). Coupling was performed using 4 eq. of the amino acid, 4 eq. HOBt, 4 eq. HBTU, and 8 eq. DIPEA as described previously [15]. ^{15}N- and ^{19}F-labeled amino acids were coupled manually with a longer coupling time (18 h) using 1.2 to 2 eq. of the labeled amino acids. Peptide purification was done using a JASCO (Groß-Umstadt, Germany) high-pressure liquid chromatography (HPLC) system. Some selected chromatograms of the purified peptides, along with their mass spectra, are shown in Figure S2. A semi-preparative Vydac C18 column was used, with a water/acetonitrile gradient containing 5 mM HCl. An LC-MS system (liquid chromatography instrument from Agilent, Waldbronn, Germany; QTOF ESI mass spectrometer from Bruker, Bremen, Germany) was used to check the identity and purity of the purified peptides, which were found to be at least 95% pure. For ^{15}N-NMR, a single ^{15}N-labeled Leu residue was incorporated into the position of Leu6. For ^{19}F-NMR, a selective CF$_3$-reporter group in the form of CF$_3$-Bpg was placed into the same position, with minimal perturbation due to its hydrophobic nature [13,24]. For all experiments, the weighed-in amount of peptide was used for preparing stock solutions and calculating the concentrations. Note that in every series of samples within any one assay, we used the same weight of peptide, i.e., a constant mass-to-mass peptide-to-lipid ratio, as described previously [15]. Only for visualization and ease of interpretation have we stated in several cases a molar peptide-to-lipid ratio, which we defined for KL14.

2.3. Circular Dichroism Spectroscopy (CD)

2.3.1. Sample Preparation

For CD measurements in liposomes, stock solutions of the lipids (5 mg/mL in CHCl$_3$/MeOH 1/1 vol/vol) and of the lyophilized peptides (1 mg/mL in H$_2$O/MeOH

1/9 vol/vol) were prepared. The final KL samples for CD measurements in pure water were obtained by diluting aliquots of the peptide stock solutions with deionized water, resulting in a peptide concentration of 0.1 mg/mL (27–113 µM, depending on the molecular weight of the peptide). POPC/POPG (1/1 mol/mol) and DMPC were chosen as lipid systems for the CD measurements in liposomes. The lipid concentration in the sample was adjusted to 1 mg/mL to get an analyzable CD signal of the peptides. Merging the peptide and lipid molecules on a molecular level without preceding aggregation of peptides was achieved by mixing suitable aliquots of the organic peptide and lipid stock solutions (co-solubilization). For KL14, a molar peptide-to-lipid ratio (P/L) of 1/20 was chosen, corresponding to a concentration of 65 µM in DMPC (or 75 µM in POPC/POPG 1:1) for the peptide and 1.3 mM or 1.5 mM for the lipid, respectively. The same corresponding peptide-to-lipid mass-to-mass ratio was used for all other KL peptides in order to keep the charge ratio constant, resulting in differing molar P/L ratios. After mixing the organic peptide/lipid solutions and subsequent evaporation of the organic solvents under N_2 gas flow, the residual peptide/lipid film was vacuum-dried for 4 h and re-dissolved in H_2O. Then, the samples were vortexed for 5 min and freeze/thaw-cycled tenfold. In a final step, the solutions were treated in an ultrasonic bath (UTR 200, Hielscher, Germany) for 16 min to generate small unilamellar vesicles (SUVs), which were stored overnight at room temperature.

Before the measurement of one series of aqueous peptide solutions, the samples were kept at the slightly acidic pH (~5–6) produced by the residual HCl from HPLC purification in the lyophilized peptides. In a second series of aqueous peptide solutions, and also for the liposome dispersions, the pH was set to 9–10 by adding a small aliquot (2–3 µL) of 0.1 M NaOH to each sample. The pH of each sample was checked with indicator paper.

2.3.2. Measurements

The CD measurements were conducted using a J-815 circular dichroism spectropolarimeter (Jasco, Groß-Umstadt, Germany). Spectra were recorded in the spectral range from 260–190 nm, at 0.1 or 0.5 nm intervals with a scanning speed of 10 nm/min and a spectral bandwidth of 1 nm as described earlier [14]. Samples were measured in a 1 mm quartz glass cuvette (110-QS, Hellma Analytics, Müllheim, Germany) at 30 °C. All KL peptides were measured in pure water and in the presence of small unilamellar vesicles (SUVs) composed of DMPC or POPC/POPG (1/1 mol/mol). Additional details are given in the Supplementary Materials.

2.4. Oriented CD

To examine the membrane alignment and conformation of the KL peptides in different lipid systems, oriented CD (OCD) experiments were performed on a Jasco J-810 spectropolarimeter with an OCD cell built in-house [6]. The samples were prepared from peptide stock solutions in MeOH/H_2O (9/1 vol/vol) with a concentration of 1 mg/mL and lipid stock solutions in MeOH/CHCl$_3$ (1/1 vol/vol) with a concentration of 5 mg/mL. The chosen lipid systems were POPC/POPG (1/1 mol/mol) and DMPC/lyso-MPC (2/1 mol/mol). The spectra were taken at wavelengths from 260–185 nm at 8 rotation angles (0°, 45°, 90°, . . . , 315°) with a scanning speed of 10 nm/min and a spectral bandwidth of 1 nm. Two spectral scans were performed at each rotation angle. These measurements at different rotation angles were done to avoid artefacts caused by linear dichroism. The final spectrum was obtained by averaging all spectra of all angles and baseline zeroing them at 260 nm. More details are given in the Supplementary Materials.

2.5. MIC (Minimum Inhibitory Concentration) Assay

The antimicrobial activity of the KL peptides was determined with a MIC assay. Peptide activity was tested against Gram-negative *Escherichia coli* (DSM 1116) and *Enterobacter helveticus* (DSM 18390) and against Gram-positive *Bacillus subtilis* (DSM 347) and *Staphylococcus xylosus* (DSM 20287) as previously reported [15]. Our standard MIC assay [14,25] was modified to avoid unnecessary exposure of peptides to the phosphate-containing growth medium, which was previously found to affect the results due to phosphate-induced aggregation of peptides [15]. Details can be found in the Supplementary Materials.

2.6. Hemolysis Assay

Hemolytic activity was examined with a serial 2-fold dilution assay as described earlier [15,26]. Details of the assay are described in the Supplementary Materials.

2.7. Vesicle Leakage Assay

For the leakage experiments [27,28], multilamellar vesicles (MLVs) were prepared in buffer containing the fluorophore ANTS, the quencher DPX, 50 mM NaCl, and 10 mM HEPES (pH 7.5). MLVs were extruded 41 times through a Nuclepore polycarbonate membrane with a pore size of 100 nm (Whatman GE Healthcare Europe, Freiburg, Germany) to obtain large unilamellar vesicles (LUV). Dye outside LUVs was removed by gel filtration through spin columns filled with Sephacryl 100-HR (Sigma-Aldrich, Taufkirchen, Germany). Before use, columns were equilibrated with elution buffer (150 mM NaCl, 10 mM HEPES, pH 7.5).

Leakage of entrapped ANTS was monitored by fluorescence dequenching of ANTS [29]. Fluorescence measurements were performed at 30 °C on a FluoroMax2 spectrofluorometer (HORIBA Jobin Yvon, Unterhaching, Germany) with an excitation wavelength of 355 nm and an emission wavelength of 515 nm. The sample was placed in a 10 mm × 10 mm quartz glass cuvette, and constant stirring was applied during the measurement. LUV and peptide solutions were diluted with elution buffer to obtain 100 µM lipids and the wanted peptide-to-lipid ratio in the sample. The level of 0% leakage corresponded to the fluorescence value immediately after the addition of vesicles, while 100% leakage was the fluorescence value obtained after the addition of Triton-X100, 10 min after the addition of vesicles. Additional details about the assay are given in the Supplementary Materials.

2.8. Leakage of FITC-Dextrans

The size of lesions in the membrane induced by the peptides was investigated using a modified leakage assay based on the fluorescence quenching of FITC-labeled dextran polymers of different sizes [30,31]. In short, POPC/POPG (1:1) vesicles were prepared with carboxyfluorescein and different size fluorescein isothiocyanate (FITC)-dextrans (FDs, Sigma-Aldrich, Taufkirchen, Germany) inside. When the FDs leak out, the fluorescence signal is quenched by anti-FITC antibodies (SouthernBiotech, Birmingham, AL, USA). Leakage was followed for 30 min, then Triton-X100 was added to completely destroy the vesicles and obtain 100% leakage. The size of the FITC-dextrans was determined by dynamic light scattering (Zetasizer Nano S, Malvern Instruments Ltd., Malvern, UK), as described in the Supplementary Materials.

2.9. Solid-State NMR

Macroscopically oriented NMR samples were prepared by co-dissolving appropriate amounts of peptides and lipids and spreading the solution onto thin glass plates. The peptide-to-lipid ratio (P/L) is given in mol/mol. For all samples and sample types, the P/L ratios were calculated for KL14, and the same peptide-to-lipid mass ratios were used for the other peptides to keep the charge ratio constant. All ^{15}N-, ^{19}F-, and ^{31}P-NMR solid-state NMR measurements were carried out on a Bruker Avance 500 or 600 MHz spectrometer (Bruker Biospin, Karlsruhe, Germany) at 308 K, as previously reported [32–37]. More experimental details are given in the Supplementary Materials.

NMR Data Analysis

^{31}P-NMR spectra give information about the phospholipids. The line shape is sensitive to the lipid phase [38], and the degree of orientation of macroscopically oriented samples can also be determined [39]. In our oriented samples, the membrane normal is usually oriented parallel to the external magnetic field B_0 (the sample tilt is 0°). In a well-oriented sample, there should then be one oriented peak around 30 ppm, but in less oriented samples, there is also a broad powder line shape with a maximum of around -15 ppm. The degree of orientation can be determined by integration.

In oriented samples, the ^{15}N-NMR chemical shift of peptides labeled with ^{15}N at the backbone amide gives information about the approximate orientation of the ^{15}N-^{1}H bond. A peak around 200 ppm indicates that the ^{15}N-^{1}H bond is oriented parallel to B_0, and a peak around 90 ppm indicates that the ^{15}N-^{1}H bond is oriented perpendicular to B_0. In a sample with a $0°$ tilt, a peak at 90 ppm indicates that the bond is in the membrane plane, and a peak at 200 ppm indicates that the bond is along the membrane normal [40,41].

In the ^{19}F-NMR spectrum of CF_3-Bpg labeled peptides, the dipolar couplings lead to a triplet signal, and the coupling strength gives information about the orientation of the C-CF_3 bond with respect to the external magnetic field B_0 [34,42]. In our oriented samples, the membrane normal is usually oriented parallel to B_0 (the sample tilt is $0°$). The sign of the dipolar couplings can also be determined from the spectrum [42]. If the coupling is close to the maximum value of $+16$ kHz, then the C-CF_3 bond is parallel to B_0. If the coupling is close to the minimum value of -8 kHz, then the C-CF_3 bond is perpendicular to B_0 [42–44]. A coupling close to -8 kHz can also be found in samples where the peptides aggregate and a powder pattern is obtained in ^{19}F-NMR; in this case, the coupling will be the same if the oriented sample is rotated so that the membrane normal is perpendicular to B_0 (the sample tilt is $90°$). If, on the other hand, the coupling is scaled by a factor -0.5 upon changing the sample tilt from $0°$ to $90°$; this is a sign of fast rotational diffusion of peptides in the membrane [34].

3. Results

3.1. Peptide Synthesis

Twenty different peptides consisting of a repetitive Lys-Leu motif were used in this study, with lengths between 6 and 26 amino acids. Peptides starting with Lys are called KLx peptides, and those starting with Leu are called LKx peptides, where x is the total number of amino acids in the sequence. Further analogues of KL10 and KL14 were synthesized for NMR analyses, either with a non-perturbing ^{15}N-NMR label at the backbone amide of Leu-6 or with Leu-6 replaced by CF_3-Bpg for ^{19}F-NMR experiments. All peptides were amidated at the C-terminus, and all sequences are listed in Table 1.

3.2. Circular Dichroism (CD)

CD spectroscopy was used to investigate the secondary structure of all KL and LK peptides in solution. In slightly acidic water at pH $\approx$ 5–6 (Figure 2A–C), the CD line shapes indicated that all peptides were unstructured (random coil spectra), with a minimum close to 197 nm. Phosphate buffer (PB) pH 7.0, which is well established in CD spectroscopy for measurements in aqueous environments, has to be avoided by all means because it will induce strong aggregation of KL into β-sheets that precipitate [15]. In the presence of anionic POPC/POPG (1/1) lipid vesicles in aqueous suspension at pH $\approx$ 9–10 and P/L = 1/20, (Figure 2D–F), almost all peptides showed a transition to a β-pleated conformation, as indicated in the spectra by a maximum at approximately 198 nm and a minimum close to 219 nm. The only exception was KL6, which was mostly disordered with a non-canonical line shape in the presence of vesicles. The intensity of the CD spectra was significantly reduced for the very long peptides (KL18 and longer), which had a strong tendency to precipitate as β-pleated aggregates. The reason is that aggregation resulted in turbid solutions, in which macroscopic particles would sediment and no longer contribute to the signal. Moreover, even for the shorter KL peptides, one cannot exclude significant differential light scattering and absorption flattening artifacts [45], which are also known to occur at increased turbidity. Therefore, no quantitative secondary structure estimation was attempted.

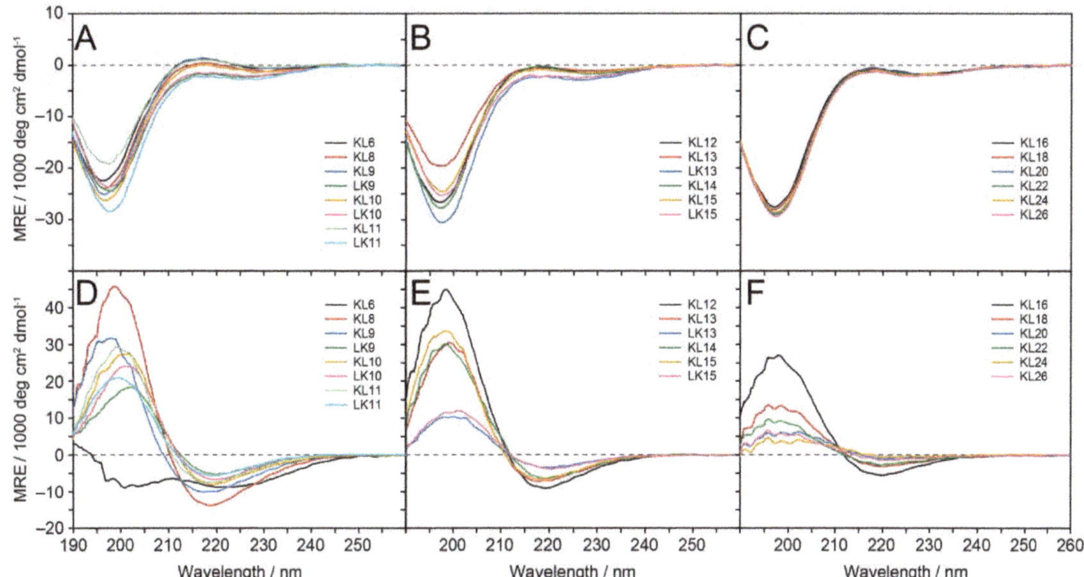

Figure 2. Circular dichroism spectra of KL and LK peptides (including KL6, KL10, KL14 and K18 from ref. [15]). (**A–C**) CD spectra in pure water at pH ≈ 5–6, with a peptide concentration of 0.1 mg/mL. (**D–F**) CD spectra in the presence of POPC/POPG (1/1) SUVs at pH ≈ 9–10 (HPLC-purified samples were made more basic by adding a small aliquot of 0.1 M NaOH). The lipid concentration was 1 mg/mL (1.3 mM), and the peptide concentration for KL14 was 0.13 mg/mL (65 μM), resulting in a peptide-to-lipid (P/L) molar ratio of 1/20, which provides enough bilayer surface for binding. The other samples were prepared with the same peptide-to-lipid mass-to-mass ratio in order to keep the total weight of peptide material as well as the charge ratio (Lys residues per anionic lipid) constant. The charge ratio of 7 cationic Lys residues per 10 anionic POPG lipids ensures that there is sufficient electrostatic attraction for all peptides to bind to the vesicles.

Further CD analyses were performed on KL and LK peptides in more alkaline water at pH ≈ 9–10 and in the presence of uncharged DMPC vesicles (cf. Figure S3, Supplementary Materials). In aqueous solution at this pH, where the N-terminal amino group is deprotonated and contributes to the polarity and total charge, the shortest peptides, KL6 and KL8, remain mostly unstructured. KL10 exhibits a mixture of random coil and β-pleated signals, while all other peptides show predominantly β-sheet structures. These spectra resemble those in small unilamellar DMPC vesicles dispersed in water at pH ≈ 9–10, where maxima at 195 and minima at 215 nm are found; i.e., there seems to be only a moderate induction of additional β-pleated structure due to the zwitterionic DMPC lipid. We note that in the negatively charged POPC/POPG vesicles discussed above, the β-sheet maxima and minima are shifted towards 198 and 219 nm, respectively, which seems to be an indication of higher aggregation tendency. In contrast, the binding and lipid-induced β-stranded conformation seems to be less pronounced in uncharged zwitterionic membranes, and the signal height is similar or somewhat increased compared to the spectra in water with the pH adjusted to 9–10 in the absence of lipids.

3.3. Oriented CD

Oriented CD (OCD) measurements of the KL peptides were performed in macroscopically aligned POPC/POPG (1/1) and DMPC/lyso-MPC (2/1) lipid systems. The idea behind these experiments was to elucidate whether changes in the OCD line shape would reveal any preferred orientation of the β-strands either parallel to the oriented lipid bilayer plane (see Figure 1A–C,F), strongly tilted (Figure 1D), or essentially parallel to the membrane normal (Figure 1E). Usually, OCD is applied only to α-helical peptides, where it gives straightforward information on the approximate helix tilt angle in oriented membrane samples [6,46,47]. Only very few attempts have been made so far to characterize the directional dependence of the CD bands of β-stranded and/or β-sheet structures in lipid bilayers. Note, moreover, that it is not straightforward to discriminate a local β-stranded conformation from an oligomeric β-sheet. Only in the case of poly[Leu-Lys] bound to phosphatidylcholine was it shown that the long-wavelength negative band is stronger in OCD than that in the isotropic CD spectrum, while the short-wavelength positive band is weaker, which was interpreted in terms of a backbone orientation parallel to the surface of the bilayers [48]. However, the β-sheet CD tensor is far more complex than that of an α-helix [49], and as there are numerous structural variations of β-pleated conformations (e.g., parallel β-sheets, antiparallel β-sheets, twisted amyloid fibrils, β-helices, etc.), no comprehensive theory or evaluation procedure exists. Nonetheless, here we tried to apply OCD to the KL peptides in the above-mentioned oriented bilayers in order to determine whether the spectral pattern of OCD spectra is different from that of the isotropic spectrum, which would at least indicate some orientational preference in the membrane, allowing to exclude some of the models in Figure 1.

For better comparison, all OCD spectra in Figure 3A,B have been scaled to the negative band of KL14 at approximately 217 nm. The lipid bilayers were oriented on a solid quartz glass support and kept hydrated at 97% relative humidity; i.e., the membranes are essentially fully hydrated, but there is no excess water present. Peptides with an even number of amino acids from 8 to 26 were examined. In the POPC/POPG lipid system, an OCD minimum was found at approximately 217 nm. A maximum is seen between 194 and 199 nm for peptides longer than KL12, indicating that they form β-pleated structures in these membranes under OCD conditions. The longer peptides have characteristic β-strand/-sheet spectra with quite broad positive bands. These exhibit distinct absorption flattening artifacts [45], indicating a high tendency to form aggregates, in which the β-strands seem to have no preferred alignment.

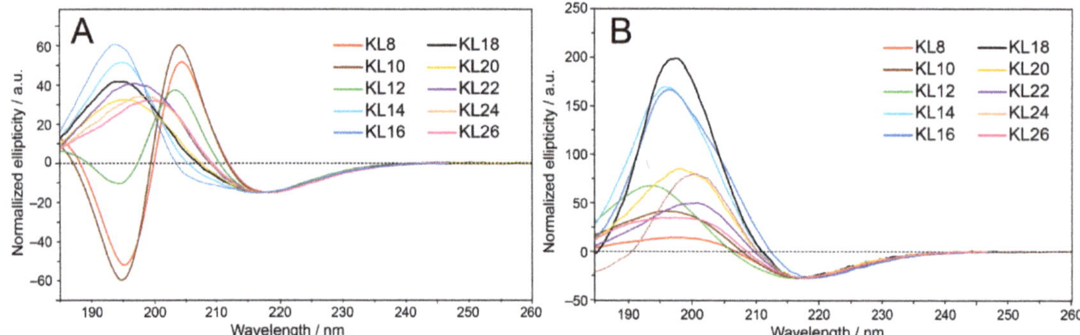

Figure 3. Oriented CD (OCD) spectra of KL peptides with an even number of residues. The P/L of KL14 was 1/20, and the same mass ratio was used for the other peptides. Spectra are normalized to the intensity of KL14 at 217 nm. (**A**) In POPC/POPG (1/1), all peptides show a minimum close to 217 nm, indicating extended β-strands or complete β-sheet formation. The shortest peptides KL8, KL10 and KL12 also have another minimum at approximately 195 nm, suggesting a partially unstructured conformation. (**B**) In DMPC/lyso-MPC (2/1), all peptides show a minimum at 217 nm and very broad positive bands, indicating β-sheet formation, given, for the longer peptides, strong absorption flattening artifacts due to aggregation.

Generally, for the KL peptides, a comparison of oriented OCD and conventional isotropic CD spectra is difficult due to the high aggregation tendency at P/L = 1/20 (lower P/L ratios could not be measured due to low sensitivity). If one considers the ratio of the positive and negative bands for KL14 and KL16, where absorption flattening artifacts are less pronounced, only small differences compared with the liposome spectra can be found. This means that no preferred alignment of the β-strands can be identified in these oriented OCD samples. Interestingly, in POPC/POPG bilayers, the shortest peptides with 8–12 amino acids exhibit an additional minimum at approximately 195 ppm, revealing that these peptides have a significant fraction of an unordered conformation under OCD conditions. Either a sub-population of these peptides is bound to the membrane as fully assembled β-sheets, while the residual fraction is not bound and very mobile with no defined structure, or all peptides are bound to the membrane but not fully extended in a straight β-stranded conformation.

Finally, a highly informative DMPC/lyso-MPC (2/1) lipid system was included, which exhibits a highly positive spontaneous curvature. The intrinsic monolayer curvature of this lyso-lipid mixture is known to drive α-helical KIA peptides into stable transmembrane pores of the toroidal wormhole type [50,51]. For the KL peptides, however, we find that the OCD spectra in DMPC/lyso-MPC are very similar to those in POPC/POPG, with a minimum at 217 nm. Despite the distinct characteristics in charge and curvature, the only difference compared to POPC/POPG can be seen for very short KL peptides, which give characteristic β-sheet signals with no minimum at 195 nm. However, the positive band around 200 nm is rather small, which can be due to a random coil contribution, which is largest for KL8 and reduced for KL10 and KL12. KL14, KL16 and KL18 have higher positive bands than in POPC/POPG, while the even longer peptides again demonstrate clear signs of aggregation, with very broad positive bands and pronounced absorption flattening artifacts reducing the band around 200 nm.

3.4. Antimicrobial Activity

The activity of KL and LK peptides against bacteria was examined using a MIC assay. In these two-fold dilution series, any values differing by a factor of two are not considered to be significantly different. We have previously shown that KL peptides, especially the longer ones, aggregate vigorously in the presence of phosphate ions [15]. We also demonstrated that the standard MIC assay that is routinely used in many laboratories is biased because peptides are exposed to the phosphate-containing medium before coming in contact with the bacteria. Therefore, here we used a modified MIC assay that was developed in our previous study [15], as described in the Methods section. In this approach, a two-fold dilution series was first prepared with peptides dissolved in water, and only thereafter were the bacteria in the medium added so that the peptides were simultaneously exposed to bacteria and phosphate-containing medium.

As seen in Table 2, there is a clear correlation between peptide length and biological activity, with the highest activity for peptide lengths of approximately 12 amino acids. KL6 was always inactive, in line with the observation above by CD that it does not form any β-strands in the presence of membranes. Regarding the even-numbered KL peptides, KL8 is somewhat antimicrobially active, KL10 and KL12 show the highest activity, while KL14 is moderately active, and KL16 and longer peptides are significantly less active. The odd-numbered KL peptides are slightly more active than the KL peptides with one less amino acid, but the values lie within a factor of two. The inverted sequence LK9 shows activity similar to that of KL9, LK10 is similar to KL10, and LK11 is similar to KL11. For the particular lengths 13 and 15, the charge-dominated KL peptides are more active (with 4–8 times lower MIC values) than the hydrophobicity-dominated LK peptides.

Table 2. MIC values for KL peptides against four bacterial strains and HC_{50} data ($\mu g/mL$).

Peptide	Gram-Negative		Gram-Positive		Hemolysis [a]
	E. coli	*E. helveticus*	*B. subtilis*	*S. xylosus*	HC_{50}
KL6	>256	128	>256	256	>256
KL8	64	16	16	8	>256
KL9	32	8	16	2	>256
LK9	16	4	4	2	145
KL10 [b]	4	2	2	2	47
LK10 [b]	8	2	4	1	256
KL11 [b]	4	2	4	1	>256
LK11	2	1	2	1	3
KL12	2	2	2	2	8
KL13	2	1	2	1	9
LK13	8	4	8	4	2
KL14	8	4	4	4	1
KL15	8	2	4	2	2
LK15	64	16	16	16	<1
KL16	32	8	16	8	2
KL18	32	16	32	8	1
KL20	32	32	32	16	<1
KL22	32	32	32	16	<1
KL24	32	16	32	8	<1
KL26	64	16	32	16	<1

[a] Note that HC_{50} values are obtained from concentration-dependent curves by interpolation (Supplementary Figure S2); therefore, these values do not correspond to powers of 2^n such as the MIC values. [b] The peptide candidates with the most promising therapeutic profile (low MIC value and high HC_{50} value) are highlighted in grey.

3.5. Hemolysis

Amphipathic antimicrobial peptides not only show membranolytic effects against bacteria but can also permeabilize eukaryotic cells such as erythrocytes. The hemolytic activities of the KL peptides were recorded for several different peptide concentrations and are summarized in Table 2 as HC_{50} values (concentration of peptide giving 50% hemolysis). The complete graphs of hemolysis as a function of peptide concentration are shown in Figure S4. We see that the KL peptides can be highly hemolytic, and the general trend is that hemolysis increases with length. Only KL6, KL8, KL9, and KL11 have very low hemolytic activity. KL10 gave 20% hemolysis at 8 $\mu g/mL$, KL11 gave almost no hemolysis, and KL12 gave 50% or more hemolysis at 8 $\mu g/mL$. Peptides with 14 or more amino acids gave 50% or more hemolysis even at 2 $\mu g/mL$. The trend seen above with regard to MIC values, i.e., that longer peptides become less active again, is clearly not observed for the HC_{50} values of hemolysis. Only the concentration-dependent curves show an apparent reduction of hemolysis at higher concentrations for the longer peptides (Figure S4) due to rapid aggregation in solution before the peptides could damage the erythrocytes.

LK peptides with an odd number of residues are found to be more hemolytic than KL peptides of the same length. This difference can be simply attributed to their increased overall hydrophobicity, which is generally known to promote hemolysis [52,53] (as odd-length LK peptides have one excess Leu residue, whereas odd-length KL peptides carry one extra Lys). Interestingly, in the Leu/Lys-balanced pair of 10 amino acids in length, the hemolytic activity turns out to be the other way around: KL10 is considerably more hemolytic than LK10. This finding seems to reflect the more pronounced amphiphilic character of an extended KL10 molecule, which carries a doubly charged Lys at the N-terminus and an entirely hydrophobic C-terminus with an amidated Leu.

3.6. Vesicle Leakage

The intriguing length-dependent biological activities raised the question of whether the window of optimal antibiotic activity for peptides KL10 to KL13 might be attributed to the concept of hydrophobic mismatch. Namely, when these peptides are fully extended as β-strands, they have a length of 37 to 48 Å, which seems perfectly matched to span the hydrophobic thickness of a membrane as a transmembrane pore, as illustrated for the two models in Figure 1D,E. It can be generally assumed that the activity of pore-forming peptides will show a distinct dependence not only on peptide length but also on membrane thickness. That is, in a toroidal wormhole and possibly even in a β-barrel, the amphiphilic unit has to be long enough to span the hydrophobic region of the bilayer; otherwise, it should remain inactive. Second, this threshold length should vary when its activity is being compared in membranes of different thicknesses. We thus set forth to determine whether hydrophobic matching between the length of the KL backbone and the bilayer thickness plays a role, as had been previously demonstrated for the amphiphilic α-helical KIA peptides [54].

For these experiments, it was imperative to vary the membrane thickness, which had not been possible in the MIC and hemolysis assays above, because the acyl chain composition of living cells obviously cannot be controlled. Therefore, we performed complementary in vitro experiments by measuring the leakage of a fluorescent dye from small unilamellar vesicles with different well-defined bilayer thicknesses (and comparable spontaneous lipid curvature). Different synthetic lipids were chosen with distinctly different acyl chain lengths (all of them being unsaturated so that they could be measured and compared in the liquid crystalline phase). In all cases, a 1/1 (mol/mol) mixture of zwitterionic phosphatidylcholine (PC) and anionic phosphatidylglycerol (PG) head groups was used to attract the cationic peptides electrostatically to the vesicles and promote complete binding. Furthermore, anionic lipids are known to be one of the main components of bacterial membranes, which in some cases contain more than 50 mol% PG, and also contain other anionic lipids like cardiolipin [55]. Vesicles were prepared in 10 mM HEPES buffer at pH = 7.5.

In POPC/POPG (1/1) vesicles at P/L = 1/20 (Figure 4A), we see hardly any leakage for KL6 to KL11, but KL12 gives almost full leakage within 10 min. KL14 is similar to KL12, and the longer peptides are slightly less active but generally give >50% leakage. Unlike the sharp threshold seen here for the KL peptides (i.e., between KL11 and KL12 in POPC/POPC with C16/C18 acyl chains), the LK peptides do not show such a pronounced length-dependent jump in activity. They also give a lower overall leakage, with a maximum of about 35% for peptides with 11 or more amino acids. Interestingly, LK11 causes essentially the same leakage as LK15, whereas KL11 gives much less leakage than KL15 and less than LK11.

In contradiction to our hydrophobic mismatch hypothesis, it turned out that in thin DMoPC/DMoPG vesicles with short acyl chains (14 carbons, 19.2 Å hydrophobic thickness [56]) and in thick DErPC/DErPG vesicles with very long acyl chains (22 carbons, 34.4 Å hydrophobic thickness [57]), almost the same extent of leakage as in POPC/POPG was found for KL peptides with an even number of amino acids (Figure 4B). In all cases, peptides with up to 10 amino acids give only very little leakage, while KL12 gives almost full leakage, and longer peptides also give a high degree of leakage. Therefore, we can conclude that leakage as a function of peptide length is essentially independent of membrane thickness. It only varies with the peptide length as such but shows no relation to the actual bilayer thickness. This result is in stark contrast to our earlier observations on α-helical KIA peptides with different lengths, where the relative length of peptides compared to the membrane thickness was critically important for function. Those KIA peptides are only active when they are long enough to span the membrane, and this threshold length varies linearly with the thickness of the lipid bilayer [54]. It, therefore, seems that the β-stranded KL-type peptides use a membrane permeabilization mechanism that is completely different from that of α-helical KIA peptides. The KIA peptides have been demonstrated to form toroidal wormhole pores with peptides in a membrane-spanning orientation [50,51], but for the KL peptides, neither this model (Figure 1E) nor the H-bonded β-barrel (Figure 1D) is supported by our leakage experiments.

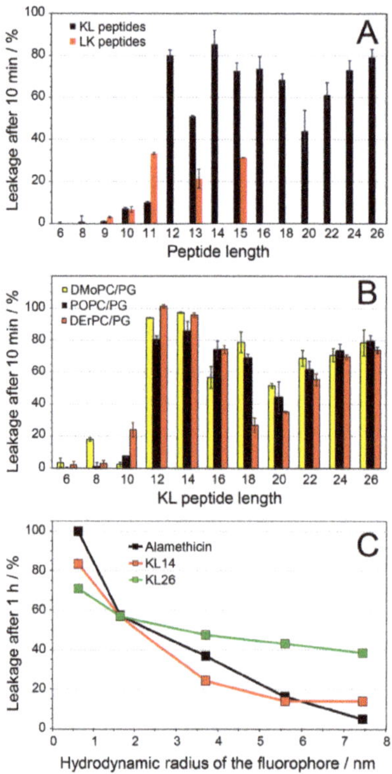

Figure 4. Fluorescent leakage assays of KL peptides in small unilamellar lipid vesicles. (**A,B**) For ANTS/DPX leakage, a P/L ratio of 1/20 was set for KL14, and the same peptide-to-lipid mass ratio was used for all other peptides. Leakage was monitored for 10 min after mixing peptides and vesicles. The 100% leakage signal was determined by the addition of Triton X-100. (**A**) Comparison of KL (black) and LK (red) peptides of different lengths in POPC/POPG (1/1) vesicles. (**B**) Leakage of even-numbered KL peptides in lipid systems with different bilayer thickness. Even though we see a length dependence, with a minimum length of 12 amino acids required for leakage to occur, a comparison of all three lipid systems shows that there is essentially no dependence on membrane thickness. (**C**) Leakage of FITC-dextrans of different sizes by KL14 and KL26 in POPC/POPG vesicles. Alamethicin (20 amino acids length) with a published pore size of 1.8 nm was used as a reference peptide. The lipid concentration was 100 μM, and the concentrations of KL14, KL26 and alamethicin were 12, 8 and 2.5 μM, respectively. These concentrations were chosen to obtain a similar leakage for FD4 for each peptide. For KL14 and KL26, the same peptide-to-lipid mass ratio was used. The hydrodynamic radii of the FITC-dextrans were determined by dynamic light scattering assuming spherical particles (see Supplementary Materials for more details).

Next, a modified leakage assay based on the fluorescence quenching of fluorescein-labeled dextran polymers with different sizes [30,31] was used to assess the diameter of the lesions in the membrane that are responsible for leakage. Here, we use the term "lesion" to describe any defects in the membrane that allow molecules to pass through without inferring any mechanism or structure. Carboxyfluorescein and FITC-dextrans with different diameters were trapped inside vesicles, and when leakage was induced by one of the KL peptides, the escaping fluorophores were quenched by specific antibodies outside the vesicles. KL14 and KL26 were used in this experiment to represent different peptide lengths, as they both showed high leakage in the ANTS/DPX assay. As a control, the helical antimicrobial peptide alamethicin was used, which is known from the literature to form water-filled barrel-stave pores with an inner diameter of 1.8 nm in DLPC and 2.6 nm in DPhPC [58,59]; similar-size pores might be expected in our POPC/POPG lipid system. As seen in Figure 4C, the leakage induced by KL14 or KL26 is reduced for the larger dextran sizes. This observation suggests that the KL peptides form lesions of limited size and do not dissolve the membrane completely. The size-dependent curve for KL14 was rather similar to that of alamethicin, indicating that lesions of similar size are formed by the two peptides. This result, however, does not imply that KL14 forms barrel-stave pores like the helical alamethicin. KL26 gave much more leakage for the larger dextrans than did KL14 or alamethicin, so we conclude that it forms larger lesions and less specific damage in the membranes than alamethicin.

3.7. Solid-State NMR

To determine the backbone orientation of the membrane-embedded KL peptides, solid-state ^{15}N- and ^{19}F-NMR experiments were performed on the KL10 peptide in macroscopically oriented samples, carrying either a selective ^{15}N-label (^{15}N-Leu) or a single ^{19}F-labeled amino acid (CF$_3$-Bpg) at the position of Leu6. These oriented samples are similar to the OCD samples in terms of composition, preparation and hydration, so we can infer from the corresponding OCD observations that the peptides must also form β-strands in the NMR samples. From previous studies of α-helical peptides, it is known that POPC/POPG (1/1), due to its negative spontaneous curvature, prefers to accommodate molecules within the headgroup region. Amphiphilic peptides thus tend to stay surface-bound in these unsaturated bilayers, maintaining an alignment parallel to the membrane surface. Bilayers with positive, spontaneous curvature, on the other hand, favor the insertion of peptides into the membrane. Therefore, we also used DMPC/lyso-MPC (2/1) for comparison. As a third lipid system with intermediate spontaneous curvature close to zero, we also included DMPC.

Oriented ^{15}N-NMR spectra are shown in Figure 5. Measurements were performed at different peptide concentrations, with P/L ratios from 1/100 to 1/20. ^{31}P-NMR spectra were measured before and after the ^{15}N-NMR experiments in order to monitor the orientational quality and stability of the samples. A ^{31}P-NMR signal at approximately 30 ppm corresponds to the well-oriented portion of the lipid bilayers, while a broad component with a maximum around -15 ppm arises from unoriented phospholipids. In DMPC (Figure 5A,B), P/L ratios of 1/50 and 1/20 were compared. The ^{31}P-NMR spectra indicate that at high concentration (P/L = 1/20, Figure 5B), the peptide strongly perturbs the membrane, as the degree of lipid orientation is massively decreased, with less than half of the lipids being arranged as well-oriented bilayers and the remainder contributing to an unoriented morphology. In POPC/POPG (Figure 5C,D), the lipid orientation at 1/20 was also very poor; hence only the spectra at 1/100 and 1/50 are shown. In DMPC/lyso-MPC (Figure 5E,F), the sample orientation was better, and even at a high peptide concentration of P/L = 1/20, the sample orientation was still quite good (in these spectra, there is a distinct peak from oriented lyso-lipids at approximately 20 ppm). In all lipid systems, a single ^{15}N-NMR peak was found close to 100 ppm, with no indication of any change in peptide orientation between the lipid systems (e.g., as a function of spontaneous curvature) or as a function of peptide concentration. The position of the ^{15}N-NMR signal represents the alignment of the chemical shift anisotropy tensor (CSA, with its main axis roughly aligned along the ^{15}N-^{1}H bond vector) of the labeled peptide bond in the oriented membrane sample. The observed ^{15}N chemical shift of 100 ppm is fully consistent with β-strands that are aligned either flat on the membrane surface (as in Figure 1A–C) or with peptides aligned in a transmembrane orientation forming some kind of β-barrel type structure (as in Figure 1D,E). On the other hand, these ^{15}N-NMR data do not support model F, in which the peptides form some kind of β-helix (Figure 1F), which can thus be ruled out with confidence.

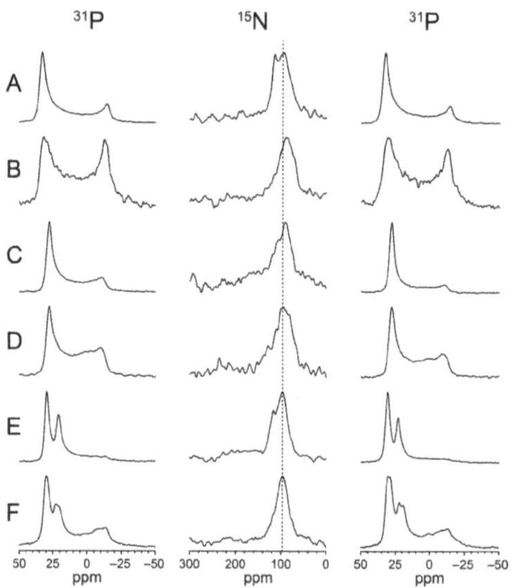

Figure 5. Solid-state NMR data on ^{15}N-labeled KL10 embedded in macroscopically aligned membranes with the sample normal aligned parallel to the static magnetic field direction. The middle column shows the ^{15}N-NMR spectra. ^{31}P-NMR spectra before (on the left side) and after ^{15}N-NMR experiments (on the right side) are also given. The lipid systems are (**A**) DMPC, P/L = 1/50; (**B**) DMPC, P/L = 1/20; (**C**) POPC/POPG (1/1), P/L = 1/100; (**D**) POPC/POPG (1/1), P/L = 1/50; (**E**) DMPC/lyso-MPC (2/1), P/L = 1/50; (**F**) DMPC/lyso-MPC (2/1), P/L = 1/20. (P/L ratios are for KL14, and the same peptide-to-lipid mass ratios were used for KL10).

While ^{15}N-NMR has allowed us to examine the peptide backbone, oriented ^{19}F-NMR experiments will next provide complementary information on the alignment of the peptide side chains, namely on the direction of the C_α-C_β bond vector. The rigid amino acid CF_3-Bpg is used as a ^{19}F-label, because its side chain (i.e., the CF_3-bond vector) extends perpendicularly to the plane of a β-sheet/-strand. This means that the relevant dipolar vector of the CF_3-group is aligned essentially orthogonally to the principal axis of the ^{15}N CSA tensor. It should therefore be possible to distinguish the different peptide architectures of models discussed in the introduction in which the peptide backbone lies flat on the membrane surface. When the C-CF_3 bond vector of an immobilized peptide points straight down into the membrane (Figure 1C), i.e., parallel to the bilayer normal, this will produce a dipolar splitting of approximately +15 kHz. On the other hand, in a flipped peptide, the C-CF_3 bond vector is oriented 90° to the membrane normal, giving a splitting of approximately −7.5 kHz, which would be expected for a β-barrel (Figure 1D), a toroidal wormhole (Figure 1E) or a β-helix (Figure 1F). Besides these orientational aspects, the ^{19}F-NMR analysis also provides important information on peptide mobility, which will help to discriminate highly dynamic monomeric peptides (as in Figure 1A,B) from immobilized oligomeric assemblies (Figure 1C) and detect uniaxial rotational averaging around the membrane normal (as expected in the discrete pore models of Figure 1D–F).

^{19}F-NMR experiments were performed on ^{19}F-labeled KL10 and KL14 in the same lipid systems as used for the ^{15}N-NMR analysis, and spectra are shown in Figure 6. Basically, two different splittings were observed. KL10 at low peptide concentrations of P/L = 1/1000 and 1/400 gave a splitting of 9–10 kHz, which at higher P/L increased to 12–15 kHz. In POPC/POPG, the larger splitting dominated already at P/L = 1/100, but in DMPC and DMPC/lyso-MPC, the smaller splitting was still observed at 1/50, co-existing with the larger splitting. In DMPC and DMPC/lyso-MPC at P/L = 1/50 and 1/100, a smaller negative splitting of −4.5 kHz was observed, which probably originates from non-oriented parts of the sample, where fast peptide motions lead to an averaging of the smaller 9 kHz splitting by a factor of $-1/2$. At P/L = 1/20 in DMPC/lyso-MPC, a splitting of about -7 kHz was seen, which indicates non-oriented immobilized peptides (a so-called "powder" signal). For KL14, the larger splitting is more dominant. In POPC/POPG, only a large splitting of 15 kHz is observed even at P/L = 1/1000. In DMPC and DMPC/lyso-MPC, the large splitting dominates at 1/100, though at 1/400, a smaller splitting of approximately 10 kHz is observed. At a high P/L, a powder splitting of -7.4 kHz was observed in all lipid systems. These ^{19}F-NMR results indicate that the KL10 and KL14 peptide backbones are aligned flat on the membrane surface, i.e., compatible with models A/B/C illustrated in Figure 1. This will be explained in detail in the discussion section below.

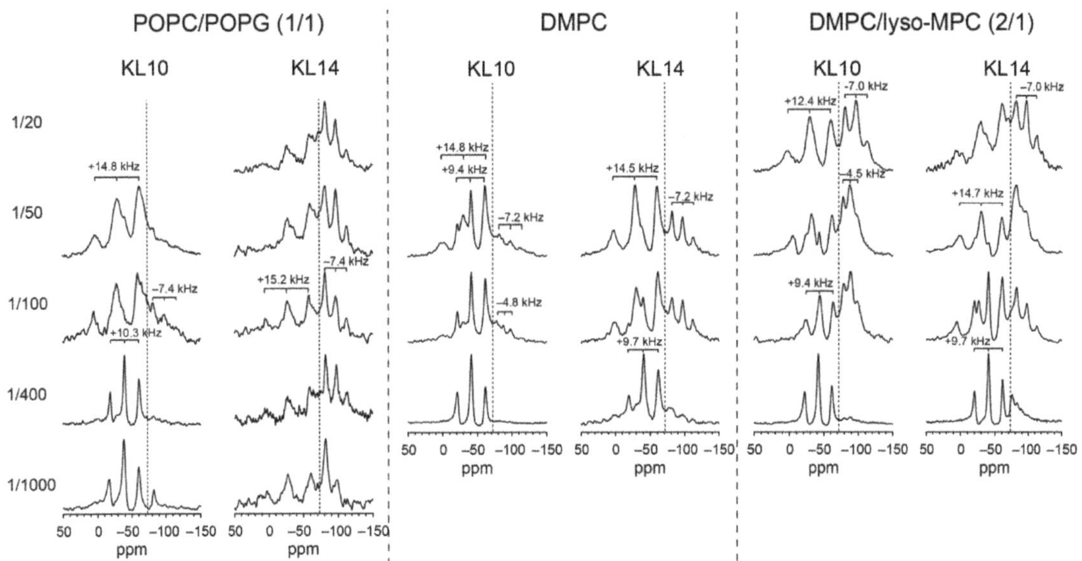

Figure 6. Solid-state ^{19}F-NMR spectra of KL10-^{19}F and KL14-^{19}F (labeled with a single CF$_3$-group at position 6) in macroscopically oriented membrane samples aligned as above. Dotted lines indicate the isotropic chemical shift (-72 ppm). Triplets arise from the dipolar coupling between the ^{19}F nuclei within the CF$_3$-group, and the splitting depends on the orientation of the C-CF$_3$ bond with regard to the membrane normal. Triplets to the left and right of the isotropic chemical shift have positive and negative signs of dipolar coupling, respectively. Three lipid systems were used, as denoted above each column, and several P/L ratios, as indicated on the left of each row of spectra (P/L values are for KL14, and the same peptide-to-lipid mass ratios were used for KL10).

4. Discussion

In this study, we have systematically explored how the length of amphipathic β-stranded peptides affects their membranolytic activity. KL model peptides with lengths from 6 to 26 amino acids were constructed from simple leucine-lysine repeats. We will first discuss and compare the activity of these different peptides against bacteria, erythrocytes and lipid vesicles and then relate these general activities with the structural data from CD and SSNMR spectroscopy. This way, the different architectures of models A–F illustrated in Figure 1 can be differentiated and discriminated for these novel KL-type peptides in order to come up with a mechanistic interpretation of antimicrobial action. Figure 7 shows an overview of the models and how well they fit with our experimental data.

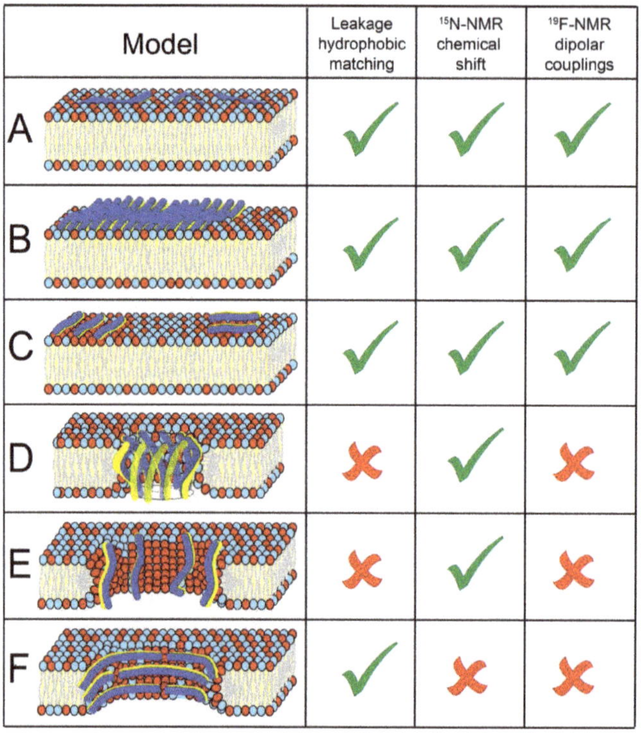

Model	Leakage hydrophobic matching	^{15}N-NMR chemical shift	^{19}F-NMR dipolar couplings
A	✓	✓	✓
B	✓	✓	✓
C	✓	✓	✓
D	✗	✓	✗
E	✗	✓	✗
F	✓	✗	✗

Figure 7. Compatibility of the different hypothetical models **A–F** from Figure 1 with our experimental data. A green tick (✓) indicates that the model is consistent with the data, while a red cross (✗) means that the model has to be rejected due to inconsistency with the data.

The activity of KL peptides against four strains of bacteria shows a clear length dependence, with maximum antimicrobial activity for peptides with 9–15 amino acids (Table 2). Shorter and longer peptides are less active. On the other hand, the hemolytic activity increases steadily with length; peptides with 11 or more amino acids are strongly hemolytic, and there is clearly no reduction in activity for longer peptides (except at high peptide concentrations, where aggregation probably occurs). This finding means that the KL peptides have an "optimum" length of 9–11 amino acids, where they have both good antimicrobial activity and low hemolytic side effects. Shorter peptides are not active at all, and longer peptides have both lower antimicrobial activity and much higher hemolytic side effects, making them less therapeutically promising. For vesicle leakage, we observe a similar length dependence as that for hemolysis, with low activity for peptides with 10 or fewer amino acids and high activity for longer peptides. The size-dependent leakage assay shows that large molecules can also leak out of vesicles, and long KL26 peptides can induce leakage of larger molecules than short KL14 peptides can (Figure 4C). However, from the results of dextran leakage, it is hard to conclude the exact size of the holes formed. The analysis above assumes dextrans to be spherical, but it has been shown that the larger dextrans can be represented as prolate ellipsoids with large axial ratios (for example, the axial ratio for a dextran with a molecular mass of 33 kDa was found to be 16) [60]. Therefore, even those dextran molecules with a nominal hydrodynamic radius that is too large to fit in the 1.8–2.6 nm diameter pore formed by alamethicin [58,59] can leak out to some extent. By comparing KL14 and KL26 with alamethicin, we can see that KL14 forms lesions of a similar size as those of alamethicin, while KL26 forms somewhat larger lesions.

These results from the antimicrobial and hemolytic assays, as well as vesicle leakage tests, indicate that long KL peptides are intrinsically active when bound to the membrane. However, their strong tendency to aggregate almost instantaneously in aqueous solution leads to losses in "active" material that can reach the membrane. This outcome is especially significant in the case of the antimicrobial assay, where phosphate in the medium enhances aggregation but cannot be avoided.

Odd-numbered LK peptides are more hemolytic and less antimicrobial than the corresponding KL peptides. The odd-numbered LK peptides (sequence L[KL]$_n$) have one more Leu than Lys, whereas the odd-numbered KL peptides (sequence [KL]$_n$K) have one more Lys than Leu, so the balance between hydrophobicity and charge is shifted. Similar to what is known from the literature on α-helical peptides, the more hydrophobic odd-numbered LK peptides are also more hemolytic [52,53,61,62]. It appears that those odd-numbered LK peptides are therefore less useful as antimicrobial agents. We conclude that the most promising therapeutic candidates are KL10, LK10 and KL11, and of these, KL11 has the least undesired hemolytic side effects.

Why are the peptides with about 10 amino acids the best? This result must be explicable in terms of the structure and dynamics of the peptides, i.e., with their membranolytic mechanism, which we have investigated with biophysical methods. From CD spectroscopy in solution, we observed that in water, all peptides were unstructured (Figure 2A–C). In the presence of anionic POPC/POPG (1/1) vesicles, all peptides (except the very short KL6) formed β-stranded structures, with a CD minimum at ~217 nm and a maximum at ~197 nm (Figure 2D–F). A quantitative structure analysis was performed by deconvolution of the CD line shapes of KL10 in water at pH ≈ 10, as well as in 10 mM phosphate buffer (where β-sheets are also formed, but no turbidity artifacts impeded the analysis). In both cases, KL10 was found to form mostly antiparallel β-sheets [15]. For longer peptides, the intensity of the CD signals is significantly lower, which can be attributed to aggregation in solution, so that some of the material precipitates and does not contribute to the signal. We had previously shown that the aggregation tendency of the KL peptides increases steadily with length [15].

Oriented CD samples are prepared by co-dissolving peptides and lipids in organic solvents and drying them to a film on a solid support. Then, the sample is hydrated in a chamber with 97% relative humidity. Hence, there is only a limited amount of water available and no bulk water phase. In this case, the peptides cannot precipitate from solution but are forced to stay in/on the membrane. In POPC/POPG (1/1) lipid bilayers, all peptides gave line shapes indicative of β-sheets, but the shorter peptides (KL8, KL10, KL12) also showed signals from unstructured conformations, with a minimum at 195 nm (Figure 3A). It appears that these peptides have a lower propensity to form aggregated β-sheets in the membrane, which correlates with their high antimicrobial activity and low hemolytic activity.

In DMPC/lyso-MPC (2/1) lipids, all peptides formed β-sheets, and the short peptides did not contribute any unstructured signals. In unsaturated POPC/POPG bilayers, which have a negative spontaneous curvature, it has been observed that membrane-active peptides in general tend to reside ON/IN the membrane surface, i.e., with their amphiphilic faces parallel to the bilayer plane. In DMPC/lyso-MPC bilayers, however, which have a positive spontaneous curvature, the peptide backbones were seen to readily tilt into the membrane once a threshold concentration was reached [35,63,64]. These previous observations fit well with our OCD data, where the shorter peptides are less structured in POPC/POPG (residing on the surface) than in DMPC/lyso-MPC (able to penetrate more deeply into the bilayer).

When ^{19}F-NMR spectroscopy was used to characterize KL10 and KL14 in oriented membranes, two distinct populations of peptides were observed (Figure 6). One population gives a ^{19}F dipolar splitting of approximately +9 kHz, and the other one gives a splitting of approximately +15 kHz. In a previous study on KIGAKI peptides, an alternating amphiphilic β-strand forming sequence similar to that of the KL peptides, two sets of splittings had also been observed [14]. The smaller splitting (in that case, +7.5 kHz) could be assigned to more mobile, less structured peptides, while the larger splitting (of +15 to +17 kHz) was associated with peptides forming immobilized β-sheets. It seems that the KL peptides are also present in these two forms. Compared to KL10, the longer KL14 showed much more spectral intensity in the signal with the large splitting, which correlates well with its higher tendency to self-assemble as β-sheets, which is, of course, also concentration dependent. In POPC/POPG (1/1), KL10 gave the small +9 kHz splitting at low peptide concentrations of P/L = 1/1000 and 1/400, while the large +15 kHz splitting dominated at 1/100 and 1/50. KL14, on the other hand, gave the large splitting exclusively, even at a high dilution of 1/1000. In DMPC and DMPC/lyso-MPC, both peptides at 1/400 gave only the small +9 kHz splitting, while at 1/100, a larger splitting was also observed, which, for KL14, had a higher intensity than for KL10. The higher proportion of aggregated peptides in POPC/POPG may be related to the presence of charged phospholipids, as it was also shown previously that KL peptides aggregate strongly in the presence of phosphate ions [15].

It is very clear that there is a pronounced correlation between aggregation behavior and biological activity. Extremely short peptides (e.g., KL6) are completely inactive, as they neither bind to membranes nor do they aggregate with themselves. Excessively long peptides aggregate vigorously in solution before they can even reach the membrane surface. Therefore, the optimal length of peptides is approximately 10 amino acids. Slightly longer peptides have even better antimicrobial activity, but beyond KL14, this beneficial feature decreases again.

From a structural point of view, the optimal window for antibiotic activity may well be related to the mechanism of action, in analogy to helical pore-forming peptides. Alamethicin [65], magainins [66–68] and KIA peptides [51,54] have all been shown to form transmembrane pores in lipid bilayers (under suitable conditions). For KIA peptides, it was found that a certain minimum length was needed for membrane activity, i.e., the α-helix must be long enough to span the bilayer to be active [54]. For KIA peptides, a length of 21 amino acids was sufficient to kill the same bacteria that have been tested here [69]. This length corresponds to an α-helix of 31.5 Å, assuming a 1.5 Å increment per residue and a completely helical sequence. For an extended β-strand in an ideal β-sheet, the increment per residue is approximately 3.7 Å. Thus, a β-strand with 9 amino acids has a length (9 × 3.7 Å = 33.3 Å) similar to that of an α-helix with 21 amino acids (21 × 1.5 Å = 31.5 Å), which should thus be long enough to span the membrane. Such an extended β-strand would obviously have to be aligned in an upright transmembrane orientation and be accompanied by other strands to form a stable oligomeric pore, as illustrated in Figure 1D,E (Figure 7D,E). This hydrophobic mismatch hypothesis fits well with our bacterial MIC data above, showing that KL peptides shorter than 9 amino acids are not active. It can be argued that the highly antibiotic KL peptides with a length of up to 14 amino acids fit well across a lipid bilayer as a pore, provided that they can adjust their angle and tilt away slightly from the bilayer normal. Finally, KL peptides with a length of 15 amino acids or more are again less active, which would actually correspond in length to an α-helix with 37 amino acids. The longest α-helical KIA peptide that had been tested contained only 28 amino acids, so it is conceivable that such long KIA peptides would also show less activity. Alternatively, the reason could be the strong aggregation tendencies of long KL peptides, which are far stronger than those of any KIA peptides.

The hypothesis of KL peptides forming pores in a lipid bilayer in an upright, transmembrane orientation was tested in two ways. If KL peptides need a minimum length to span the membrane in order to be active (Figure 1D,E, Figure 7D,E), they should show different leakage thresholds in lipid vesicles that are composed of bilayers with different thicknesses, as had been observed for KIA peptides [54,70]. Fluorescent leakage experiments of KL peptides of different lengths were performed with thin membranes (DMoPC/DMoPG, hydrophobic thickness 19.2 Å [56]), intermediate membranes (POPC/POPG, hydrophobic thickness 27.1 Å [57] and 27.8 Å [71]), and thick membranes (DErPC/DErPG, hydrophobic thickness 34.4 Å [57]). Remarkably, in all cases, the same length dependence was found: low leakage for peptides up to 10 amino acids and high leakage for longer peptides (Figure 4B). This result clearly contradicts the hypothesis of membrane-spanning pores.

The second test of the pore hypothesis was performed using solid-state NMR. To obtain information about the orientation of α-helical peptides in membranes, ^{15}N-NMR is a straightforward and very useful method. In macroscopically oriented samples, the chemical shift of the narrow peak from a single ^{15}N-labeled amide group in the middle of the helical region provides the approximate tilt angle of the peptide [51,72]. For β-sheet-forming peptides such as the KL series, however, there is no such simple theory that can yield the corresponding information from ^{15}N-labeled peptides. Nonetheless, the main reason for performing these experiments was the hope that a shift in the NMR signal would be observed for different KL lengths, at different concentrations, or in different bilayers, which could then be correlated with further ^{19}F-NMR results. However, all experiments under all conditions showed more or less the same peak at approximately 100 ppm, so the ^{15}N-NMR experiments were rather inconclusive. We can only exclude the unusual β-helical model in Figure 1F (Figure 7F) with some confidence because, in this case, a ^{15}N-NMR chemical shift closer to 200 ppm would have been expected, as visualized in the corresponding vector diagrams. Note that the chemical shift of an oriented peptide depends on both CSA tensor orientation (and thereby segmental orientation or possibly even whole-body orientation of the peptide backbone) as well as dynamics (local segmental wobble and any larger-scale fluctuations). Therefore, in a potentially wobbly β-stranded peptide (monomer, oligomer, aggregate, or fibril), the position of the NMR signal cannot be translated into the peptide alignment, as it is routinely possible in the case of α-helical peptides.

^{19}F-NMR experiments, finally, gave a clear result. If KL peptides form β-strands on the membrane surface as in Figure 1A–C (Figure 7A–C), then the CF_3 label points down into the membrane, almost parallel to the bilayer normal (which is aligned with B_0 of the NMR magnet). It should give a large splitting of about +15 kHz when the peptide is immobile (and slightly less in the case of some local or global motional averaging). If, on the other hand, peptides are placed upright in a transmembrane orientation as in Figure 1D,E (Figure 7D,E), the CF_3 label should point almost perpendicular to the membrane normal and give half the splitting of -7.5 kHz when immobilized (or less in the case of dynamics). The ^{19}F-NMR spectra in Figure 6 always gave large splittings corresponding to peptides oriented flat on the membrane surface. Additionally, in DMPC/lyso-MPC (2/1), in which it is generally very favorable for KIA peptides and other amphiphilic helices to assume a transmembrane orientation, no experimental evidence for an upright insertion was found for KL peptides. Therefore, we conclude that KL peptides do not form traditional pores in a transmembrane orientation, neither in the form of β-barrels consisting of H-bonded β-sheets (Figures 1D and 7D), nor from monomers in a toroidal wormhole style (Figures 1E and 7E). Additionally, β-helical pores with the peptide backbones oriented parallel to the membrane surface can also be excluded by the ^{19}F-NMR data (Figures 1F and 7F) since the CF_3 groups would also, in this case, be oriented perpendicular to the membrane normal and give splittings of approximately -7.5 kHz, which is not observed.

It thus seems clear that KL peptides bind to the membrane surface and use some type of carpet mechanism to permeabilize the bilayer and induce leakage, forming defects of a limited size. Monomeric, flexible β-strands as in Figure 1A (Figure 7A) will obviously prevail at low peptide concentration, especially for short KL peptides with a low tendency to aggregate. Longer KL peptides form larger aggregates with less mobility. As seen above, there is also a lipid dependence, with, for example, more immobilized peptides seen in POPC/POPG than in DMPC/lyso-MPC. Aggregation into amyloid-like fibrils consisting of H-bonded β-sheets as in Figure 1B (Figure 7B) cannot be discriminated, at this point in time, from the kind of lateral self-assembly depicted in Figure 1C (Figure 7C), where the cationic peptides are segregated in some immobilized phase that is enriched with anionic lipids. Future solid-state NMR techniques might be able to provide more information about the detailed aggregation behavior.

5. Conclusions

Membrane-active antimicrobial peptides are typically cationic and amphipathic. Here, we have shown that very simple KL (and LK) model peptides, forming amphipathic β-strands, can have high antimicrobial activity and low hemolytic side effects at the same time. However, these KL peptides are therapeutically promising only with a length of 10 or 11 amino acids. Shorter KL peptides actually bind to the membrane [15], but they are not membranolytic, most likely because they do not form sufficiently large aggregates that are needed to permeabilize the lipid bilayer. Longer peptides, on the other hand, are extremely hemolytic. These long peptides aggregate vigorously already in solution, and therefore, few of them reach the membrane, especially in the antimicrobial MIC assay where phosphate ions are present in the buffer. Nonetheless, once the long peptides are bound to the membrane, they are as active as the intermediate peptides (on a mass-to-mass ratio). For any therapeutic or biotechnological applications, KL peptides should thus be long enough to be able to form aggregates in the membrane but be short enough to have some flexibility and not pre-aggregate in solution.

This conclusion indicates that the mechanism of action observed in bacteria and in red blood cells involves membrane permeabilization. Using ^{19}F-NMR spectroscopy, we have shown that the KL peptides do not form pores in a transmembrane orientation, despite the intriguing length-dependent optimum in antimicrobial activity. Vesicle leakage data using different membrane thicknesses also indicate that these β-stranded peptides, unlike many α-helical antimicrobial peptides, do not have to span the membrane to form some sort of pore to be active. Instead, a carpet mechanism seems to be operating here, where the peptides transiently accumulate in the outer leaflet and cause such an imbalance in lateral pressure that the lipid bilayer forms local lesions. Remarkably, our size-dependent leakage experiments have shown that the diameter of these lesions is, nonetheless, neither indefinitely large nor small but rather comparable to the typical pore diameters of alamethicin and other α-helical antibiotics.

Supplementary Materials: The following Supplementary Material can be downloaded at https://www.mdpi.com/article/10.3390/biomedicines10092071/s1, Method descriptions with additional details; Table S1: Size of the FITC-dextrans according to dynamic light scattering measurements, assuming spherical particles. The size of carboxyfluorescein (CF) is taken from [13]; Table S2: Hemolysis (in %) at different KL peptide concentrations (μg/mL). Values are compared with the values after addition of triton-X, which leads to 100% hemolysis. For KL14, KL16 and longer peptides the hemolysis is no longer increasing continuously with peptide concentration, probably due to aggregation; Figure S1: Results of size determination of FITC-dextrans by dynamic light scattering; Figure S2: LC-MS chromatograms showing absorbance at 220 nm (A,C,E,G) and mass spectra (B,D,F,H) of selected peptides used in this study. In all cases there is a single peak in the chromato-gram and the expected mass is found for each peptide. (A,B) KL10, (C,D) KL10-19F, (E,F) KL14, (G,H) KL18; Figure S3: Circular dichroism spectra of KL and LK peptides. (A–C) CD spectra in pure water, at pH ≈ 9–10. The peptide concentration was 0.1 mg/mL. (D–F) CD spectra in the presence of DMPC SUVs at pH ≈ 9–10. The lipid concentration was 1 mg/mL (1.5 mM) and the peptide concentration for KL14 was 0.15 mg/mL (75 μM) resulting in a peptide-to-lipid (P/L) molar ratio of 1/20. The other samples were prepared in the same peptide-to-lipid mass-to-mass ratio to keep the charge ratio constant; Figure S4: Hemolytic activity of KL and LK peptides. (A) Peptides with length 6–11 amino acids. (B) Peptides with length 12–15 amino acids. (C) Peptides with length 16–26 amino acids. In general, longer peptides are more active. For very long peptides, activity goes down at higher concentration, probably due to peptide aggregation. The dotted line in each panel indicates 50% hemolysis, and the HC50 value is determined from the crossing of the hemolysis curves with this line. References [6,14,15,26–37,73–81] are cited in the Supplementary Materials.

Author Contributions: Conceptualization, E.S., P.W., J.B. and A.S.U.; Formal analysis, E.S., P.W. and J.B.; Funding acquisition, A.S.U.; Investigation, F.S., E.S., P.W., J.R., J.B., H.L.P.C. and L.B.; Methodology, P.W., J.R. and J.B.; Supervision, E.S., P.W. and J.B.; Visualization, F.S. and E.S.; Writing—original draft, E.S.; Writing—review & editing, F.S., E.S., P.W., J.R., J.B., H.L.P.C., L.B. and A.S.U. All authors have read and agreed to the published version of the manuscript.

Funding: This research was funded by the Helmholtz Association Program BIF-TM and by the German Research Foundation (DFG) by grants UL127/7-1 and INST 121384/58-1 FUGG. We acknowledge financial support from the KIT-Publication Fund of the Karlsruhe Institute of Technology.

Data Availability Statement: The experimental raw data are available from the authors upon request.

Acknowledgments: We thank Andrea Eisele and Kerstin Scheubeck for help with peptide synthesis, Kerstin Scheubeck for help with biological assays, Siegmar Roth and Bianca Posselt for help with CD measurements, and Stephan Grage and Markus Schmitt for help with NMR infrastructure.

Conflicts of Interest: The authors declare no conflict of interest.

References

1. Pinheiro da Silva, F.; Machado, M.C. Antimicrobial peptides: Clinical relevance and therapeutic implications. *Peptides* **2012**, *36*, 308–314. [CrossRef]
2. Wimley, W.C.; Hristova, K. Antimicrobial peptides: Successes, challenges and unanswered questions. *J. Membr. Biol.* **2011**, *239*, 27–34. [CrossRef] [PubMed]
3. Arias, C.A.; Murray, B.E. Antibiotic-resistant bugs in the 21st century—A clinical super-challenge. *N. Engl. J. Med.* **2009**, *360*, 439–443. [CrossRef] [PubMed]
4. Brogden, K.A. Antimicrobial peptides: Pore formers or metabolic inhibitors in bacteria? *Nat. Rev. Microbiol.* **2005**, *3*, 238–250. [CrossRef]
5. Boman, H.G. Antibacterial peptides: Basic facts and emerging concepts. *J. Intern. Med.* **2003**, *254*, 197–215. [CrossRef]
6. Bürck, J.; Wadhwani, P.; Fanghänel, S.; Ulrich, A.S. Oriented circular dichroism: A method to characterize membrane-active peptides in oriented lipid bilayers. *Acc. Chem. Res.* **2016**, *49*, 184–192. [CrossRef]
7. Strandberg, E.; Ulrich, A.S. NMR methods for studying membrane-active antimicrobial peptides. *Concepts Magn. Reson. A* **2004**, *23A*, 89–120. [CrossRef]
8. Strandberg, E.; Ulrich, A.S. Solid-state ^{19}F-NMR analysis of peptides in oriented biomembranes. In *Modern Magnetic Resonance*, 2nd ed.; Webb, G., Ed.; Springer International Publishing: Cham, Switzerland, 2017; pp. 1–18. [CrossRef]
9. Strandberg, E.; Ulrich, A.S. Solid-state NMR for studying peptide structures and peptide-lipid interactions in membranes. In *Modern Magnetic Resonance*, 2nd ed.; Webb, G.A., Ed.; Springer International Publishing: Cham, Switzerland, 2017; pp. 1–13. [CrossRef]
10. Wadhwani, P.; Strandberg, E. Structure analysis of membrane-active peptides using ^{19}F-labeled amino acids and solid-state NMR. In *Fluorine in Medicinal Chemistry and Chemical Biology*; Ojima, I., Ed.; Blackwell Publishing: London, UK, 2009; pp. 463–493.
11. Strandberg, E.; Ulrich, A.S. Flow charts for the systematic solid-state ^{19}F/^{2}H-NMR structure analysis of membrane-bound peptides. *Annu. Rep. NMR Spectrosc.* **2020**, *99*, 80–117. [CrossRef]
12. Blazyk, J.; Wiegand, R.; Klein, J.; Hammer, J.; Epand, R.M.; Epand, R.F.; Maloy, W.L.; Kari, U.P. A novel linear amphipathic β-sheet cationic antimicrobial peptide with enhanced selectivity for bacterial lipids. *J. Biol. Chem.* **2001**, *276*, 27899–27906. [CrossRef]
13. Wadhwani, P.; Reichert, J.; Strandberg, E.; Bürck, J.; Misiewicz, J.; Afonin, S.; Heidenreich, N.; Fanghänel, S.; Mykhailiuk, P.K.; Komarov, I.V.; et al. Stereochemical effects on the aggregation and biological properties of the fibril-forming peptide [KIGAKI]$_3$ in membranes. *Phys. Chem. Chem. Phys.* **2013**, *15*, 8962–8971. [CrossRef]
14. Wadhwani, P.; Strandberg, E.; Heidenreich, N.; Bürck, J.; Fanghänel, S.; Ulrich, A.S. Self-assembly of flexible β-strands into immobile amyloid-like β-sheets in membranes as revealed by solid-state ^{19}F NMR. *J. Am. Chem. Soc.* **2012**, *134*, 6512–6515. [CrossRef] [PubMed]
15. Strandberg, E.; Schweigardt, F.; Wadhwani, P.; Bürck, J.; Reichert, J.; Cravo, H.L.P.; Burger, L.; Ulrich, A.S. Phosphate-dependent aggregation of [KL]$_n$ peptides affects their membranolytic activity. *Sci. Rep.* **2020**, *10*, 12300. [CrossRef]
16. Grage, S.L.; Afonin, S.; Ieronimo, M.; Berditsch, M.; Wadhwani, P.; Ulrich, A.S. Probing and manipulating the lateral pressure profile in lipid bilayers using membrane-active peptides: A solid-state ^{19}F NMR study. *Int. J. Mol. Sci.* **2022**, *23*, 4544. [CrossRef] [PubMed]
17. Silvius, J.R.; Gagné, J. Calcium-induced fusion and lateral phase separations in phosphatidylcholine-phosphatidylserine vesicles–Correlation by calorimetric and fusion measurements. *Biochemistry* **1984**, *23*, 3241–3247. [CrossRef]
18. Serra-Batiste, M.; Ninot-Pedrosa, M.; Bayoumi, M.; Gairi, M.; Maglia, G.; Carulla, N. Aβ42 assembles into specific β-barrel pore-forming oligomers in membrane-mimicking environments. *Proc. Natl. Acad. Sci. USA* **2016**, *113*, 10866–10871. [CrossRef]
19. Jang, H.; Arce, F.T.; Ramachandran, S.; Capone, R.; Lal, R.; Nussinov, R. β-Barrel topology of Alzheimer's β-amyloid ion channels. *J. Mol. Biol.* **2010**, *404*, 917–934. [CrossRef]
20. Kandel, N.; Matos, J.O.; Tatulian, S.A. Structure of amyloid β(25-35) in lipid environment and cholesterol-dependent membrane pore formation. *Sci. Rep.* **2019**, *9*, 2689. [CrossRef]
21. Mani, R.; Cady, S.D.; Tang, M.; Waring, A.J.; Lehrer, R.I.; Hong, M. Membrane-dependent oligomeric structure and pore formation of a β-hairpin antimicrobial peptide in lipid bilayers from solid-state NMR. *Proc. Natl. Acad. Sci. USA* **2006**, *103*, 16242–16247. [CrossRef]
22. Ketchem, R.R.; Hu, W.; Cross, T.A. High-resolution conformation of gramicidin A in a lipid bilayer by solid-state NMR. *Science* **1993**, *261*, 1457–1460. [CrossRef]

23. Carpino, L.A.; Han, G.Y. 9-Fluorenylmethoxycarbonyl amino-protecting group. *J. Org. Chem.* **1972**, *37*, 3404–3409. [CrossRef]

24. Afonin, S.; Mikhailiuk, P.K.; Komarov, I.V.; Ulrich, A.S. Evaluating the amino acid CF$_3$-bicyclopentylglycine as a new label for solid-state ^{19}F-NMR structure analysis of membrane-bound peptides. *J. Pept. Sci.* **2007**, *13*, 614–623. [CrossRef] [PubMed]

25. Ruden, S.; Hilpert, K.; Berditsch, M.; Wadhwani, P.; Ulrich, A.S. Synergistic interaction between silver nanoparticles and membrane-permeabilizing antimicrobial peptides. *Antimicrob. Agents Chemother.* **2009**, *53*, 3538–3540. [CrossRef] [PubMed]

26. Strandberg, E.; Tiltak, D.; Ieronimo, M.; Kanithasen, N.; Wadhwani, P.; Ulrich, A.S. Influence of C-terminal amidation on the antimicrobial and hemolytic activities of cationic α-helical peptides. *Pure Appl. Chem.* **2007**, *79*, 717–728. [CrossRef]

27. Duzgunes, N.; Wilschut, J. Fusion assays monitoring intermixing of aqueous contents. *Methods Enzymol.* **1993**, *220*, 3–14. [CrossRef]

28. Steinbrecher, T.; Prock, S.; Reichert, J.; Wadhwani, P.; Zimpfer, B.; Bürck, J.; Berditsch, M.; Elstner, M.; Ulrich, A.S. Peptide-lipid interactions of the stress-response peptide TisB that induces bacterial persistence. *Biophys. J.* **2012**, *103*, 1460–1469. [CrossRef]

29. Ellens, H.; Bentz, J.; Szoka, F.C. H$^+$- and Ca^{2+}-induced fusion and destabilization of liposomes. *Biochemistry* **1985**, *24*, 3099–3106. [CrossRef]

30. Ladokhin, A.S.; Selsted, M.E.; White, S.H. Sizing membrane pores in lipid vesicles by leakage of co-encapsulated markers: Pore formation by melittin. *Biophys. J.* **1997**, *72*, 1762–1766. [CrossRef]

31. Stutzin, A. A fluorescence assay for monitoring and analyzing fusion of biological membrane-vesicles in vitro. *FEBS Lett.* **1986**, *197*, 274–280. [CrossRef]

32. Grage, S.L.; Strandberg, E.; Wadhwani, P.; Esteban-Martin, S.; Salgado, J.; Ulrich, A.S. Comparative analysis of the orientation of transmembrane peptides using solid-state ^{2}H- and ^{15}N-NMR: Mobility matters. *Eur. Biophys. J.* **2012**, *41*, 475–482. [CrossRef]

33. Müller, S.D.; De Angelis, A.A.; Walther, T.H.; Grage, S.L.; Lange, C.; Opella, S.J.; Ulrich, A.S. Structural characterization of the pore forming protein TatA$_d$ of the twin-arginine translocase in membranes by solid-state ^{15}N-NMR. *Biochim. Biophys. Acta* **2007**, *1768*, 3071–3079. [CrossRef]

34. Glaser, R.W.; Sachse, C.; Dürr, U.H.N.; Afonin, S.; Wadhwani, P.; Strandberg, E.; Ulrich, A.S. Concentration-dependent realignment of the antimicrobial peptide PGLa in lipid membranes observed by solid-state ^{19}F-NMR. *Biophys. J.* **2005**, *88*, 3392–3397. [CrossRef] [PubMed]

35. Strandberg, E.; Zerweck, J.; Wadhwani, P.; Ulrich, A.S. Synergistic insertion of antimicrobial magainin-family peptides in membranes depends on the lipid spontaneous curvature. *Biophys. J.* **2013**, *104*, L9–L11. [CrossRef] [PubMed]

36. Heinzmann, R.; Grage, S.L.; Schalck, C.; Bürck, J.; Banoczi, Z.; Toke, O.; Ulrich, A.S. A kinked antimicrobial peptide from *Bombina maxima*. II. Behavior in phospholipid bilayers. *Eur. Biophys. J.* **2011**, *40*, 463–470. [CrossRef]

37. Walther, T.H.; Grage, S.L.; Roth, N.; Ulrich, A.S. Membrane alignment of the pore-forming component TatA$_d$ of the twin-arginine translocase from *Bacillus subtilis* resolved by solid-state NMR spectroscopy. *J. Am. Chem. Soc.* **2010**, *132*, 15945–15956. [CrossRef] [PubMed]

38. Cullis, P.R.; de Kruijff, B. Lipid polymorphism and the functional roles of lipids in biological membranes. *Biochim. Biophys. Acta* **1979**, *559*, 399–420. [CrossRef]

39. Wadhwani, P.; Tremouilhac, P.; Strandberg, E.; Afonin, S.; Grage, S.L.; Ieronimo, M.; Berditsch, M.; Ulrich, A.S. Using fluorinated amino acids for structure analysis of membrane-active peptides by solid-state ^{19}F-NMR. In *Current Fluoroorganic Chemistry: New Synthetic Directions, Technologies, Materials, and Biological Applications*; Soloshonok, V.A., Mikami, K., Yamazaki, T., Welch, J.T., Honek, J.F., Eds.; ACS Symposium Series 949; American Chemical Society: Washington, DC, USA, 2007; pp. 431–446.

40. Teng, Q.; Cross, T.A. The *in situ* determination of the ^{15}N chemical-shift tensor orientation in a polypeptide. *J. Magn. Reson.* **1989**, *85*, 439–447. [CrossRef]

41. Bechinger, B.; Zasloff, M.; Opella, S.J. Structure and orientation of the antibiotic peptide magainin in membranes by solid-state nuclear magnetic resonance spectroscopy. *Protein Sci.* **1993**, *2*, 2077–2084. [CrossRef]

42. Glaser, R.W.; Sachse, C.; Dürr, U.H.N.; Wadhwani, P.; Ulrich, A.S. Orientation of the antimicrobial peptide PGLa in lipid membranes determined from ^{19}F-NMR dipolar couplings of 4-CF$_3$-phenylglycine labels. *J. Magn. Reson.* **2004**, *168*, 153–163. [CrossRef]

43. Dürr, U.H.N.; Grage, S.L.; Witter, R.; Ulrich, A.S. Solid state ^{19}F NMR parameters of fluorine-labeled amino acids. Part I: Aromatic substituents. *J. Magn. Reson.* **2008**, *191*, 7–15. [CrossRef]

44. Grage, S.L.; Durr, U.H.; Afonin, S.; Mikhailiuk, P.K.; Komarov, I.V.; Ulrich, A.S. Solid state ^{19}F NMR parameters of fluorine-labeled amino acids. Part II: Aliphatic substituents. *J. Magn. Reson.* **2008**, *191*, 16–23. [CrossRef]

45. Miles, A.J.; Wallace, B.A. Circular dichroism spectroscopy of membrane proteins. *Chem. Soc. Rev.* **2016**, *45*, 4859–4872. [CrossRef] [PubMed]

46. Bürck, J.; Roth, S.; Wadhwani, P.; Afonin, S.; Kanithasen, N.; Strandberg, E.; Ulrich, A.S. Conformation and membrane orientation of amphiphilic helical peptides by oriented circular dichroism. *Biophys. J.* **2008**, *95*, 3872–3881. [CrossRef] [PubMed]

47. Huang, H.W.; Wu, Y. Lipid-alamethicin interactions influence alamethicin orientation. *Biophys. J.* **1991**, *60*, 1079–1087. [CrossRef]

48. Bazzi, M.D.; Woody, R.W. Interaction of amphipathic polypeptides with phospholipids: Characterization of conformations and the CD of oriented β-sheets. *Biopolymers* **1987**, *26*, 1115–1124. [CrossRef]

49. Woody, R.W. The circular dichroism of oriented β sheets: Theoretical predictions. *Tetrahedron Asymmetry* **1993**, *4*, 529–544. [CrossRef]

50. Strandberg, E.; Bentz, D.; Wadhwani, P.; Ulrich, A.S. Chiral supramolecular architecture of stable transmembrane pores formed by an α-helical antibiotic peptide in the presence of lyso-lipids. *Sci. Rep.* **2020**, *10*, 4710. [CrossRef]

51. Grau-Campistany, A.; Strandberg, E.; Wadhwani, P.; Rabanal, F.; Ulrich, A.S. Extending the hydrophobic mismatch concept to amphiphilic membranolytic peptides. *J. Phys. Chem. Lett.* **2016**, *7*, 1116–1120. [CrossRef]
52. Strandberg, E.; Kanithasen, N.; Bürck, J.; Wadhwani, P.; Tiltak, D.; Zwernemann, O.; Ulrich, A.S. Solid state NMR analysis comparing the designer-made antibiotic MSI-103 with its parent peptide PGLa in lipid bilayers. *Biochemistry* **2008**, *47*, 2601–2616. [CrossRef]
53. Dathe, M.; Wieprecht, T.; Nikolenko, H.; Handel, L.; Maloy, W.L.; MacDonald, D.L.; Beyermann, M.; Bienert, M. Hydrophobicity, hydrophobic moment and angle subtended by charged residues modulate antibacterial and haemolytic activity of amphipathic helical peptides. *FEBS Lett.* **1997**, *403*, 208–212. [CrossRef]
54. Grau-Campistany, A.; Strandberg, E.; Wadhwani, P.; Reichert, J.; Bürck, J.; Rabanal, F.; Ulrich, A.S. Hydrophobic mismatch demonstrated for membranolytic peptides, and their use as molecular rulers to measure bilayer thickness in native cells. *Sci. Rep.* **2015**, *5*, 9388. [CrossRef]
55. Epand, R.F.; Savage, P.B.; Epand, R.M. Bacterial lipid composition and the antimicrobial efficacy of cationic steroid compounds (Ceragenins). *Biochim. Biophys. Acta* **2007**, *1768*, 2500–2509. [CrossRef] [PubMed]
56. Marsh, D. Energetics of hydrophobic matching in lipid-protein interactions. *Biophys. J.* **2008**, *94*, 3996–4013. [CrossRef] [PubMed]
57. Kucerka, N.; Tristram-Nagle, S.; Nagle, J.F. Structure of fully hydrated fluid phase lipid bilayers with monounsaturated chains. *J. Membr. Biol.* **2006**, *208*, 193–202. [CrossRef]
58. He, K.; Ludtke, S.J.; Worcester, D.L.; Huang, H.W. Neutron scattering in the plane of membranes: Structure of alamethicin pores. *Biophys. J.* **1996**, *70*, 2659–2666. [CrossRef]
59. Qian, S.; Wang, W.C.; Yang, L.; Huang, H.W. Structure of the alamethicin pore reconstructed by x-ray diffraction analysis. *Biophys. J.* **2008**, *94*, 3512–3522. [CrossRef]
60. Bohrer, M.P.; Deen, W.M.; Robertson, C.R.; Troy, J.L.; Brenner, B.M. Influence of molecular configuration on the passage of macromolecules across the glomerular capillary wall. *J. Gen. Physiol.* **1979**, *74*, 583–593. [CrossRef]
61. Chen, Y.X.; Mant, C.T.; Farmer, S.W.; Hancock, R.E.W.; Vasil, M.L.; Hodges, R.S. Rational design of alpha-helical antimicrobial peptides with enhanced activities and specificity/therapeutic index. *J. Biol. Chem.* **2005**, *280*, 12316–12329. [CrossRef]
62. Zamora-Carreras, H.; Strandberg, E.; Mühlhäuser, P.; Bürck, J.; Wadhwani, P.; Jiménez, M.Á.; Bruix, M.; Ulrich, A.S. Alanine scan and ^{2}H NMR analysis of the membrane-active peptide BP100 point to a distinct carpet mechanism of action. *Biochim. Biophys. Acta* **2016**, *1858*, 1328–1338. [CrossRef]
63. Strandberg, E.; Ulrich, A.S. AMPs and OMPs: Is the folding and bilayer insertion of β-stranded outer membrane proteins governed by the same biophysical principles as for α-helical antimicrobial peptides? *Biochim. Biophys. Acta* **2015**, *1848*, 1944–1954. [CrossRef]
64. Strandberg, E.; Tiltak, D.; Ehni, S.; Wadhwani, P.; Ulrich, A.S. Lipid shape is a key factor for membrane interactions of amphipathic helical peptides. *Biochim. Biophys. Acta* **2012**, *1818*, 1764–1776. [CrossRef]
65. Leitgeb, B.; Szekeres, A.; Manczinger, L.; Vagvolgyi, C.; Kredics, L. The history of alamethicin: A review of the most extensively studied peptaibol. *Chem. Biodivers.* **2007**, *4*, 1027–1051. [CrossRef]
66. Matsuzaki, K. Magainins as paradigm for the mode of action of pore forming polypeptides. *Biochim. Biophys. Acta* **1998**, *1376*, 391–400. [CrossRef]
67. Yang, L.; Weiss, T.M.; Lehrer, R.I.; Huang, H.W. Crystallization of antimicrobial pores in membranes: Magainin and protegrin. *Biophys. J.* **2000**, *79*, 2002–2009. [CrossRef]
68. Ludtke, S.J.; He, K.; Heller, W.T.; Harroun, T.A.; Yang, L.; Huang, H.W. Membrane pores induced by magainin. *Biochemistry* **1996**, *35*, 13723–13728. [CrossRef]
69. Strandberg, E.; Bentz, D.; Wadhwani, P.; Bürck, J.; Ulrich, A.S. Terminal charges modulate the pore forming activity of cationic amphipathic helices. *BBA-Biomembranes* **2020**, *1862*, 183243. [CrossRef]
70. Gagnon, M.C.; Strandberg, E.; Grau-Campistany, A.; Wadhwani, P.; Reichert, J.; Bürck, J.; Rabanal, F.; Auger, M.; Paquin, J.F.; Ulrich, A.S. Influence of the length and charge on the activity of α-helical amphipathic antimicrobial peptides. *Biochemistry* **2017**, *56*, 1680–1695. [CrossRef]
71. Pan, J.; Heberle, F.A.; Tristram-Nagle, S.; Szymanski, M.; Koepfinger, M.; Katsaras, J.; Kucerka, N. Molecular structures of fluid phase phosphatidylglycerol bilayers as determined by small angle neutron and X-ray scattering. *Biochim. Biophys. Acta* **2012**, *1818*, 2135–2148. [CrossRef] [PubMed]
72. Bechinger, B.; Gierasch, L.M.; Montal, M.; Zasloff, M.; Opella, S.J. Orientations of helical peptides in membrane bilayers by solid state NMR spectroscopy. *Solid State Nucl. Magn. Reson.* **1996**, *7*, 185–191. [CrossRef]
73. Rance, M.; Byrd, R.A. Obtaining high-fidelity spin-1/2 powder spectra in anisotropic media–Phase-cycled Hahn echo spectroscopy. *J. Magn. Reson.* **1983**, *52*, 221–240. [CrossRef]
74. Levitt, M.H.; Suter, D.; Ernst, R.R. Spin dynamics and thermodynamics in solid-state NMR cross polarization. *J. Chem. Phys.* **1986**, *84*, 4243–4255. [CrossRef]
75. Fung, B.M.; Khitrin, A.K.; Ermolaev, K. An improved broadband decoupling sequence for liquid crystals and solids. *J. Magn. Reson.* **2000**, *142*, 97–101. [CrossRef] [PubMed]
76. Zhang, S.; Wu, X.L.; Mehring, M. Elimination of ringing effects in multiple-pulse sequences. *Chem. Phys. Lett.* **1990**, *173*, 481–484. [CrossRef]

77. Bennett, A.E.; Rienstra, C.M.; Auger, M.; Lakshmi, K.V.; Griffin, R.G. Heteronuclear decoupling in rotating solids. *J. Chem. Phys.* **1995**, *103*, 6951–6958. [CrossRef]
78. Reichert, J.; Grasnick, D.; Afonin, S.; Bürck, J.; Wadhwani, P.; Ulrich, A.S. A critical evaluation of the conformational requirements of fusogenic peptides in membranes. *Eur. Biophys. J.* **2007**, *36*, 405–413. [CrossRef]
79. Sigma-Aldrich. Fluorescein Isothiocyanate-Dextran. Available online: https://www.sigmaaldrich.com/technical-documents/protocols/biology/fluorescein-isothiocyanate-dextran.html#ref (accessed on 14 July 2022).
80. Wooten, M.K.C.; Koganti, V.R.; Zhou, S.S.; Rankin, S.E.; Knutson, B.L. Synthesis and nanofiltration membrane performance of oriented mesoporous silica thin films on macroporous supports. *ACS Appl. Mater. Interfaces* **2016**, *8*, 21806–21815. [CrossRef]
81. Mayer, L.D.; Hope, M.J.; Cullis, P.R. Vesicles of variable sizes produced by a rapid extrusion procedure. *Biochim. Biophys. Acta* **1986**, *858*, 161–168. [CrossRef]

Article

Antibacterial and Anti-Inflammatory Effects of Apolipoprotein E

Manoj Puthia [1,†], Jan K. Marzinek [2,†], Ganna Petruk [1], Gizem Ertürk Bergdahl [3], Peter J. Bond [2,4] and Jitka Petrlova [1,*]

[1] Division of Dermatology and Venereology, Institution of Clinical Sciences, Lund University, SE-22184 Lund, Sweden; manoj.puthia@med.lu.se (M.P.); ganna.petruk@med.lu.se (G.P.)
[2] Bioinformatics Institute (A*STAR), Singapore 138671, Singapore; marzinekj@bii.a-star.edu.sg (J.K.M.); peterjb@bii.a-star.edu.sg (P.J.B.)
[3] Division of Infection Medicine, Institution of Clinical Sciences, Lund University, SE-22184 Lund, Sweden; gizemertrk@yahoo.com
[4] Department of Biological Sciences, National University of Singapore, 14 Science Drive 4, Singapore 117543, Singapore
* Correspondence: jitka.petrlova@med.lu.se; Tel.: +46-733604082
† These authors contributed equally to this work.

Abstract: Apolipoprotein E (APOE) is a lipid-transport protein that functions as a key mediator of lipid transport and cholesterol metabolism. Recent studies have shown that peptides derived from human APOE display anti-inflammatory and antimicrobial effects. Here, we applied in vitro assays and fluorescent microscopy to investigate the anti-bacterial effects of full-length APOE. The interaction of APOE with endotoxins from *Escherichia coli* was explored using surface plasmon resonance, binding assays, transmission electron microscopy and all-atom molecular dynamics (MD) simulations. We also studied the immunomodulatory activity of APOE using in vitro cell assays and an in vivo mouse model in combination with advanced imaging techniques. We observed that APOE exhibits anti-bacterial activity against several Gram-negative bacterial strains of *Pseudomonas aeruginosa* and *Escherichia coli*. In addition, we showed that APOE exhibits a significant binding affinity for lipopolysaccharide (LPS) and lipid A as well as heparin. MD simulations identified the low-density lipoprotein receptor (LDLR) binding region in helix 4 of APOE as a primary binding site for these molecules via electrostatic interactions. Together, our data suggest that APOE may have an important role in controlling inflammation during Gram-negative bacterial infection.

Keywords: apolipoprotein E; antimicrobial peptides; Gram-negative bacteria; host defense; innate immunity; aggregation

Citation: Puthia, M.; Marzinek, J.K.; Petruk, G.; Bergdahl, G.E.; Bond, P.J.; Petrlova, J. Antibacterial and Anti-Inflammatory Effects of Apolipoprotein E. *Biomedicines* **2022**, *10*, 1430. https://doi.org/10.3390/biomedicines10061430

Academic Editor: Shaker A. Mousa

Received: 24 May 2022
Accepted: 15 June 2022
Published: 17 June 2022

Publisher's Note: MDPI stays neutral with regard to jurisdictional claims in published maps and institutional affiliations.

1. Introduction

Apolipoprotein E (APOE) is a 299-residue glycoprotein that functions as a key regulator of cholesterol and lipid levels in the blood plasma and brain. APOE is an exchangeable apolipoprotein that is a protein component of high-density lipoproteins (HDL), very low-density lipoproteins (VLDL), and low density (LDL) remnant particles, which all have atherogenic effects. Multifunctional APOE has been reported to affect cholesterol efflux, coagulation, macrophage function, oxidative processes, central nervous system physiology, cell signaling, and inflammation [1].

In humans, APOE exhibits polymorphism, with the three alleles of APOE genes located at chromosome 19: APOE2, APOE3 and APOE4. These three isoforms only differ in amino acids composition at positions 112 and 158. The most frequent isoform APOE3 (78% of human population) has cysteine at residue 112 and arginine at residue 158; in contrast, APOE2 (8%) has cysteine at both sites, and APOE4 (14%) has arginine at both sites [2,3].

Multiple lysine and arginine amino acids in the 130–150 residues region of APOE are essential for the binding of low-density lipoprotein receptors (LDLR), as well as heparin [4,5]. The N-terminal domain of APOE forms an amphipathic four-helix bundle with LDLR- and heparin-binding sites within helix 4. The four helices are organized in an antiparallel fashion. The C-terminal domain of APOE is arranged as amphipathic α-helices, which are responsible for host lipid/lipoprotein-binding capability and protein–protein interactions [3].

It has been shown that APOE = driven peptides from the receptor- and heparin-binding region of APOE exhibit antibacterial, antiviral and immunomodulatory effects [6–9]. Moreover, we have previously reported that full-length APOE has antibacterial activity against Gram-negative bacteria, both in vitro and in vivo [10].

Although the role of lipid-poor/free APOE in the skin is not fully known, we hypothesized that the APOE may have a neutralizing effect on Gram-negative bacterial infection. In this study, we demonstrated the antibacterial activity of APOE against multiple Gram-negative bacterial strains in vitro. We also detected specific interaction and the formation of APOE aggregates upon lipopolysaccharide (LPS) challenge, which were also confirmed by molecular dynamics (MD) simulations. Moreover, we showed an immunomodulatory effect of APOE in cultured cells and animal models during LPS challenge.

2. Materials and Methods

2.1. Bacterial Strains

E. coli (25922 and 700928) and *P. aeruginosa* (27853) were purchased from American Type Culture Collection (ATCC). *P. aeruginosa* clinical strains 15159 and *P. aeruginosa* (PAO1) were kindly provided by Dr. B. Iglewski (University of Rochester).

2.2. Endotoxins

LPS from *E. coli* (serotype 0111:B4, cat# L3024), Lipoteichoic acid (LTA) from *S. aureus* (cat# tlrl-pslta) and LPS-EB Biotin from *E. coli* (cat# tlrl-lpsbiot) were purchased from Sigma-Aldrich. Lipid A from *E. coli* (serotype R515, cat# ALX-581-200-L002) was purchased from AH Diagnostics.

2.3. Proteins and Peptides

Human plasma APOE (cat# IHUAPOE) and human plasma APOA1 (cat# IRHPL0059) were purchased from Innovative Research. The thrombin-derived peptide TCP-25 (GKYG-FYTHVFRLKKWIQKVIDQFGE) (97% purity, acetate salt) was synthesized by Ambiopharm (Madrid, Spain). Both APOE and Thrombin-derived C-terminal Peptide (TCP-25) contain amphipathic regions with a strong positive charge (K and L amino acid rich) and hydrophobicity (L, V, and I amino acid rich). Those two structural features are well-known to play important roles in antimicrobial, anti-inflammatory, and immunomodulatory activities of host defense molecules.

2.4. Cells

THP-1-XBlue-CD14 reporter monocytes (InvivoGen) were cultured in RMPI 1640-GlutaMAX-1 (Gibco, Life Technology ltd., Renfrew, UK) and the media was supplemented with 10% (v/v) heat-inactivated FBS (FBSi, Invitrogen, Waltham, MA, USA) and 1% (v/v) antibiotic-antimycotic solution (AA, Invitrogen) at 37 °C in 5% CO_2.

2.5. Animals

BALB/c tg(NF-κB-RE-Luc)-Xen reporter male mice (10–12 weeks old), purchased from Taconic Biosciences, Rensselaer, NY, USA, were used for all experiments. The animals were housed under standard conditions of light and temperature and had free access to standard laboratory chow and water.

2.6. Viable Count Assay (VCA)

The potential antibacterial activity of APOE on *E. coli* and *P. aeruginosa* strains was explored by incubating one colony overnight in 5 mL of Todd-Hewitt (TH) medium. The next morning, the bacterial culture was refreshed and grown to a mid-logarithmic phase (OD_{620nm} 0.4). The bacteria were then centrifuged, washed, and diluted 1:1000 in 10 mM Tris buffer at pH 7.4 to obtain an approximate concentration of bacteria amounting to 2×10^6 cfu/mL. Next, 50 µL of bacterial suspension was incubated with 2 µM of APOE, 2 µM of TCP-25 (used as a positive control), or buffer control (10 mM Tris buffer at pH 7.4) for 2 h at 37 °C. After 2 h, serial dilutions of the samples were plated on TH agar plates, incubated overnight at 37 °C, and followed by colony counting the next day [10,11].

2.7. NF-κB Activity Assay

NF-κB/AP-1 activation in THP-1-XBlue-CD14 reporter monocytes was determined after 20–24 h of incubation according to the manufacturer's protocol (InvivoGen). Briefly, 1×10^6 cells/mL in RPMI were seeded in 96-well plates (180 µL) and incubated with peptides (TCP-25 1 µM; APOE 0.5, 1 and 2 µM; APOAI 1 µM), LPS (10 ng/mL) or both overnight at 37 °C, 5% CO_2 in a total volume of 200 µL. The following day, the activation of NF-κB/AP-1 was analyzed as the secretion of embryonic alkaline phosphatase (SEAP). The supernatant (20 µL) from the cells was transferred to 96-well plates, and 180 µL of Quanti-Blue was added. The plates were incubated for 2 h at 37 °C, and the absorbance was measured at 600 nm in a VICTOR3 Multilabel Plate Counter spectrofluorometer [12].

2.8. MTT Viability Assay

Sterile filtered MTT (3-(4,5-dimethylthiazolyl)-2,5-diphenyltetrazolium bromide; Sigma-Aldrich) solution (5 mg/mL in PBS) was stored in the dark at −20 °C until usage. We added 20 µL of MTT solution to the remaining overnight culture of THP-1-XBlue-CD14 reporter monocytes from the above NF-κB activity assay in 96-well plates, which were incubated at 37 °C (see above). After 2 h of incubation at 37 °C, we removed the supernatant and dissolved the blue formazan product generated in cells by the addition of 100 µL of DMSO (100%) in each well. The plates were then gently shaken for 10 min at room temperature to dissolve the precipitates. The absorbance was measured at 550 nm in a VICTOR3 Multilabel Plate Counter spectrofluorometer.

2.9. Mouse Model of Subcutaneous Inflammation and In Vivo Imaging

A mouse model of subcutaneous inflammation using NF-κB-RE-Luc reporter mice was used to study the immunomodulatory effects of APOE. The reporter mice carry a transgene containing six NF-κB-responsive elements and a modified firefly luciferase cDNA. The reporter gene is inducible by LPS and helps in in vivo studies of transcriptional regulation of the NF-κB gene. Mice were anesthetized using a mixture of 4% isoflurane and oxygen. Using a trimmer, hair was shaved from the back of the mouse and cleaned. Overnight culture of *P. aeruginosa* (PAO1) was refreshed and grown to mid-logarithmic phase in TH media. Bacteria were washed (5.6 × 1000 rpm, 15 min) and heat-killed for 30 min at 80 °C. Heat-killed bacteria (1×10^6 cfu/mouse) and APOE (2 µM) were mixed and injected immediately without preincubation. A total of 200 µL of the mixture was injected subcutaneously either on the left or the right side of the dorsum. In vivo inflammation was then longitudinally measured by imaging bioluminescence with IVIS Spectrum (PerkinElmer Life Sciences, Boston, MA, USA). Fifteen minutes before the in vivo imaging, mice were intraperitoneally given 100 µL of D-luciferin (150 mg/kg body weight). The IVIS imaging was performed 3 and 6 h after the subcutaneous injection. The data were acquired and analyzed using Living Image 4.0 Software (PerkinElmer). Five or six mice per treatment group were used. The animal model was previously described in [12].

2.10. Fluorescence Microscopy
Live/Dead Bacteria

E. coli and *P. aeruginosa* viability in the aggregates was assessed by using LIVE/DEAD® BacLight™ Bacterial Viability Kit (Invitrogen, Molecular Probes, Carlsbad, CA, USA). Bacterial suspensions were prepared as described above for VCA. Bacterial strains were treated with 2 µM APOE, 2 µM TCP-25, or 10 mM Tris at pH 7.4. After a 1 h incubation time at 37 °C, samples were mixed 1:1 with the dye mixture, followed by incubation for 15 min in the dark at room temperature. The dye mixture was prepared according to the manufacturer's protocol, i.e., 1.5 µL of component A (SYTO-9 green-fluorescent nucleic acid stain) and 1.5 µL of B (red-fluorescent nucleic acid stain propidium iodide) were dissolved in 1 mL of 10 mM Tris at pH 7.4. Five µL of stained bacterial suspension were trapped between a slide and an 18 mm square coverslip. Ten view fields (1 × 1 mm) were examined from three independent sample preparations using a Zeiss AxioScope A.1 fluorescence microscope (objectives: Zeiss EC Plan-Neofluar 40×; camera: Zeiss AxioCam MRm; acquisition software: Zeiss Zen 2.6 [blue edition]) [10].

2.11. Thioflavin T Assays (ThT)

Amyloid formation was determined using the dye Thioflavin T (ThT). Thioflavin T preferentially binds to the β-sheet structures of amyloidogenic proteins/peptides. For examination of the concentration dependence of the aggregation, we incubated APOE (2 µM) and LPS from *E. coli* (100 µg/mL) in buffer (10 mM Tris, pH 7.4) for 30 min at 37 °C before measurements. Two hundred microliters of the materials were incubated with 100 µM ThT for 15 min in the dark (ThT stock was 1 mM stored in the dark at 4 °C). We measured ThT fluorescence using a VICTOR3 Multilabel Plate Counter spectrofluorometer (PerkinElmer, Boston, MA, USA) at an excitation of 450 nm, with excitation and emission slit widths of 10 nm. The baseline (10 mM Tris pH 7.4 and LPS) was subtracted from the signal of each sample [12].

2.12. Transmission Electron Microscopy (TEM)

APOE was visualized using TEM (Jeol Jem 1230; Jeol, Tokyo, Japan) in combination with negative staining after incubation with LPS and lipid A or buffer. Images of endotoxins (100 µg/mL) in the presence or absence of APOE (2 µM), were taken after incubation for 30 min at 37 °C. For the mounted samples, ten view fields were examined on the grid (pitch 62 µm) from three independent sample preparations. Samples were adsorbed onto carbon-coated grids (Copper mesh, 400) for 60 s and stained with 7 µL of 2% uranyl acetate for 30 s. The grids were rendered hydrophilic via glow discharge at low air pressure. The size of aggregates was analyzed as the mean of gray value/µm ± standard deviation (SD) by ImageJ 1.52k, after all the images were converted to 8-bit and the threshold was manually adjusted [10].

2.13. Biacore Analysis

Binding experiments were carried out by using Biacore X100 (Cytiva Life Sciences, Uppsala, Sweden) with control software version of v.2.0. Sensor chip CM5 or SA (Cytiva Life Sciences, Uppsala, Sweden) were used as the gold surface for the immobilization and all the assays were carried out at 25 °C. An amine coupling kit (Cytiva Life Sciences, Uppsala, Sweden) which contained EDC [1-Ethyl-3-(3-dimethylamino-propyl)carbodiimide] (75 mg/mL), NHS (N-hydroxysuccinimide) (11.5 mg/mL) and ethanolamine (1 M, pH 8.5) or an biotin CAPture kit (Cytiva Life Sciences, Uppsala, Sweden) were used for the covalent immobilization of the ligands (APOE and LPS-biotin) on the gold surface.

Before starting the immobilization procedure, the chip (CM5 or SA) was docked into the instrument and the chip surface was activated following the EDC/NHS or/SA protocol with HBS-P or -EP buffer (0.01 M HEPES pH 7.4; 0.15 M NaCl; 0.005% *v/v* Surfactant P20) or as the running buffer. The ligand at a concentration of 0.01 mg/mL (in 10 mM acetate buffer, pH 5.0) was injected for 7 min (flow rate: 10 µL/min) followed by a 7 min (flow rate:

10 µL/min) injection of 1.0 M ethanolamine in order to deactivate excess reactive groups. The immobilization procedure was completed after the targeted immobilization level ($\approx$1200 RU for APOE and $\approx$600 RU for LPS-biotin) was reached. Only the flow channel_2 (active channel) was used for the ligand immobilization, while the flow channel_1 (reference channel) was used as a reference to investigate non-specific binding. The subtracted channel (flow channel_2–flow channel_1) was used to evaluate the results of the analysis.

Analytes (LPS/LipA/LTA or apolipoproteins) were injected into the active (Fc_2) and reference channels (Fc_1) in concentration ranges between 0 mg/mL and 2 mg/mL, respectively. Triplicate injections were carried out for each concentration. The association time was set to 120 s while the dissociation time was kept for 600 s. The flow rate was set to 10 µL/min and 10 mM glycine-HCl (pH 2.5) was used as the regeneration buffer. To evaluate the analysis, the parameters were determined using Biacore Evaluation Software v.2.0 in equilibrium binding analysis, which was performed by plotting the RU values measured in the plateau for each concentration and fitting the data to the steady state affinity [13,14].

2.14. Blue Native-Polyacrylamide Gel Electrophoresis and Western Blot

Twenty-one µL of APOE (2 µM) was mixed with either 10 mM Tris as control, endotoxins (100 µg/mL) or heparin sodium salt (500 µg/mL, Sigma-Aldrich, St. Louis, MO, USA). Samples were incubated for 30 min at 37 °C before mixing with loading buffer (4 $\times$ Loading Buffer Native Gel, cat#BN2003, Life Technologies, Carlsbad, CA, USA), and, subsequently, 28 µL was loaded onto 4–16% Bis-Tris Native Gels (cat#BN1002BOX, Life Technologies). Samples were run in parallel with a marker (Native Marker Unstained Protein Standard, cat#LC0725, Life Technologies) at 150 V for 100 min. Gels were run in duplicates for each experiment: one for gel analysis after de-staining from Coomassie and subsequent staining with Gel Code Blue Safe Protein (cat# 1860983, Thermo Scientific, Waltham, MA, USA), while the other was transferred to a 0.2 µm Polyvinylidene fluoride (PVDF) membranes (Trans Blot Transfer Pack, cat #1704156, Bio-Rad, Hercules, CA, USA) via a Trans Turbo Blot system (Bio-Rad). Thereafter, the membrane was de-stained with 70% ethanol and blocked with 5% milk in 1$\times$ PBS-Tween (PBS-T) for 30 min at room temperature. The membrane was incubated with mouse mAb anti-human APOE (cat#ab1906, Abcam, Cambridge, UK) at a concentration of 1 µg/mL, diluted in 1% fat-free milk in 1$\times$ PBS-T overnight at 4 °C. APOE and its high-molecular weight complexes were then detected using a secondary rabbit anti-mouse polyclonal antibody that was conjugated to horseradish peroxidase conjugate (HRP) (cat#P0260, Dako, Glostrup, Denmark) (diluted 1:1000 in 1$\times$ PBS-T complemented with 5% milk) after incubation for 60 min at room temperature. PBS-T was used to wash the membrane after each step (3 $\times$ 10 min), and the last wash after the secondary antibody was performed five times. The bands were revealed by incubating the membrane in the developing substrate (Super Signal West Pico PLUS Chemiluminescent Substrate, cat#34580, Thermo Scientific). The signal was acquired by a Chemi-Doc (Bio-Rad) system. All the experiments were performed at least three times [10].

2.15. Slot-Blot Assay

We used a slot-blot assay to detect the interaction between APOE and LPS. APOE (1 µg per well) was bound to the nitrocellulose membrane (Hybond-C, GE Healthcare (Chicago, IL, USA), Biosciences) after pre-soaking in 10 mM Tris, pH 7.4. The membrane was incubated in the blocking solution (2% BSA in PBS-T, pH 7.4) for 1 h at room temperature and subsequently incubated with 20 µg/mL biotinylated LPS (LPS-EB Biotin, InvivoGen, San Diego, CA, USA) for 1 h at room temperature. Next, the membranes were washed three times for 10 min in PBS-T and incubated with streptavidin-HRP (Thermo Scientific, Rockford, IL, USA). Binding was detected using peroxide solution and a luminol/enhancer solution (1:1 *v/v*) (SuperSignal West Pico Chemiluminescent Substrate, Thermo Scientific). To test for competitive inhibition of peptide binding to LPS, we also performed binding studies in the presence of unlabeled heparin (6 mg/mL). Signal was acquired by a Chemi-Doc (Bio-Rad) system. All the experiments were performed at least four times [15].

2.16. Blood Stimulation Assay and ELISA

Fresh venous blood was collected in the presence of lepirudin (50 mg/mL) from healthy donors. The blood was diluted 1:4 in RPMI-1640-GlutaMAX-I (Gibco) and 1 mL of this solution was transferred to 24-well plates and stimulated with 0.05 or 0.1 ng/mL LPS in the presence or the absence of APOE (10, 50 and 100 nM), APOAI (100 nM) or TCP-25 (1000 nM). After 24 h of incubation at 37 °C in 5% CO_2, the plate was centrifuged for 5 min at $1000 \times g$ and then the supernatants were collected and stored at −80 °C before analysis. The experiment was performed at least four times by using blood from different donors each time.

The cytokines TNF-α and IL-1β were measured in human plasma obtained after the blood stimulation experiment described above. The assay was performed by using human inflammation DuoSet® ELISA Kit (R&D Systems, Minneapolis, MN, USA) specific for each cytokine, according to the manufacturer's instructions. Absorbance was measured at a wavelength of 450 nm. Data shown are mean values ± SEM obtained from at least four independent experiments all performed in duplicate.

2.17. Molecular Dynamics Simulations

The initial structure of APOE was obtained from the protein data bank (PDB: 2L7B [16]). All systems were prepared using the CHARMM-GUI web server [17]. Either the full-length proteins (residues 1–299) or the truncated variants (residues 1–167) were modelled. Hydrogen atoms were added to the protein, while assuming neutral pH. The CHARMM36m [18] forcefield with TIP3P explicit water model [19] were used. In all systems, the protein was placed in the center of a cubic box of dimensions $9.5 \times 9.5 \times 9$ nm^3. Approximately 25,000 water molecules were added to each system; magnesium chloride salt was included to a concentration of ~150 mM, whilst neutralizing the overall system charge. For simulations in the presence of multiple lipid or heparin molecules, these were randomly placed around the surface of APOE, corresponding to twenty, five, or five molecules of heparin, lipid A or LPS, respectively. Blind docking of a single molecule of heparin or LPS was also performed using Vina-Carb [20], while AutoDock Vina 1.0 was used for lipid A [21], and, in each case, the top scoring pose was chosen for subsequent simulations. The heparin molecule was built as a tetrasaccharide 2-O-sulfo-±-L-iduronic acid 2-deoxy-2-sulfamido-±-D-glucopyranosyl-6-O-sulfate [22] using CHARMM-GUI Glycan Reader and Modeler [23]. Lipid A and LPS corresponded to the most frequently occurring species present in *E. coli*.

Energy minimization was performed for each system using steepest descents for ≤5000 steps with a 0.01 nm step size. Equilibration protocols were performed using simulations in the NVT followed by NPT ensembles for a total 50 ns with position restraints applied to protein backbone atoms. All unrestrained production simulations were run for 2 µs each in the NPT ensemble, using GROMACS2018 [24]. Simulations of full length APOE3 were run in triplicate. Equations of motion were integrated via the Verlet leapfrog algorithm with a 2 fs time step. All bonds connected to hydrogens were constrained with the LINCS algorithm. The cutoff distance was 1.2 nm for the short-range neighbor list and van der Waals interactions. The Particle Mesh Ewald method [25] was applied for long-range electrostatic interactions with a 1.2 nm real-space cutoff. The velocity rescaling thermostat was used to maintain the temperature at 310 K. Pressure was maintained at 1 bar using the Parrinello-Rahman barostat [26]. Simulations were performed on an in-house Linux cluster composed of 8 nodes containing 2 GPUs (Nvidia GeForce RTX 2080 Ti) and 24 CPUs (Intel® Xeon® Gold 5118 CPU @ 2.3 GHz) each. All snapshots were generated using VMD [27].

2.18. Statistical Analysis

The graphs of VCA, K_D constants (from SPR), ThT assay, TEM analysis, SPR analyses, BN gel analyses and slot blot analyses are presented as the mean ± SEM from at least three independent experiments. We assessed differences in these assays using one-way ANOVA with Dunnett's multiple comparison tests or *t*-test. All data were analyzed using GraphPad Prism (GraphPad Software, Inc., La Jolla, CA, USA). Additionally, *p*-values less than 0.05 were considered to be statistically significant (* $p < 0.05$, ** $p < 0.01$, *** $p < 0.001$, and **** $p < 0.0001$).

3. Results

3.1. Antibacterial Activity of APOE In Vitro

The antimicrobial activity of APOE (2 µM) against several bacterial strains of *P. aeruginosa* or *E. coli* was analyzed using a viable count assay (VCA). We detected a significant reduction in bacterial growth, which was more than 100 folds for all *P. aeruginosa* strains (27853, 15159 and PAO1) (Figure 1A). Additionally, we observed approximately 50% of the bacterial killing of both *E. coli* strains (25922 and 700928) (Figure 1B). The bacterial killing efficiencies of APOE against *P. aeruginosa* and *E. coli* were compared to the thrombin-derived antimicrobial peptide TCP-25, which was used here as a positive control [28].

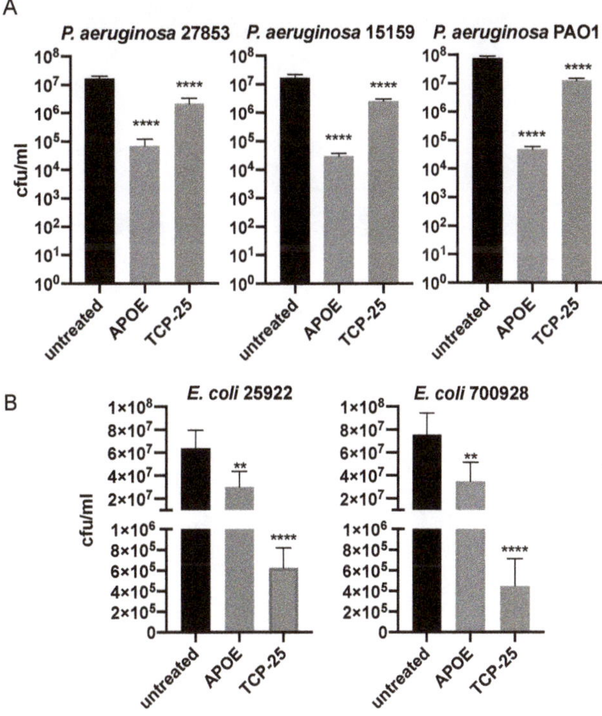

Figure 1. Antimicrobial activity of APOE in vitro. *P. aeruginosa* (**A**) and *E. coli* (**B**) were incubated with 2 µM APOE for 2 h and then viable count assay was performed. Thrombin-derived peptide 2 µM TCP-25 was used as a positive control, whereas the untreated bacteria were used as negative control. Data are presented as the mean ± SEM of four independent experiments (*n* = 4). Statistical analysis was performed using a one-way ANOVA with Dunnett's multiple comparison tests, ** = $p \leq 0.01$ and **** = $p \leq 0.0001$.

The data from VCA were confirmed by utilizing a live/dead imaging assay. The results revealed aggregation and dead bacteria upon treatment with 2 µM of APOE (Figure 2).

Dead bacteria with disturbed membrane integrity were stained red and live bacteria with intact membranes were stained green.

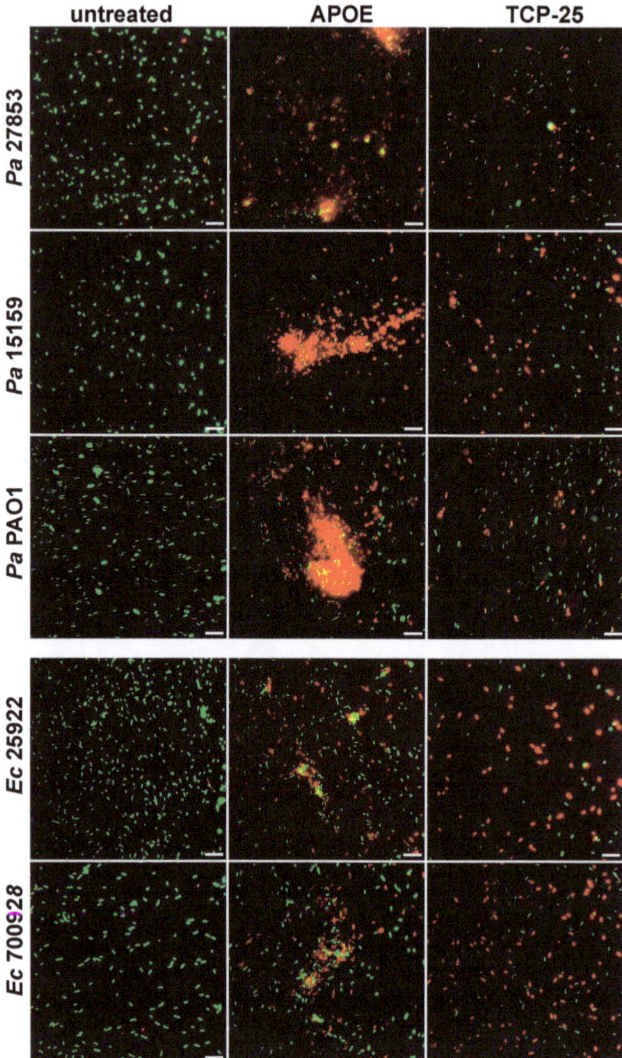

Figure 2. Visualization of *E. coli* and *P. aeruginosa* viability. Live/Dead Viability Assay of Gramnegative bacteria stimulated with 2 μM APOE, 10 mM Tris buffer at pH 7.4 (negative control) or 2 μM TCP-25 (positive control). Representative images from three independent experiments are presented (n = 4). At least ten individual fields were acquired per experiment. Live bacteria were stained with green SYTO 9 nucleic acid fluorescent dye, and dead bacteria were stained using red propidium iodine dye. Scale bar represents 5 μm.

3.2. Interaction of APOE with Bacterial Products

Next, we investigated the interaction of APOE with LPS and lipid A from *E. coli* by using the surface plasmon resonance system from Biacore. Lipid A is a lipid component of LPS, which anchors the endotoxin to the outer membrane of Gram-negative bacteria. The molecular interactions between APOE-immobilized sensor chip and LPS/lipid A were analyzed from binding curves (response vs. time) and K_D constants were calculated by the Biacore Evaluation Software v.2.0. The K_D constant for LPS was 53.7 $\pm$ 5.1 μM and for lipid A was 96.17$\pm$ 17.8 μM (Figure 3A,B) The calibration curves corresponding to the interaction of APOE with LPS or lipid A are shown in Supplementary Figure S1A. Moreover, we validated the binding specificity of APOE by using LTA, which is the main constituent of the cell wall of Gram-positive bacteria *S. aureus*, as a negative control. No interaction was detected between APOE and LTA under the same conditions as in the previous experiment with LPS and lipid A (Figure S1B).

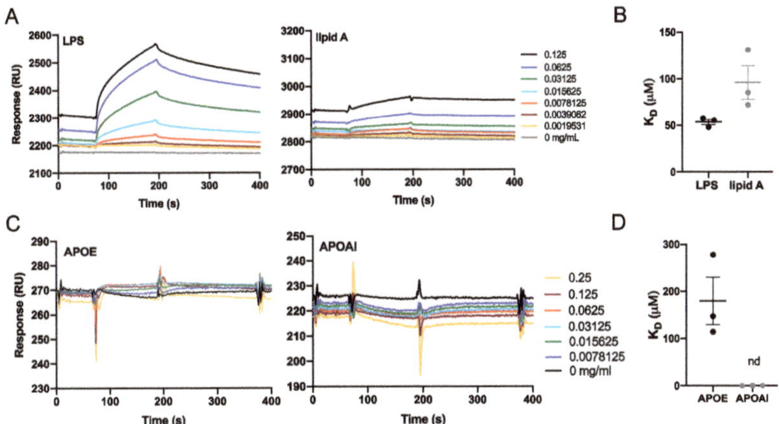

Figure 3. Biacore binding analysis. Binding interactions between the analytes and immobilized APOE protein or LPS. (**A,B**) Sensorgrams that show the response unit plotted as a function of time for LPS and lipid A or for APOE and APOAI. (**C,D**) Dissociation constants (K_D) of protein–ligand interactions were calculated from the sensorgrams (nd = not detected).

To further verify the specific interaction of APOE with LPS, we immobilized LPS on the sensor chip and investigated the interaction of LPS with APOE and apolipoprotein AI (APOAI). APOAI is the major protein component of high-density lipoprotein particles, which play important roles in lipid metabolism. No binding interaction was observed between LPS and APOAI. The K_D constant of APOE was calculated from the binding curves (K_D 179.7 $\pm$ 50.1 μM). On the other hand, the K_D of APOAI could not be calculated due to the lack of affinity between the binding partners (Figure 3C,D).

3.3. Bacterial Products Induce the Formation of APOE Aggregates

To further investigate the binding capability of APOE to LPS or lipid A from *E. coli*, we used TEM. Analysis of the TEM images revealed that aggregate-like complexes of APOE were formed after exposure to both LPS and lipid A (Figure 4A,B). To verify the protein specificity of APOE, we carried out the TEM experiment with APOAI under the same conditions as above. TEM images analysis confirmed that APOAI did not form aggregate-like complexes in the presence of LPS (Figure S2). Furthermore, we observed an increase in APOE aggregation upon LPS challenge by detecting a significant increase in ThT fluorescence, which is suggestive of an increase in beta-sheet structural features typical of aggregating proteins (Figure 4C).

3.4. Heparin Blocks Interaction between APOE and LPS

Additionally, we confirmed that the interaction between APOE and LPS can be blocked by including heparin in the mixture, which bonded to the similar receptor-binding region of APOE as LPS. The blockage of the interaction was verified by blue Native gel, which was followed by a Western blot assay. We performed an image analysis of the most pronounced protein band on the WB membrane, which corresponded to the dimeric form of APOE (above 66 kDa). We observed that dimeric APOE is released from the APOE-LPS complex in the presence of heparin (Figure 5A). Moreover, we used a slot-blot assay to confirm that heparin was able to block APOE and LPS interaction. We observed a significant decrease in LPS-band intensity on the membrane additionally treated with heparin (Figure 5B).

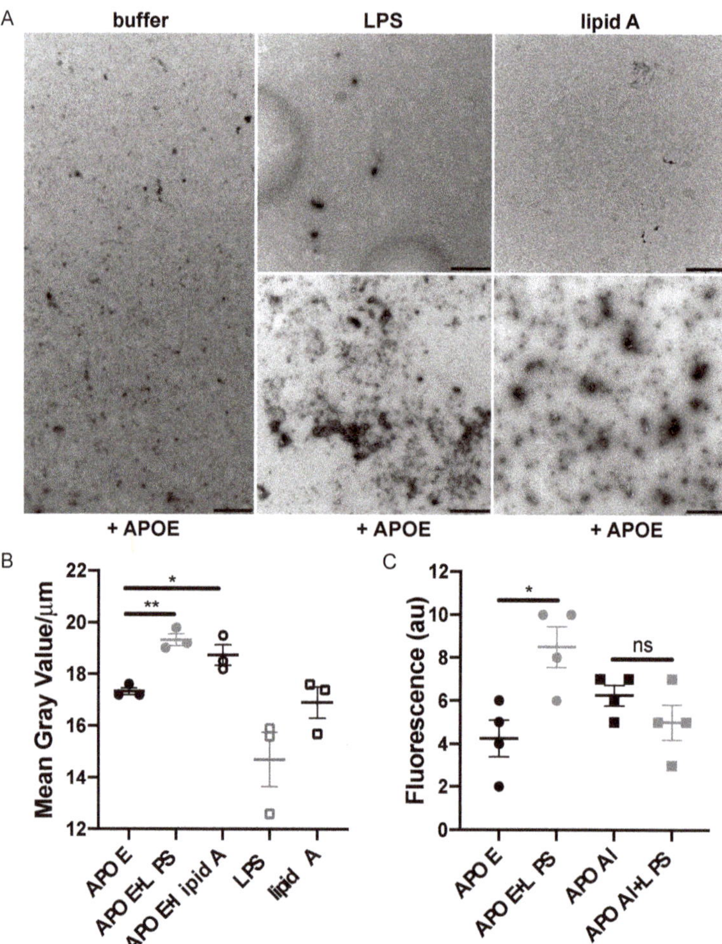

Figure 4. Detection of macromolecular complexes of APOE and bacterial products. (**A**) APOE (2 μM) was incubated with 100 μg/mL LPS or with lipid A from *E. coli* for 30 min at 37 °C. At the end of incubation, the macromolecular complexes were visualized by TEM. One representative image for each condition from three independent experiments is shown ($n = 3$). Scale bar represents 1 μm. (**B**) Analysis of the complexes of APOE and ligands following TEM. Quantification was performed using ImageJ 1.52k after all the images were converted to 8-bit, and the threshold was adjusted. The complexes of APOE and ligands are expressed as the mean of gray value/μm ± SEM. In the graph, each point represents at least ten pictures per each experiment ($n = 3$). Statistical analysis was performed using a one-way ANOVA with Dunnett's multiple comparison tests, * = $p \leq 0.05$ and ** = $p \leq 0.01$. (**C**) ThT assay demonstrates aggregation of APOE in the presence of LPS from *E. coli* ($n = 4$). Statistical analysis was performed using a one-way ANOVA with Dunnett's multiple comparison tests, * = $p \leq 0.05$ and ns = not significant. Black dot = APOE, gray dot = APOE + LPS, empty circle = APOE + lipid A, empty gray square = LPS, empty black square = lipid A, black square = APOAI and gray square = APOAI + LPS.

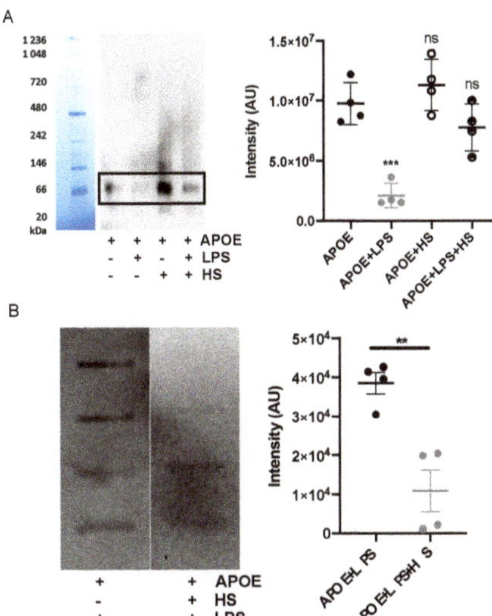

Figure 5. Heparin blocks interaction of APOE with LPS. (**A**) APOE (2 μM) was mixed with 10 mM Tris, 100 μg/mL LPS with or without heparin. Samples were incubated for 30 min at 37 °C and then run on Blue Native gel, which was followed by Western blot. One representative image from four independent experiments is shown (*n* = 4). The intensity of the bands representing dimer of APOE was measured. Statistical analysis was performed using a one-way ANOVA with Dunnett's multiple comparison tests, *** = $p \leq 0.001$ and ns = not significant. (**B**) A slot-blot assay demonstrated blocked binding of APOE (1 μg) to biotin-labeled LPS after the addition of heparin. Statistical analysis was performed using a *t*-test, ** = $p \leq 0.01$.

3.5. Molecular Dynamics Simulations of APOE and Its Interactions with Heparin, Lipid A and LPS

In order to gain molecular insights into the dynamics of APOE, as well as its interactions with heparin, lipid A and LPS binding, we performed a series of explicitly solvated all-atom MD simulations. We first carried out 2 μs simulations of isolated full-length wild type APOE, performed in triplicate to enhance conformational sampling. The C-terminal domain, which partially shields the N-terminal domain LDLR binding site in crystallographic structures, was observed to be highly dynamic, with its backbone root-mean square deviation (RMSD) fluctuating between ~1 and 2 nm (Figure S3A). This reflected the dissociation of the C-terminal domain from the N-terminal domain across all replicas, consistently leading to increased exposure of the highly positively charged LDLR receptor-binding region (Figure S3B), located on helix 4 containing multiple lysine and arginine amino acids (residues ~130–150). This was quantified by a measurement of the solvent accessible surface area (SASA) of this region, which rapidly increased from an initial value of ~2.5 nm^2 to a relatively constant value of ~10 nm^2 over the remainder of the simulation (Figure S3C).

In light of the above observations, and to improve sampling of ligand interactions with the N-terminal domain, all subsequent simulations with heparin, lipid A or LPS molecules were run with truncated APOE (residues 1–167). A series of 2 μs simulations were next performed in which different ligands were placed randomly in bulk water around APOE, in order to observe spontaneous assembly with functional regions of the protein surface. For comparison, blind docking of a single molecule of each ligand was also performed, and in each case the most frequently occurring pose involved the receptor binding domain; this pose was subsequently also used to initiate 2 μs simulations.

In the case of heparin, NMR studies showed that porcine heparin from intestinal mucosa is mostly composed of the trisulfated disaccharide: 2-O-sulfo-±-L-iduronic acid 2-deoxy-2-sulfamido-±-D-glucopyranosyl-6-O-sulfate [22], so a tetrasaccharide composed of two blocks of this were modeled. Over the course of the simulation, several heparin molecules spontaneously contacted the protein surface, with three of these tightly bound to the receptor binding region (Figure 6A). APOE residues which were involved in stable interactions with heparin for over 80% of the simulation time are shown in Figure 6B. These residues correspond to multiple basic amino acids located on: i) helix 4, with R134-R150 located in the receptor binding region along with K157-R158; ii) helix 3, including R90, R92, K95, R103 and R114; and iii) R167 at the C-terminus. In the alternative simulation initiated from a single docked heparin molecule, electrostatic interactions of heparin sulfate groups with the basic residues of helix 4 and R114 of helix 3 were observed to dominate (Figure S4A).

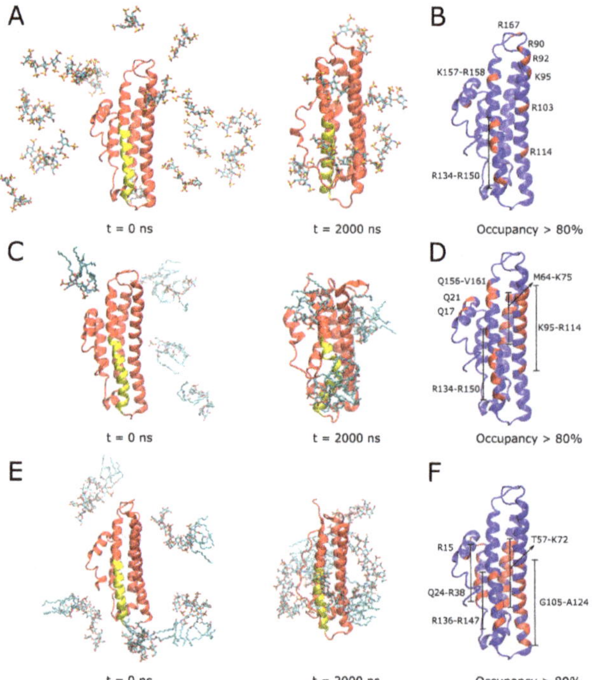

Figure 6. MD simulations to investigate binding modes of heparin, lipid A and LPS with APOE. In (**A,C,E**), initial and final simulation snapshots of systems containing truncated APOE with heparin, lipid A, and full LPS molecules are shown, respectively. In (**B,D,F**), APOE is shown in blue with a per-residue ligand occupancy (<0.3 nm cutoff distance) for >80% of total simulation sampling highlighted in red, for heparin, lipid A, and LPS systems, respectively. In (**A,C,E**), N-terminal domain is colored in red with receptor binding region on helix 4 colored in yellow. Ligand molecules are shown in CPK sticks representation (cyan—carbon; red—oxygen; blue—nitrogen; brown—phosphorus; sulfur—yellow). Simulation time points are indicated below snapshots, reported in nanoseconds (ns).

In the case of lipid A (Figure 6C, Movie S1) and LPS (Figure 6E, Movie S2), binding simulations, electrostatic interactions between lipid phosphates and basic residues of APOE were dominant, further supported by hydrophobic interactions with lipid acyl tails wrapping around the protein surface. APOE residues involved in stable interactions with lipid A for over 80% of the simulation time are shown in Figure 6D. Similarly to heparin, these included multiple basic amino acids located on helix 4 (residues 134–150 of the receptor binding region and Q156-V161) and helix 3 (K95-R114), in addition to helix 2 (M64-K75). The interaction region including multiple basic residues was even more extensive in the case of the larger full LPS molecule (Figure 6F), including helix 4 (residues R136-R147), helix 3

(G105-A124), helix 2 (R15, T57-K72) as well as helix 1 (Q24-R38). Simulations of a single lipid A (Figure S4B) or LPS (Figure S4C) docked to the APOE receptor binding region further supported the role of basic residues from this region, binding to ligand phosphates, along with some nonpolar amino acid sidechains stabilizing lipid A tails. It should be noted that the reported simulations aimed to provide a structural description of the interaction between APOE and bacterial products such as LPS molecules in isolation to complement the corresponding surface plasmon resonance experiments and are indicative of the protein's "scavenging" role and propensity for lipid-associated aggregation. Nevertheless, in the context of its antimicrobial activity, the observed dominance of electrostatics in our simulations is indicative of a potential binding mode to the Gram-negative outer membrane involving the initial approach of the receptor binding region to the anionic LPS moieties on the lipid bilayer surface.

3.6. Immunomodulatory Activity of APOE In Vitro

We next used reporter THP-1 monocytes to detect the effects on LPS-signaling by APOE. APOE (0.5–2 μM) significantly reduced the activation of NF-κB/AP-1 triggered by *E. coli* LPS (Figure 7A). The MTT viability assay did not show any significant cytotoxic effect of APOE on THP-1 cells, which suggests that the reduction in the NF-κB/AP-1 activation was due to the neutralizing effect of APOE on LPS and not by any APOE-mediated toxic effects on the cells (Figure 7A). TCP-25 peptide was used here as a positive control. Furthermore, we confirmed that APOAI did not reduce NF-κB activation and did not cause cell toxicity under similar experimental conditions as in the above experiment described for APOE (Figure S5).

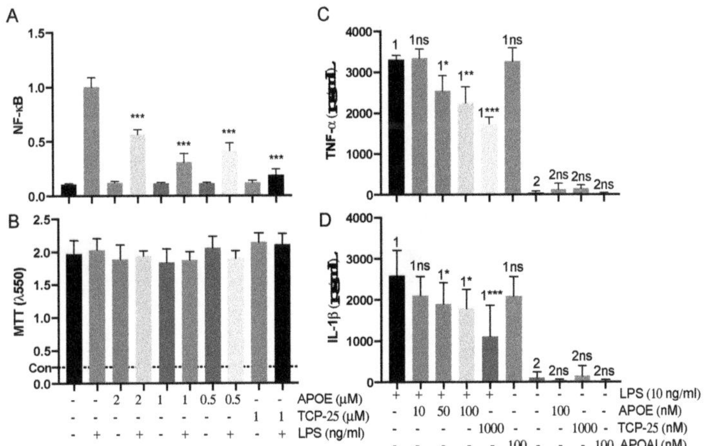

Figure 7. Anti-endotoxic effect of APOE in vitro. (**A**) THP-1-XBlue-CD14 cells were treated with APOE (0.5, 1 and 2 μM), LPS (10 ng/mL) from *E. coli*, or a combination of both. APOE yielded a significant reduction of activation of NF-κB/AP-1. *** $p \leq 0.001$. (**B**) MTT viability assay for analysis of toxic effects of APOE on THP-1 cells. The dotted line (con) represents positive control of dead cells. The mean values of five measurements ± their SEM are shown (*n* = 5). *p* values were determined using one-way ANOVA with Dunnett's multiple comparison test. Cytokine analysis of TNF-α (**C**) and IL-1ß (**D**) the blood collected from healthy donors at 24 h after treatment with 10 ng/mL LPS together with 10–50–100 nM APOE. Untreated blood was used as a baseline control. Blood treated with 10 ng/mL LPS together with 100 nM APOAI or 1000 nM TCP-25 were used for comparison. The mean ± SEM values of four independent experiments performed in duplicate are shown (*n* = 4). * = $p \leq 0.05$, ** = $p \leq 0.01$ and *** = $p \leq 0.001$, ns = not significant, determined using two-way ANOVA with Sidak's multiple comparisons test.

Additionally, we performed a blood stimulation assay to validate the immunomodulatory effect of APOE on cytokine release by human immune cells in the blood. Here, we confirmed that APOE significantly decreased proinflammatory cytokines levels of both TNF-α and IL-1ß. Moreover, we observed that APOAI did not significantly reduce the release of both TNF-α and IL-1ß cytokines under similar conditions (Figure 7B).

3.7. Immunomodulatory Activity of APOE In Vivo

We next investigated whether APOE could suppress local inflammation triggered by bacteria in vivo. For this experiment, we utilized the (NF-κB-RE-Luc)-Xen reporter mouse model and studied the effects of APOE on subcutaneous inflammation induced by heat-killed bacteria *P. aeruginosa* (PAO1). We used killed bacteria to exclude the possible confounding effects of bacterial growth in vivo after subcutaneous deposition of live bacteria. Heat-killed bacteria (1×10^6 cfu/mouse) were subcutaneously injected into the left or right side of mouse dorsum, either with or without 200 μL of APOE (2 μM). After injection of luciferin substrate, the luminescent signal, which corresponded to NF-κB activation, was detected by an in vivo bioimaging system (IVIS Spectrum) (Figure 8). IVIS image analysis showed a significant reduction in NF-κB activation after 3 and 6 h at the site challenged by both APOE and heat-killed bacteria, which was compared to heat killed bacteria-treatment alone. APOE alone did not yield any significant increase in NF-κB activation (Figure S6).

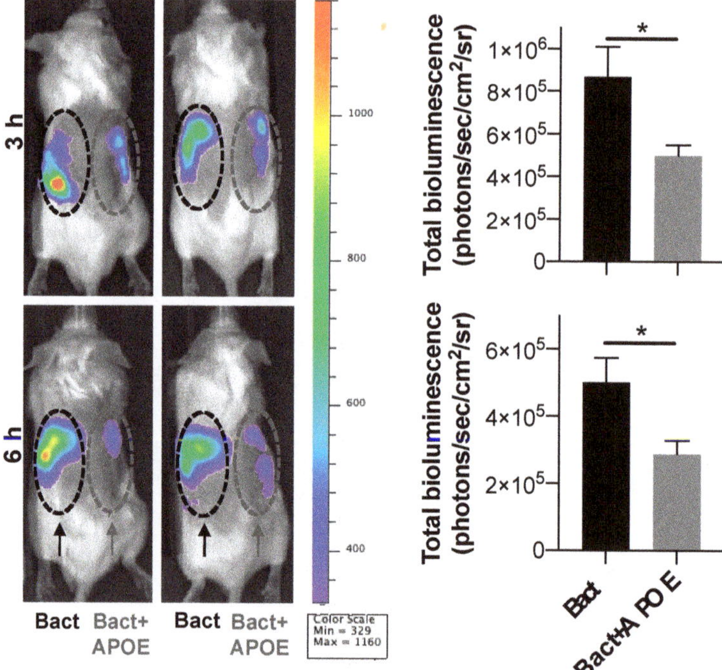

Figure 8. Anti-endotoxic effect of APOE in vivo. NF-κB activation in the NF-κB-RE-luc random transgenic mouse model was analyzed by the IVIS imaging. Heat-killed bacteria PAO1 (1×10^6 cfu) (left side of the dorsum) or heat-killed bacteria (1×10^6 cfu) treated with APOE (2 μM) (right side of the dorsum) were injected subcutaneously and the NF-κB response was longitudinally imaged 3 and 6 h after the subcutaneous injection. Representative images show bioluminescence at 3 and 6 h after subcutaneous injection. Bar charts show bioluminescence measured from these reporter mice. Data are presented as the mean ± SEM (n = 5). p value was determined using an unpaired t-test. * = $p \leq 0.05$.

4. Discussion

In this study, we extended our previous work on the antimicrobial effect of full-length APOE. We have reported that APOE has a strong killing ability against only Gram-negative bacteria in vitro and in vivo [10]. Here, we confirmed the aggregation ability and antimicrobial activity of APOE against several Gram-negative laboratory or clinical bacterial strains of *P. aeruginosa* and *E. coli* at physiological plasma levels of protein, which are in healthy subjects between 3 and 7 mg/mL (proximately 0.9–2 μM) [29].

APOE is secreted from many cells throughout the human body, such as hepatic parenchymal cells, monocytes, macrophages, adipocytes, muscle and skin cells [30–32]. Although most of plasma APOE is lipid-bound, APOE also exists in lipid-free or lipid-poor forms. Lipid-free/poor apolipoproteins are generated in vivo by dissociation from the surface of lipoproteins or new protein synthesis [33]. Additionally, APOE is expressed in skin cells and possibly plays an important part in our innate defense system [30]. Although the exact role of lipid-poor/free APOE in the skin is not fully known, our previously published findings suggest that APOE may have a neutralizing effect against Gram-negative bacteria and their endotoxins [10].

It has been reported that peptides derived from the receptor-binding region of apolipoprotein E have an antimicrobial effect against Gram-positive and Gram-negative bacteria at the physiological protein level [34]. The antibacterial [9,35], anti-inflammatory [8], and LPS-binding [31] effects of APOE-derived peptides were described in the literature and summarized in our previous publication [10]. We extended our previous report by detecting the direct affinity of APOE to Gram-negative bacterial endotoxins using surface plasma resonance, which was complemented by MD to understand the interaction of APOE and endotoxin on a molecular level. Data obtained from our MD simulations support the role of multiple positively charged residues in the N-terminal domain in binding heparin and LPS, particularly located in the receptor binding region within helix 4 which are accessible on the APOE surface and not involved in intramolecular salt bridges.

Both antimicrobial and aggregating protein regions are usually rich in hydrophobic residues. Protein aggregation results in the self-assembly of misfolded proteins, which exhibit changes in the secondary structure. Aggregation of proteins has been implicated in many diseases such as amyloidosis, Alzheimer's, and Parkinson's disease. The mechanism behind protein aggregation is not still fully understood. Multiple lines of evidence point to inflammation, which may be curtailed in amyloidogenic processes [36]. APOE could be connected to the formation of aggregates, which is triggered by inflammatory activity, as a response to bacterial infection [37].

Toll-like receptor 4 (TLR4) agonists such as LPS and lipid A interact with apolipoprotein E by hydrophobic and electrostatic interactions, which cause protein aggregation. Our data suggest that protein aggregation is a key element in blocking LPS binding to TLR hydrophobic pockets, which leads to reduced LPS-triggered inflammatory signaling. APOE reduced NF-κB activation in human monocytes in vitro and in NF-κB-RE-luc mice in vivo, which indicates scavenging of LPS. Moreover, the specific and strong LPS-binding affinity of APOE compared to APOAI can be explained by previous reports showing that APOAI binds directly to LPS only via weak hydrophobic interactions. Additionally, the indirect LPS-binding affinity of APOAI was shown to be via LPS binding proteins [38].

Antimicrobial resistance is a serious threat to global public health that requires the acute need for the development of new therapeutic strategies [39–41]. The endogenous mechanism, by which pro-aggregation proteins such as APOE promote the containment of LPS-induced inflammation, is illustrated in Figure 9. We have previously described a similar mechanism of endogenous protein aggregation of thrombin C-terminal peptides trigged by bacterial endotoxins [12,15].

Here, we show that APOE (mostly APOE3 isoform) is able to scavenge LPS and dampen proinflammatory cytokines released from the human immune cell line, primary blood cells and animal models. Our observation is in good agreement with previously published in vivo studies on APOE-deficient mice, which show increased susceptibility to infection caused by Gram-negative bacteria or fungi [42]. We also showed apolipoprotein E specificity to Gram-negative bacteria and endotoxins compared to apolipoprotein AI.

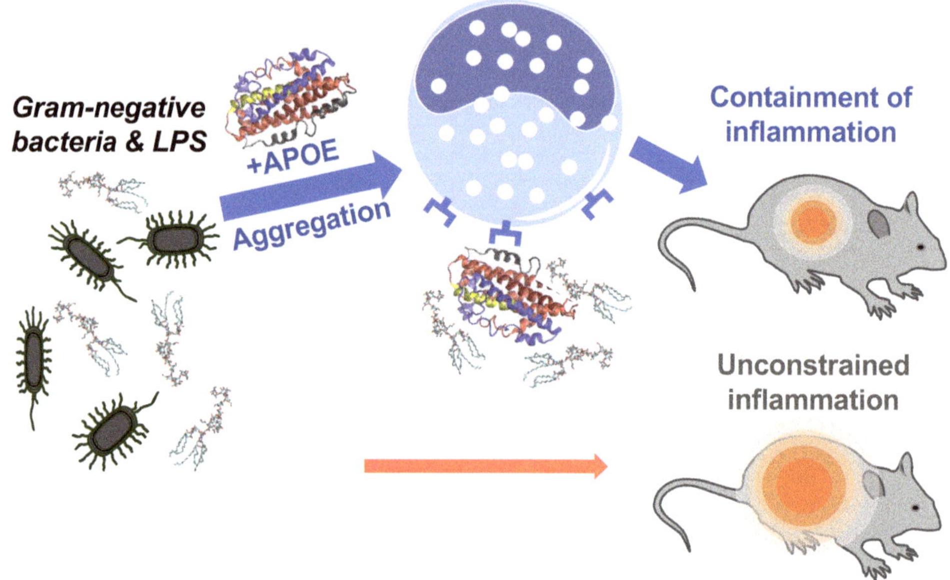

Figure 9. Summary of APOE function. Neutralizing effect of APOE on Gram-negative bacterial infection in vivo.

5. Conclusions

In conclusion, we have demonstrated that lipid-poor/free APOE has antimicrobial activity against several Gram-negative bacterial strains of *E. coli* and *P. aeruginosa*. Additionally, we showed that APOE has a stronger killing ability for all strains of *P. aeruginosa*, whereas *E. coli* was less sensitive to the antimicrobial actions of APOE. The apolipoprotein interaction with endotoxins was specific to APOE protein and Gram-negative bacteria. The further investigation of the APOE interaction with endotoxins from Gram-negative bacteria led to the discovery of protein aggregation. Taken together, these results suggest that multifunctional APOE has an additional role in innate immunity during bacterial infection.

6. Patents

The peptide TCP-25 and variants are patent protected.

Supplementary Materials: The following supporting information can be downloaded at: https://www.mdpi.com/article/10.3390/biomedicines10061430/s1, Figure S1: title Biacore binding curves and negative control; Figure S2: Visualization of macromolecular complexes of APOAI and LPS; Figure S3: Molecular dynamics simulations of full-length APOE in explicit solvent; Figure S4: Molecular dynamics simulations of a docked ligand to a truncated APOE; Figure S5: Anti-endotoxic effect of APOAI in vitro; Figure S6: Immunomodulatory effect of APOE in vivo; Movie S1: MD simulations of five lipid A molecules binding truncated APOE; Movie S2: MD simulations of five LPS molecules binding truncated APOE.

Author Contributions: J.P. and P.J.B. conceived the project and designed the experiments. J.P. performed the in vitro antimicrobial experiments, biding assays, cells assay, fluorescence microscopy, SPR experiment and electron microscopy. M.P. performed and analyzed the in vivo experiments. J.K.M. performed in silico analysis. G.P. performed ELISA assay. G.E.B. performed the SPR analyses.

J.P. and M.P. wrote the manuscript, with contributions from J.K.M. and P.J.B. All of the authors discussed the results and commented on the final manuscript. All authors have read and agreed to the published version of the manuscript.

Funding: This work was supported by grants from the Swedish Research Council (projects 2012-1883 and 2017-02341); the Welander-Finsen, Crafoord, Österlund, Johanssons, Hedlunds, Kockska, LEO and Söderberg Foundations; the Swedish Foundation for Strategic Research; the Knut and Alice Wallenberg Foundation; The Swedish Government Funds for Clinical Research (ALF); and the Medical Faculty of Lund University. P.J.B. and J.K.M. were supported by BII (A*STAR) core funds.

Institutional Review Board Statement: All animal experiments were performed according to the Swedish Animal Welfare Act SFS 1988:534 and were approved by the Animal Ethics Committee of Malmö/Lund, Sweden (permit numbers M5934-19 and M8871-19). Animals were kept under standard conditions of light and temperature and water ad libitum.

Informed Consent Statement: Not applicable.

Data Availability Statement: Data is contained within the article and Supplementary Material.

Acknowledgments: We thank Artur Schmidtchen for donating TCP-25 peptide. We thank Ann-Charlotte Strömdahl for excellent technical assistance. We thank Rolf Lood for letting us use his Biacore instrument. We acknowledge the Lund University Bioimaging Centrum (LBIC) for access to electron microscopy facilities.

Conflicts of Interest: The authors declare no conflict of interest. The funders had no role in the design of the study; in the collection, analyses, or interpretation of data; in the writing of the manuscript, or in the decision to publish the results.

References

1. White, C.R.; Garber, D.W.; Anantharamaiah, G.M. Anti-inflammatory and cholesterol-reducing properties of apolipoprotein mimetics: A review. *J. Lipid Res.* **2014**, *55*, 2007–2021. [CrossRef] [PubMed]
2. Petrlova, J.; Hong, H.S.; Bricarello, D.A.; Harishchandra, G.; Lorigan, G.A.; Jin, L.W.; Voss, J.C. A differential association of Apolipoprotein E isoforms with the amyloid-beta oligomer in solution. *Proteins* **2011**, *79*, 402–416. [CrossRef] [PubMed]
3. Phillips, M.C. Apolipoprotein E isoforms and lipoprotein metabolism. *IUBMB Life* **2014**, *66*, 616–623. [CrossRef] [PubMed]
4. Laskowitz, D.T.; Thekdi, A.D.; Thekdi, S.D.; Han, S.K.; Myers, J.K.; Pizzo, S.V.; Bennett, E.R. Downregulation of microglial activation by apolipoprotein E and apoE-mimetic peptides. *Exp. Neurol.* **2001**, *167*, 74–85. [CrossRef]
5. Saito, H.; Dhanasekaran, P.; Nguyen, D.; Baldwin, F.; Weisgraber, K.H.; Wehrli, S.; Phillips, M.C.; Lund-Katz, S. Characterization of the heparin binding sites in human apolipoprotein E. *J. Biol. Chem.* **2003**, *278*, 14782–14787. [CrossRef]
6. Azuma, M.; Kojimab, T.; Yokoyama, I.; Tajiri, H.; Yoshikawa, K.; Saga, S.; Del Carpio, C.A. A synthetic peptide of human apoprotein E with antibacterial activity. *Peptides* **2000**, *21*, 327–330. [CrossRef]
7. Croy, J.E.; Brandon, T.; Komives, E.A. Two apolipoprotein E mimetic peptides, ApoE(130–149) and ApoE(141–155)2, bind to LRP1. *Biochemistry* **2004**, *43*, 7328–7335. [CrossRef]
8. Laskowitz, D.T.; Fillit, H.; Yeung, N.; Toku, K.; Vitek, M.P. Apolipoprotein E-derived peptides reduce CNS inflammation: Implications for therapy of neurological disease. *Acta Neurol. Scand. Suppl.* **2006**, *185*, 15–20. [CrossRef]
9. Pane, K.; Sgambati, V.; Zanfardino, A.; Smaldone, G.; Cafaro, V.; Angrisano, T.; Pedone, E.; Di Gaetano, S.; Capasso, D.; Haney, E.F.; et al. A new cryptic cationic antimicrobial peptide from human apolipoprotein E with antibacterial activity and immunomodulatory effects on human cells. *FEBS J.* **2016**, *283*, 2115–2131. [CrossRef]
10. Petruk, G.; Elven, M.; Hartman, E.; Davoudi, M.; Schmidtchen, A.; Puthia, M.; Petrlova, J. The role of full-length apoE in clearance of Gram-negative bacteria and their endotoxins. *J. Lipid Res.* **2021**, *62*, 100086. [CrossRef]
11. Puthia, M.; Butrym, M.; Petrlova, J.; Stromdahl, A.C.; Andersson, M.A.; Kjellstrom, S.; Schmidtchen, A. A dual-action peptide-containing hydrogel targets wound infection and inflammation. *Sci. Transl. Med.* **2020**, *12*, eaax6601. [CrossRef] [PubMed]
12. Petrlova, J.; Petruk, G.; Huber, R.G.; McBurnie, E.W.; van der Plas, M.J.A.; Bond, P.J.; Puthia, M.; Schmidtchen, A. Thrombin-derived C-terminal fragments aggregate and scavenge bacteria and their proinflammatory products. *J. Biol. Chem.* **2020**, *295*, 3417–3430. [CrossRef] [PubMed]
13. Chowdhury, S.; Khakzad, H.; Bergdahl, G.E.; Lood, R.; Ekstrom, S.; Linke, D.; Malmstrom, L.; Happonen, L.; Malmstrom, J. Streptococcus pyogenes Forms Serotype- and Local Environment-Dependent Interspecies Protein Complexes. *mSystems* **2021**, *6*, e0027121. [CrossRef] [PubMed]
14. Khakzad, H.; Happonen, L.; Karami, Y.; Chowdhury, S.; Bergdahl, G.E.; Nilges, M.; Tran Van Nhieu, G.; Malmstrom, J.; Malmstrom, L. Structural determination of Streptococcus pyogenes M1 protein interactions with human immunoglobulin G using integrative structural biology. *PLoS Comput. Biol.* **2021**, *17*, e1008169. [CrossRef] [PubMed]

15. Petrlova, J.; Hansen, F.C.; van der Plas, M.J.A.; Huber, R.G.; Morgelin, M.; Malmsten, M.; Bond, P.J.; Schmidtchen, A. Aggregation of thrombin-derived C-terminal fragments as a previously undisclosed host defense mechanism. *Proc. Natl. Acad. Sci. USA* **2017**, *114*, E4213–E4222. [CrossRef]

16. Chen, J.; Li, Q.; Wang, J. Topology of human apolipoprotein E3 uniquely regulates its diverse biological functions. *Proc. Natl. Acad. Sci. USA* **2011**, *108*, 14813–14818. [CrossRef]

17. Jo, S.; Kim, T.; Iyer, V.G.; Im, W. CHARMM-GUI: A web-based graphical user interface for CHARMM. *J. Comput. Chem.* **2008**, *29*, 1859–1865. [CrossRef]

18. Huang, J.; Rauscher, S.; Nawrocki, G.; Ran, T.; Feig, M.; de Groot, B.L.; Grubmuller, H.; MacKerell, A.D., Jr. CHARMM36m: An improved force field for folded and intrinsically disordered proteins. *Nat. Methods* **2017**, *14*, 71–73. [CrossRef]

19. Jorgensen, W.L.; Chandrasekhar, J.; Madura, J.D.; Impey, R.W.; Klein, M.L. Comparison of Simple Potential Functions for Simulating Liquid Water. *J. Chem. Phys.* **1983**, *79*, 926–935. [CrossRef]

20. Nivedha, A.K.; Thieker, D.F.; Makeneni, S.; Hu, H.; Woods, R.J. Vina-Carb: Improving Glycosidic Angles during Carbohydrate Docking. *J. Chem. Theory Comput.* **2016**, *12*, 892–901. [CrossRef]

21. Trott, O.; Olson, A.J. AutoDock Vina: Improving the speed and accuracy of docking with a new scoring function, efficient optimization, and multithreading. *J. Comput. Chem.* **2010**, *31*, 455–461. [CrossRef] [PubMed]

22. Aquino, R.S.; Pereira, M.S.; Vairo, B.C.; Cinelli, L.P.; Santos, G.R.; Fonseca, R.J.; Mourao, P.A. Heparins from porcine and bovine intestinal mucosa: Are they similar drugs? *Thromb. Haemost.* **2010**, *103*, 1005–1015. [CrossRef] [PubMed]

23. Park, S.J.; Lee, J.; Qi, Y.; Kern, N.R.; Lee, H.S.; Jo, S.; Joung, I.; Joo, K.; Lee, J.; Im, W. CHARMM-GUI Glycan Modeler for modeling and simulation of carbohydrates and glycoconjugates. *Glycobiology* **2019**, *29*, 320–331. [CrossRef] [PubMed]

24. Van Der Spoel, D.; Lindahl, E.; Hess, B.; Groenhof, G.; Mark, A.E.; Berendsen, H.J. GROMACS: Fast, flexible, and free. *J. Comput. Chem.* **2005**, *26*, 1701–1718. [CrossRef]

25. Simmonett, A.C.; Brooks, B.R. A compression strategy for particle mesh Ewald theory. *J. Chem. Phys.* **2021**, *154*, 054112. [CrossRef] [PubMed]

26. Parrinello, M.; Rahman, A. Polymorphic Transitions in Single-Crystals—A New Molecular-Dynamics Method. *J. Appl. Phys.* **1981**, *52*, 7182–7190. [CrossRef]

27. Humphrey, W.; Dalke, A.; Schulten, K. VMD: Visual molecular dynamics. *J. Mol. Graph.* **1996**, *14*, 33–38. [CrossRef]

28. Papareddy, P.; Rydengard, V.; Pasupuleti, M.; Walse, B.; Morgelin, M.; Chalupka, A.; Malmsten, M.; Schmidtchen, A. Proteolysis of human thrombin generates novel host defense peptides. *PLoS Pathog.* **2010**, *6*, e1000857. [CrossRef]

29. Kaneva, A.M.; Bojko, E.R.; Potolitsyna, N.N.; Odland, J.O. Plasma levels of apolipoprotein-E in residents of the European North of Russia. *Lipids Health Dis.* **2013**, *12*, 43. [CrossRef]

30. Barra, R.M.; Fenjves, E.S.; Taichman, L.B. Secretion of apolipoprotein E by basal cells in cultures of epidermal keratinocytes. *J. Investig. Dermatol.* **1994**, *102*, 61–66. [CrossRef]

31. Zhu, Y.; Kodvawala, A.; Hui, D.Y. Apolipoprotein E inhibits toll-like receptor (TLR)-3- and TLR-4-mediated macrophage activation through distinct mechanisms. *Biochem. J.* **2010**, *428*, 47–54. [CrossRef] [PubMed]

32. Gordon, D.A.; Fenjves, E.S.; Williams, D.L.; Taichman, L.B. Synthesis and secretion of apolipoprotein E by cultured human keratinocytes. *J. Investig. Dermatol.* **1989**, *92*, 96–99. [CrossRef] [PubMed]

33. Crouchet, E.; Lefevre, M.; Verrier, E.R.; Oudot, M.A.; Baumert, T.F.; Schuster, C. Extracellular lipid-free apolipoprotein E inhibits HCV replication and induces ABCG1-dependent cholesterol efflux. *Gut* **2017**, *66*, 896–907. [CrossRef] [PubMed]

34. Zanfardino, A.; Bosso, A.; Gallo, G.; Pistorio, V.; Di Napoli, M.; Gaglione, R.; Dell'Olmo, E.; Varcamonti, M.; Notomista, E.; Arciello, A.; et al. Human apolipoprotein E as a reservoir of cryptic bioactive peptides: The case of ApoE 133–167. *J. Pept. Sci.* **2018**, *24*, e3095. [CrossRef]

35. Dobson, C.B.; Sales, S.D.; Hoggard, P.; Wozniak, M.A.; Crutcher, K.A. The receptor-binding region of human apolipoprotein E has direct anti-infective activity. *J. Infect. Dis.* **2006**, *193*, 442–450. [CrossRef]

36. Kunjithapatham, R.; Oliva, F.Y.; Doshi, U.; Pérez, M.; Avila, J.; Muñoz, V. Role for the alpha-helix in aberrant protein aggregation. *Biochemistry* **2005**, *44*, 149–156. [CrossRef]

37. Moir, R.D.; Lathe, R.; Tanzi, R.E. The antimicrobial protection hypothesis of Alzheimer's disease. *Alzheimers Dement.* **2018**, *14*, 1602–1614. [CrossRef]

38. Ma, J.; Liao, X.L.; Lou, B.; Wu, M.P. Role of apolipoprotein A-I in protecting against endotoxin toxicity. *Acta Biochim. Biophys. Sin.* **2004**, *36*, 419–424. [CrossRef]

39. Laupland, K.B.; Church, D.L. Population-based epidemiology and microbiology of community-onset bloodstream infections. *Clin. Microbiol. Rev.* **2014**, *27*, 647–664. [CrossRef]

40. Marston, H.D.; Dixon, D.M.; Knisely, J.M.; Palmore, T.N.; Fauci, A.S. Antimicrobial Resistance. *JAMA* **2016**, *316*, 1193–1204. [CrossRef]

41. Llewelyn, M.J.; Fitzpatrick, J.M.; Darwin, E.; SarahTonkin, C.; Gorton, C.; Paul, J.; Peto, T.E.A.; Yardley, L.; Hopkins, S.; Walker, A.S. The antibiotic course has had its day. *BMJ* **2017**, *358*, j3418. [CrossRef] [PubMed]

42. Vonk, A.G.; De Bont, N.; Netea, M.G.; Demacker, P.N.; van der Meer, J.W.; Stalenhoef, A.F.; Kullberg, B.J. Apolipoprotein-E-deficient mice exhibit an increased susceptibility to disseminated candidiasis. *Med Mycol.* **2004**, *42*, 341–348. [CrossRef] [PubMed]

Article

High Level Expression and Purification of Cecropin-like Antimicrobial Peptides in *Escherichia coli*

Chih-Lung Wu [1,†], Ya-Han Chih [1,†], Hsin-Ying Hsieh [1], Kuang-Li Peng [1], Yi-Zong Lee [2], Bak-Sau Yip [1,3], Shih-Che Sue [2,*] and Jya-Wei Cheng [1,*]

[1] Department of Medical Science, Institute of Biotechnology, National Tsing Hua University, Hsinchu 300, Taiwan; s103080578@m103.nthu.edu.tw (C.-L.W.); s9980517@m99.nthu.edu.tw (Y.-H.C.); s109080583@m109.nthu.edu.tw (H.-Y.H.); richard850210@gapp.nthu.edu.tw (K.-L.P.); g15004@hch.gov.tw (B.-S.Y.)
[2] Institute of Bioinformatics and Structural Biology, National Tsing Hua University, Hsinchu 300, Taiwan; yzlee@gapp.nthu.edu.tw
[3] Department of Neurology, National Taiwan University Hospital Hsinchu Branch, Hsinchu 300, Taiwan
[*] Correspondence: scsue@life.nthu.edu.tw (S.-C.S.); jwcheng@life.nthu.edu.tw (J.-W.C.); Tel.: +886-3-5742763 (J.-W.C.); Fax: +886-3-5715934 (J.-W.C.)
[†] These authors contributed equally to this work.

Abstract: Cecropins are a family of antimicrobial peptides (AMPs) that are widely found in the innate immune system of Cecropia moths. Cecropins exhibit a broad spectrum of antimicrobial and anticancer activities. The structures of Cecropins are composed of 34–39 amino acids with an N-terminal amphipathic α-helix, an AGP hinge and a hydrophobic C-terminal α-helix. KR12AGPWR6 was designed based on the Cecropin-like structural feature. In addition to its antimicrobial activities, KR12AGPWR6 also possesses enhanced salt resistance, antiendotoxin and anticancer properties. Herein, we have developed a strategy to produce recombinant KR12AGPWR6 through a salt-sensitive, pH and temperature dependent intein self-cleavage system. The His6-Intein-KR12AGPWR6 was expressed by *E. coli* and KR12AGPWR6 was released by the self-cleavage of intein under optimized ionic strength, pH and temperature conditions. The molecular weight and structural feature of the recombinant KR12AGPWR6 was determined by MALDI-TOF mass, CD, and NMR spectroscopy. The recombinant KR12AGPWR6 exhibited similar antimicrobial activities compared to the chemically synthesized KR12AGPWR6. Our results provide a potential strategy to obtain large quantities of AMPs and this method is feasible and easy to scale up for commercial production.

Keywords: antimicrobial peptide; expression; intein; self-cleavage; cecropin-like

Citation: Wu, C.-L.; Chih, Y.-H.; Hsieh, H.-Y.; Peng, K.-L.; Lee, Y.-Z.; Yip, B.-S.; Sue, S.-C.; Cheng, J.-W. High Level Expression and Purification of Cecropin-like Antimicrobial Peptides in *Escherichia coli. Biomedicines* **2022**, *10*, 1351. https://doi.org/10.3390/biomedicines10061351

Academic Editor: Jitka Petrlova

Received: 18 May 2022
Accepted: 6 June 2022
Published: 8 June 2022

Publisher's Note: MDPI stays neutral with regard to jurisdictional claims in published maps and institutional affiliations.

1. Introduction

Antimicrobial peptides (AMPs) normally consist of 12 to 50 amino acids and can be classified as α-helices [1], β-sheets [2], extended [3], and looped peptides [4,5]. Most AMPs exert their antimicrobial activities through the incorporation and permeabilization of microbial membranes, hence the death of microbial cells [6–8]. AMPs can work alone or in combination with antibiotics to diminish antibiotic-resistant pathogens and reduce the amount of antibiotics that are needed [9,10]. Recent progress of AMPs conjugated with antibiotics also demonstrated an enhanced killing effect on drug-resistant bacterial strains [11,12]. Moreover, many AMPs possess lipopolysaccharide (LPS) neutralization as well as anticancer activities [13–15]. Recent studies also summarized the immunomodulatory activities of AMPs in medical uses [15].

Cecropins are a family of AMPs that are widely found in the innate immune systems of Cecropia moths and are composed of 34 to 39 amino acids [16]. The structure of cecropins includes an N-terminal amphipathic α-helix, a hinge motif, and a hydrophobic C-terminal α-helix. This unique structural feature causes Cecropins or Cecropin-like peptides to possess various biological functions, such as antimicrobial activities [17], anti-inflammatory activities [18,19], and anticancer activities [16]. Previously, we developed a strategy to develop AMPs using a Cecropin-like hydrophilic helix—AGP-hydrophobic helix structural feature [20]. One of the leading peptides, KR12AGPWR6 (Ac-KRIVQRIKDFLR-AGP-RRWWRW-NH$_2$) has been found to display superior antimicrobial activities, salt resistance, and LPS neutralizing activities [20].

There are two methods to produce AMPs, chemical synthesis and biosynthesis [21]. Chemical synthesis, such as solid-phase synthesis can provide efficiency and flexibility in the development of AMPs at a laboratory scale [22]. However, for AMPs with longer sequences (>25 amino acids), the yield and purity of chemical synthesized peptides will be reduced. Therefore, the cost of chemical synthesis has hindered the development of AMPs for industrial uses [23,24]. On the other hand, the biosynthesis of recombinant AMPs has several advantages, such as low cost, high yields, and a short production period. Among the expression systems, *Escherichia coli* is the most commonly used owing to its low cost of culture medium, high expression yield, and reduced production time [25]. The expression of AMPs has several challenges. Problems such as the toxicity of AMPs to host cells, their susceptibility to proteolytic degradation, and difficult purification procedures were found to be the major obstacles of biosynthesis. An effective strategy to overcome these limitations is to fuse the target AMP with a carrier protein and then release the peptide by enzymatic or chemical cleavage [24,26]. However, enzymatic cleavages such as thrombin [27], factor Xa [28], and enterokinase [29] were less efficient than chemical agents. On the other hand, chemical cleavage by CNBr or hydroxylamine has disadvantages, such as hazardousness and side-chain modifications [30–32]. Recently, we developed a salt-sensitive and self-cleavage *Nostoc punctiforme* (*Npu*) DnaE intein expression system [21]. The self-cleavage can be controlled by ionic strength, pH, and temperature. This intein expression system has several advantages, including high efficiency for self-cleavage, a quick purification process, no proteases or chemical reagents required for peptide release, and no residue in the final product [33–35]. Herein, we used this intein self-cleavage system to express and purify KR12AGPWR6. The antimicrobial activity of the biosynthesized KR12AGPWR6 was evaluated by MIC assay, and its structural feature was characterized by CD and NMR spectroscopy.

2. Materials and Methods

2.1. Chemicals and Reagents

The ECOS™ BL21 (DE3) competent Cells were purchased from Yeasterm Biotech Co., Ltd. (Taipei, Taiwan). Lysogeny broth (LB) was purchased from Cyrusbioscience (New Taipei City, Taiwan). Isopropyl β-D-1-thiogalactopyranoside (IPTG), Triton X-100, guanidine hydrochloride (GdnHCl), trifluoroacetic acid, D$_2$O, 2,2-dimethyl-2-siapentane-5-sulfonate (DSS), and ampicillin were purchased from Sigma-Aldrich (St. Louis, MO, USA). ^{15}N-labeled ammonium chloride was purchased from Cambridge Isotope Laboratories, Inc. (Tewksbury, MA, USA). Phenylmethylsulfonyl fluoride (PMSF) was purchased from Thermo Fisher Scientific (Waltham, MA, USA). Coomassie Brilliant Blue G-250 was purchased from J.T. Baker Chemical Co. (Phillipsburg, NJ, USA). Sodium azide was purchased from Merck (Millipore, Burlington, MA, USA).

2.2. Construction of the Expression Plasmid

Six histidines were attached in the N-terminus of intein, and the sequence of KR12AGPWR6 was attached to the C-terminus of intein (Figure 1A). The amino acid sequences and optimized DNA sequences are shown in Figure 1B. The pET11b-His6-intein construct was synthesized from Protech Technology Enterprise Co., Ltd. (Taipei, Taiwan).

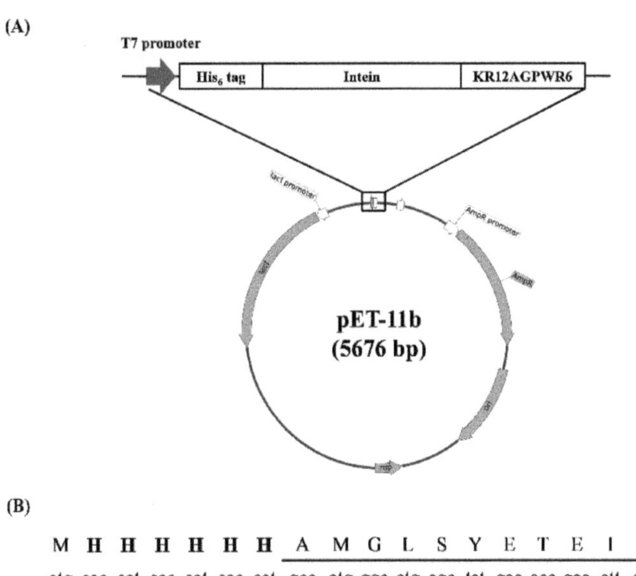

(B)

M	H	H	H	H	H	H	A	M	G	L	S	Y	E	T	E	I	L	T	V
atg	cac	cat	cac	cat	cac	cat	gcc	atg	ggc	ctg	agc	tat	gaa	acc	gaa	att	ctg	acc	gtg
E	Y	G	L	L	P	I	G	K	I	V	E	K	R	I	E	C	T	V	Y
gaa	tat	ggc	ctg	ctg	ccg	att	ggt	aaa	att	gtg	gaa	aaa	cgc	atc	gaa	tgt	act	gtt	tat
S	V	D	N	N	G	N	I	Y	T	Q	P	V	A	Q	W	H	D	R	G
agc	gtt	gat	aat	aat	ggc	aat	att	tat	acc	cag	ccg	gtg	gca	cag	tgg	cac	gat	cgc	ggc
E	Q	E	V	F	E	Y	C	L	E	D	G	S	L	I	R	A	T	K	D
gaa	cag	gaa	gtg	ttt	gaa	tat	tgt	ctg	gaa	gat	ggt	agc	ctg	att	cgc	gca	acc	aaa	gac
H	K	F	M	T	V	D	G	Q	M	L	P	I	D	E	I	F	E	R	E
cat	aaa	ttt	atg	act	gtt	gat	ggt	cag	atg	ctg	ccg	att	gat	gaa	att	ttt	gaa	cgt	gaa
L	D	L	M	R	V	D	N	L	P	N	I	K	I	A	T	R	K	Y	L
ctg	gat	ctg	atg	cgc	gtt	gat	aat	ctg	ccg	aat	atc	aaa	att	gcg	acc	cgt	aaa	tat	ctg
G	K	Q	N	V	Y	G	I	G	V	E	R	D	H	N	F	A	L	K	N
ggc	aaa	cag	aat	gtg	tat	ggc	att	ggc	gtt	gaa	cgc	gac	cat	aat	ttt	gcg	ctg	aaa	aat
G	F	I	A	S	N	**K**	**R**	**I**	**V**	**Q**	**R**	**I**	**K**	**D**	**F**	**L**	**R**	**A**	**G**
ggc	ttc	att	gct	tct	aat	aaa	cgt	atc	gtt	cag	cgt	atc	aaa	gac	ttc	ctg	cgt	gcg	ggt
P	**R**	**R**	**W**	**W**	**R**	**W**	Stop												
ccg	cgt	cgt	tgg	tgg	cgt	tgg	**tag**												

Figure 1. Gene map of the recombinant plasmid pET11b-His6-intein-KR12AGPWR6. (**A**) The illustration of inserted His6-intein-KR12AGPWR6 sequence. (**B**) Optimized DNA sequence and codon map. His6-tag are in bold; the sequence of intein is underlined; the sequence of KR12AGPWR6 is shaded.

2.3. Expression of His6-Intein-KR12AGPWR6

The constructed expression vector was inserted into *E. coli* BL21 (DE3) competent cells for expression. A single colony was selected and incubated in 100 mL LB broth containing ampicillin (0.1 mg/mL) at 37 °C with 150 rpm shaking overnight. A total of 25 mL incubated cell culture was then inoculated into 1000 mL LB broth containing ampicillin (0.1 mg/mL) until the OD_{600} reached 0.6 to 0.8. The protein was subsequently induced with 0.4 mM IPTG for 24 h at 20 °C. After induction, the cells were harvested by centrifugation at 6000× *g* for 20 min at 4 °C. The supernatant was discarded and the pellets were stocked at −20 °C.

2.4. Purification of His6-Intein-KR12AGPWR6

The cell pellets were resuspended in lysis buffer (20 mM sodium phosphate, 150 mM NaCl, 0.5% Triton X-100, 1 mM PMSF, pH 8.0) and lysed with a High-Pressure Homogenizer (AVESTIN EmulsiFlex C3, Mannheim, Germany). The cell lysates were separated with high-speed centrifugation at 4 °C (12,500× g, 30 min), and the supernatant was filtered with 0.2 μm filter before being applied to a Ni-NTA resin (QIAGEN, Hilden, Germany) equilibrated with lysis buffer. The target protein sample was separated under different concentrations of imidazole. The imidazole concentration was 40 mM at the washing steps and 400 mM at the elution steps, respectively. The eluted samples were concentrated using Amicon® Stirred Cells with 3 kDa membrane (Merck Millipore, Burlington, MA, USA), and followed by dialysis using the cellulose tubular membrane (3.5 kDa MWCO, Cellu·Sep T1 Membrane, Membrane Filtration Products, Inc., Seguin, TX, USA) within 20 mM phosphate buffer (pH 8.0) at 4 °C for 24 h.

2.5. Intein Self-Cleavage

Self-cleavage of purified His6-Intein-KR12AGPWR6 was performed in high pH buffer (20 mM phosphate buffer, pH 10.0) at 55 °C for 72 h. After intein self-cleavage, the rKR12AGPWR6 was precipitated in the pellet. Then, the rKR12AGPWR6 peptide was obtained by high-speed centrifugation (12,500× g, 30 min, 4 °C), and the supernatant was discarded. The rKR12AGPWR6 peptide was further resuspended by 6 M guanidine hydrochloride before the purification by RP-HPLC [36].

2.6. Purification of rKR12AGPWR6 by RP-HPLC

All of the samples were purified by using C18 reversed-phase high-performance liquid chromatography (RP-HPLC) on the Prominence HPLC System (Shimadzu, Kyoto, Japan) and using a COSMOSIL C_{18}-AR-II column (Nacalai Tesque, Kyoto, Japan). The column was equilibrated with ddH_2O containing 0.1% (v/v) trifluoroacetic acid (TFA) and eluted with a gradient step from 15 to 100% (v/v) methanol containing 0.1% (v/v) TFA for 75 min at a flow rate of 1 mL/min. Signals were detected by UV 220 nm. The protein samples were collected and lyophilized, then resuspended in water, and the concentrations were determined by the bicinchoninic acid assay (BCA) method (GeneCopoeia™, Rockville, MD, USA). The molecular mass of KR12AGPWR6 was verified by matrix-assisted laser desorption ionization-time of flight (MALDI-TOF) mass spectrometry.

2.7. SDS-PAGE Analysis

The samples were mixed with sample buffer (0.2 M Tris-HCl, pH 6.8, 30% glycerol, 10% sodium dodecyl sulfate (SDS), 10 mM DTT, 0.05% bromophenol blue). The samples were loaded to 12% (wt/v) SDS-polyacrylamide gel electrophoresis (SDS-PAGE) and the gel ran at 120 V for 80 min. Protein bands were detected by Coomassie Brilliant Blue G-250 staining.

2.8. Antimicrobial Activity Assays

Staphylococcus aureus ATCC 25923, *Escherichia coli* ATCC 25922, *Pseudomonas aeruginosa* ATCC 27853, and *Acinetobacter baumannii* 14B0100, were purchased from Bioresources Collection & Research Center (BCRC, FIRDI, Hsinchu, Taiwan). The antibacterial activities of peptides were determined by the standard broth microdilution method according to the guidelines of the Clinical and Laboratory Standards Institute (CLSI) [37]. Briefly, the bacteria were incubated in MHB overnight at 37 °C. The cell cultures were regrown to mid-log phase and subsequently diluted to a final concentration of 5×10^5 CFU/mL. The peptides were loaded into each well at the final concentration of 64, 32, 16, 8, 4, 2, 1 μg/mL, then the microbes were loaded into each well of a polypropylene 96-well plate. After 16 h of incubation at 37 °C, the MIC values of peptides were determined by inspecting the visible growth of bacteria. The MIC values were defined as the lowest concentration of an antimicrobial agent that inhibits the visible growth of a microorganism. All experiments were repeated three times independently.

2.9. Circular Dichroism Spectroscopy

Circular dichroism spectra were recorded in the far-UV spectral region (190 to 260 nm) at 25 °C using a 0.1 cm path-length cuvette on an AVIV CD spectrometer (Aviv Biomedical Inc., Lakewood, NJ, USA). The peptide concentration was 60 μM in 20 mM sodium phosphate buffer or in 30% TFE buffer at pH 7.4.

2.10. Nuclear Magnetic Resonance Spectroscopy (NMR)

The NMR experiments were performed using 0.6 mM of ^{15}N-labeled samples. The NMR samples were prepared in NMR buffer (20 mM sodium phosphate buffer, 10 mM NaN$_3$, pH 4.5) with 10% D$_2$O (*v/v*) for field/frequency lock. ^{15}N-edited 2D HSQC (Heteronuclear Single-Quantum Correlation) spectroscopy; NOESY (Nuclear Overhauser Effect Spectroscopy); and TOCSY (Total Correlation Spectroscopy) experiments were recorded at 298 K on a Bruker Avance 600-MHz NMR spectrometer (Bruker, Billerica, MA, USA). ^{1}H NMR data were referenced to ^{1}H resonance frequency of DSS (2,2-dimethyl-2-siapentane-5-sulfonate). Quadrature detection in the indirect dimensions was determined by using the States-TPPI (time-proportional phase incrementation) method. Signals from H$_2$O were suppressed through low power presaturation (pulse program: zgpr). An analysis of the spectra was conducted using the Sparky software (T.D. Goddard and D.G. Kneller, SPARKY3, University of California, San Francisco, CA, USA).

3. Results and Discussion

3.1. Construction of the Recombinant Plasmid

The pET-11b plasmid was used as a template and the designed construct His6-intein-KR12AGPWR6 was subcloned into the expression vector (Figure 1A). The N-terminal consecutive histidine served as purification tags. The amino acid sequence and the optimized DNA sequence were shown in Figure 1B. A DNA sequence analysis demonstrated that the His6-intein-KR12AGPWR6 sequence was correct.

3.2. Expression, Extraction and Purification of Recombinant His6-Intein-KR12AGPWR6

E. coli BL21 (DE3) cells containing the pET11b-His6-intein-KR12AGPWR6 plasmid were successfully induced by IPTG, and the expression of His6-Intein-KR12AGPWR6 was analyzed by SDS-PAGE (Figure 2). As shown in Figure 2A, lanes 3 and 4 represented expression of His6-Intein-KR12AGPWR6 at 20 °C for 4, 8, and 24 h, respectively. The bands of His6-Intein-KR12AGPWR6 were observed, and the protein levels increased along with the induction time. As shown in Figure 2B, the relative intensity of lane 5 (24 h induction with IPTG) displayed the highest level of protein expression. The enhanced background impurities in lane 5 corresponding to 24 h induction with IPTG were removed by Ni-NTA purification (Figure 3, lane 4–6). Lanes 6 to 8 showed protein expression levels without IPTG induction and smaller amounts of recombinant proteins were observed (Figure 2A). Subsequently, the pellets were resuspended in lysis buffer and lysed with a high-pressure homogenizer. Ni-NTA resin was used for protein purification. As can be seen in Figure 3, the targeted proteins were examined in cell lysate (lane 2). No target proteins could be observed in the solution flow through (lane 3). The recombinant proteins were washed three times by 40 mM imidazole (lanes 4–6) to remove non-specific targets, and most of the non-specific targets were washed out the first time. Intein may perform its self-cleavage activity at low salt concentrations. In our previous study, we proposed a 'prohibition condition' to inhibit the self-cleavage activity of intein before using Ni-NTA column to purify the target protein [38]. The 'prohibition condition' in this intein system includes a condition with salt concentration > 300 mM and pH < 7. Therefore, we chose 400 mM imidazole to prevent intein self-cleavage before the purification steps. The target proteins were then eluted by 400 mM imidazole (lane 7–10) and the protein band of recombinant His6-Intein-KR12AGPWR6 (MW 19.7 kDa) was observed on SDS-PAGE at about 20 kDa (Figure 3). Imidazole was removed from the eluted sample using cellulose tubular membrane (3.5 kDa MWCO) in dialysis buffer at 4 °C for 24 h before intein self-cleavage.

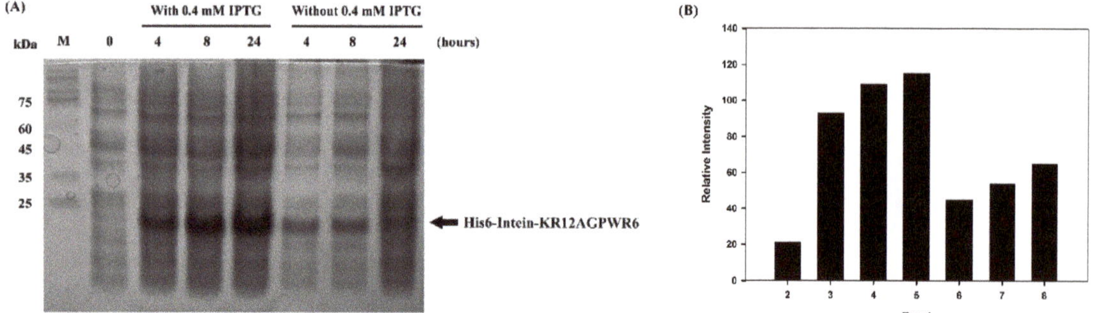

Figure 2. SDS-PAGE analysis of recombinant His6-Intein-KR12AGPWR6 in *E. coli* BL21. (**A**) SDS-PAGE of recombinant His6-Intein-KR12AGPWR6 expressed in *E. coli* BL21 with or without 0.4 mM IPTG at 20 °C for 4, 8, 24 h. Lane 1: protein molecular weight markers (kDa). Lane 2: before induction. Lanes 3–5: with IPTG induction at 20 °C for 4, 8, 24 h. Lanes 6–8: without IPTG induction at 20 °C for 4, 8, 24 h. (**B**) Relative intensities of the His6-Intein-KR12AGPWR6 bands. Proteins were visualized using Coomassie blue staining.

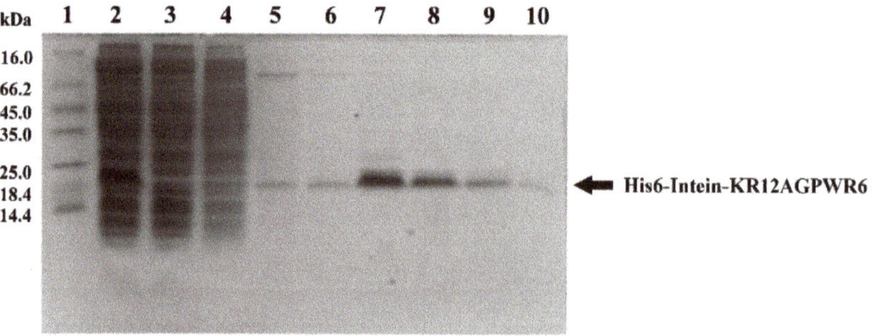

Figure 3. SDS-PAGE analysis of recombinant His6-Intein-KR12AGPWR6 expressed in *E. coli* at different steps of purification procedure. Lane 1: protein molecular weight markers (kDa); lane 2: supernatant of cell lysate; lane 3: flow through; lane 4–6: washing by 40 mM imidazole; lane 7–10: elution by 400 mM imidazole. Proteins were visualized using Coomassie blue staining.

3.3. Optimization of Intein's Self-Cleavage

Reaction time, pH, and temperature conditions were used to optimize intein's self-cleavage. After intein's self-cleavage, the mixtures were analyzed by SDS-PAGE (Figure 4A). Three major bands belonging to His6-intein-KR12AGPWR6, His6-intein, and KR12AGPWR6 were seen, and their molecular weights were 19.7 kDa, 16.9 kDa, and 2.8 kDa, respectively (Figure 4A, lane 3–8). During intein self-cleavage, we observed that KR12AGPWR6 precipitated in the pellet. Thus, it is difficult to use the quantification of KR12AGPWR6 on SDS-PAGE to find the optimized intein self-cleavage condition. On the other hand, His-intein-KR12AGPWR6 and His6-intein dissolved in the supernatant. Therefore, we used the quantification of His-intein-KR12AGPWR6 and His6-intein on SDS-PAGE to determine the best self-cleavage conditions of intein. The band of KR12AGPWR6 was the same as synthetic KR12AGPWR6 (MW 2.8 kDa) (lane 10), which suggested that KR12AGPWR6 was released via intein's self-cleavage. To optimize intein's self-cleavage efficacy, purified His6-intein-KR12AGPWR6 was kept in various pH buffers at 55 °C for 18 and 72 h. As shown in Figure 4A,B, the self-cleavage rate of intein at 72 h was higher than 18 h under different pH conditions. Further, a higher pH (pH = 10) exhibited better intein self-cleavage activity. Similarly, intein self-cleavage activity at 72 h was higher than 18 h under different temperatures. The best intein cleavage condition for His6-Intein-KR12AGPWR6 was pH 10, 55 °C, for 72 h. Moreover, intein lost its self-cleavage activity at 4 °C, while the self-cleavage activity recovered at 37 °C. These results suggested that the most optimized conditions for the self-cleavage of His6-intein-KR12AGPWR6 were under high pH, high temperature, and longer reaction time.

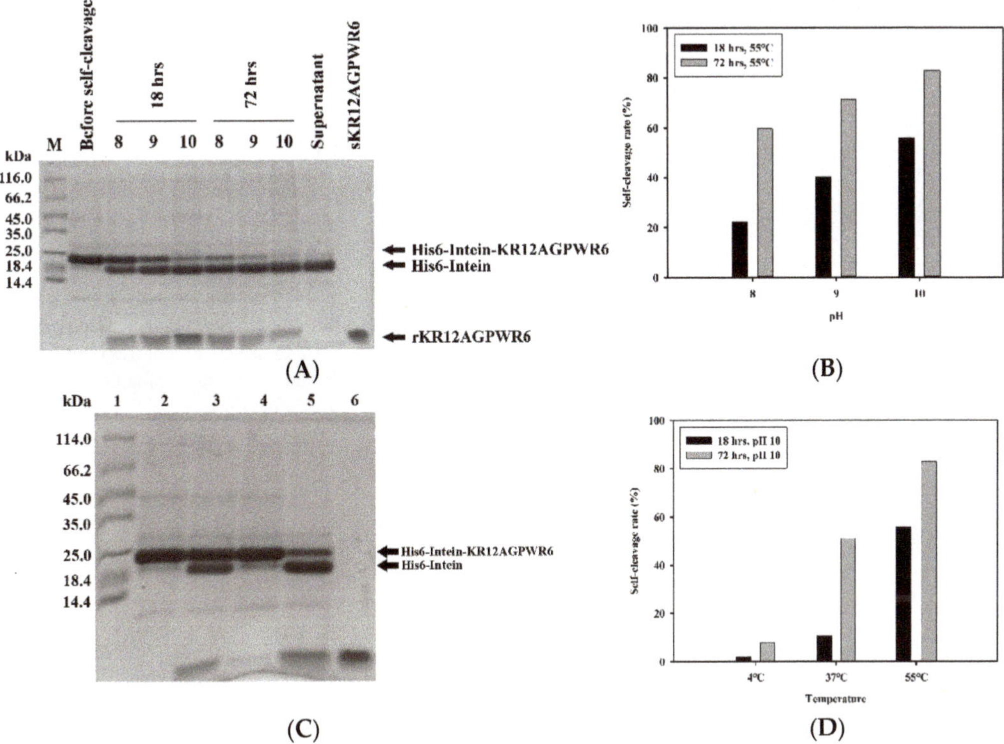

Figure 4. SDS-PAGE analysis of pH, temperature, and time optimization for intein's self-cleavage. (**A**) SDS-PAGE analysis of pH optimization for intein's self-cleavage. Lane 1: protein molecular weight markers (kDa); lane 2: protein in pH 10 buffer at 4 °C before self-cleavage; lanes 3–5: protein in different pH at 55 °C for 18 h; lanes 6–8: protein in different pH at 55 °C for 72 h; lane 9: the supernatant of protein in pH 10 at 55 °C for 72 h; lane 10: chemical synthesized sKR12AGPWR6 was used as a control. (**B**) Quantification of intein self-cleavage rates. (**C**) SDS-PAGE analysis of temperature optimization for intein's self-cleavage. Lane 1: protein molecular weight markers (kDa). Lane 2 and 3: the targeted proteins were cleaved in pH 10 buffer for 18 h at 4 °C and 37 °C, respectively. Lane 4 and 5: targeted proteins were cleaved in pH 10 buffer for 72 h at 4 °C and 37 °C, respectively. Lane 6: synthetic KR12AGPWR6 was used as a control. (**D**) Quantification of intein self-cleavage rates. Proteins were observed using Coomassie blue staining.

3.4. Purification of KR12AGPWR6

KR12AGPWR6 was redissolved by using 6 M guanidine hydrochloride before RP-HPLC purification. Reversed-phase HPLC was used to purify KR12AGPWR6 with a gradient of water/methanol containing 0.1% trifluoroacetic acid (TFA). The prolife of HPLC chromatograms is displayed in Figure 5. The retention time of KR12AGPWR6 was found at 51 min. The yield of KR12AGPWR6 was 2.41 ± 0.33 mg/L. The molecular weight of the eluted KR12AGPWR6 was determined to be 2823.664 Da by MALDI-TOF MS (Figure 6), which was close to the theoretical MW (2824.3 Da).

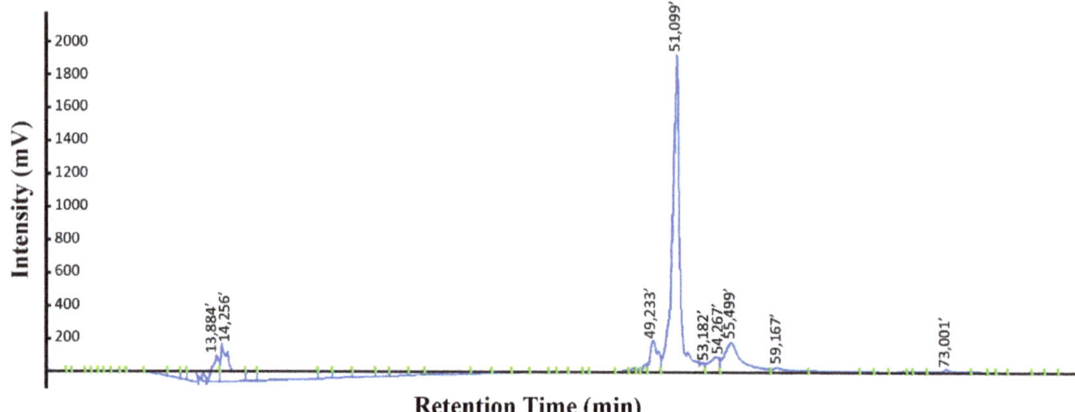

Retention Time (min)

Figure 5. Reversed-phase HPLC purification of KR12AGPWR6 after intein's self-cleavage. The samples were resuspended with 6 M guanidine hydrochloride and purified by reversed-phase HPLC on the Prominence HPLC System equipped with a C18 column. The column was equilibrated with ddH$_2$O containing 0.1% (*v/v*) trifluoroacetic acid (TFA) and the gradient ranging from 15 to 100% (*v/v*) methanol containing 0.1% (*v/v*) TFA for 75 min at a flow rate of 1 mL/min. Signals were detected by UV 280 nm.

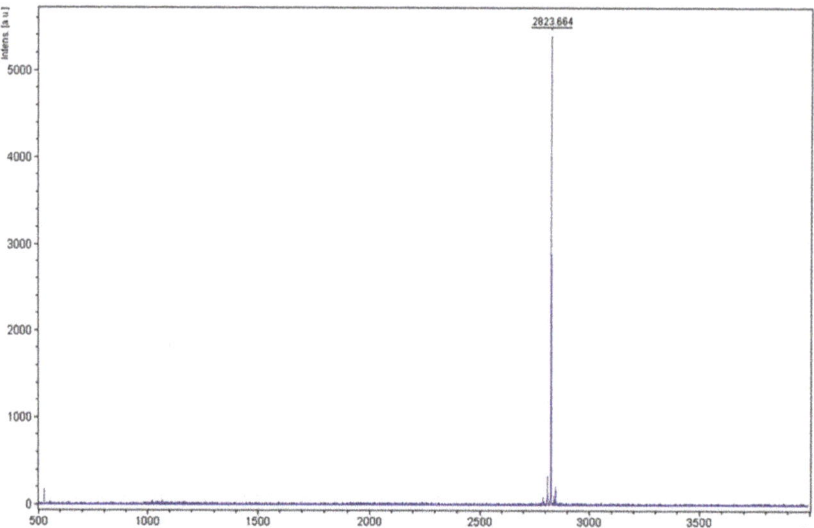

Figure 6. Mass analysis of KR12AGPWR6 from RP-HPLC. The molecular weight of the purified KR12AGPWR6 was found to be 2823.664 Da. The theoretical MW of KR12AGPWR6 was calculated to be 2824.3 Da.

3.5. Antimicrobial Activity

The antimicrobial activities of the recombinant rKR12AGPWR6 and chemically synthesized sKR12AGPWR6 against *S. aureus* ATCC 25923, *E. coli* ATCC 25922, *P. aeruginosa* ATCC 27853, and *A. baumannii* BCRC 14B0100 were evaluated by MIC assay. As shown in Table 1, the MIC value of rKR12AGPWR6 was the same as chemically synthesized sKR12AGPWR6 against *S. aureus* ATCC 25,923 (2 µg/mL) and *A. baumannii* BCRC 14B0100 (1 µg/mL). The recombinant rKR12AGPWR6 had a MIC of 4 µg/mL against *E. coli* ATCC25922 and *P. aerug-*

inosa ATCC 27853. The results showed that rKR12AGPWR6 exhibited similar antimicrobial activities as the chemically synthesized sKR12AGPWR6 against both Gram-positive and Gram-negative bacteria.

Table 1. MICs of sKR12AGPWR6 and rKR12AGPWR6 against Gram-positive and Gram-negative bacteria.

Bacterial Strains	Minimum Inhibitory Concentration (µg/mL)	
	sKR12AGPWR6	rKR12AGPWR6
S. aureus ATCC 25923	2	2
E. coli ATCC 25922	2	4
P. aeruginosa ATCC 27853	2	4
A. baumannii BCRC 14B0100	1	1

3.6. Characterization of the Recombinant rKR12AGPWR6 by CD and NMR

CD spectroscopy was used to compare the structures of the recombinant rKR12A-GPWR6 and the chemically synthesized sKR12AGPWR6 (Figure 7). The CD results indicated that both the recombinant and chemically synthesized KR12AGPWR6 adopted a typical α-helical structure under 30% TFE buffer. In order to achieve backbone assignments of rKR12AGPWR6, the 2D TOCSY and NOESY spectra of KR12AGPWR6 in 20 mM phosphate buffer were obtained (Figure S1). In addition, we successfully assigned the ^{1}H and ^{15}N backbone resonance peaks of rKR12AGPWR6 in buffer. A well-resolved ^{1}H-^{15}N HSQC spectrum of the fingerprint region of rKR12AGPWR6 is shown in Figure 8.

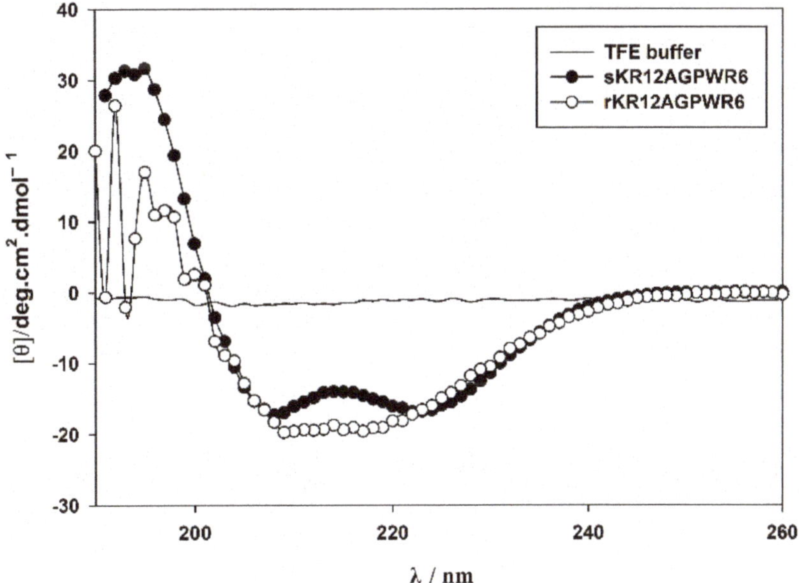

Figure 7. Circular dichroism spectra of synthetic and recombinant KR12AGPWR6 in different environments. Circular dichroism spectra of 60 µM synthetic and recombinant KR12AGPWR6 in 30% TFE buffer in pH7.4 at 25 °C.

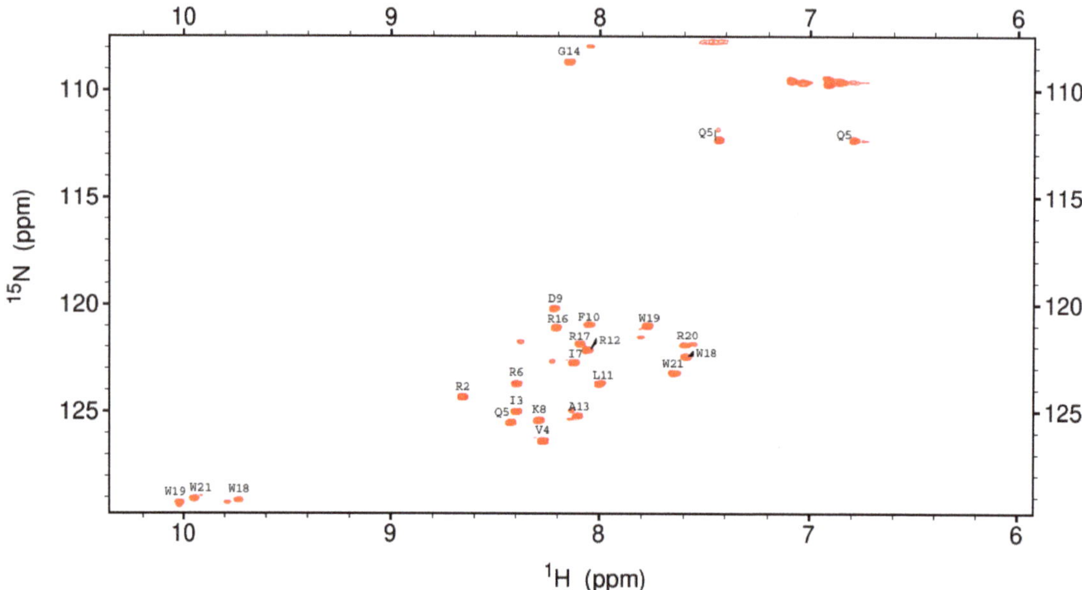

Figure 8. ^{1}H-^{15}N HSQC spectra of 0.6 mM rKR12AGPWR6 in 20 mM sodium phosphate buffer, pH 4.5, 298 K. The ^{15}N-labeled samples were expressed in *E. coli* BL21 (DE3) cells grown in M9 medium containing ^{15}N-labeled ammonium chloride (1 g/L), and the purification of peptides as mentioned above. The cross peaks of rKR12AGPWR6 were shown in red.

3.7. Design of New Cecropin-like Peptides

Cecropin analogues [38], Cecropin A/Cecropin B hybrids [39], and Cecropin A, LL-37, and Magainin hybrids exhibited strong antimicrobial and anticancer activities [40]. Many other AMPs, such as chicken cathelicidin fowlicidin-2, MSI-594, SMAP-29, Pardaxin, Cecropin A, and Papiliocin, with similar Cecropin-like structural features also possessed both antimicrobial and LPS neutralizing abilities [19,41–45].

Recently, Malmsten and co-workers developed a strategy to increase the salt resistance of short antimicrobial peptides by adding tryptophan and/or phenylalanine end-tags [46–48]. End-tagging was also found to promote other biological effects, such as anti-cancer and receptor binding activities [49,50]. We modified this strategy by adding β-naphthylalanine (Nal) to the termini of the short antimicrobial peptide S1 (Ac-KKWRKWLAKK-NH$_2$) to boost its salt resistance, serum proteolytic stability, and antiendotoxin activities [51,52]. We used solution NMR and paramagnetic relaxation enhancement techniques to study the structural differences of S1 and S1-Nal-Nal in LPS micelles [53]. The three-dimensional structures of S1 and S1-Nal-Nal in complex with LPS clearly provided an explanation for the differences in their antiendotoxin activities. Based on these structural results and the above-mentioned anti-LPS AMPs, we proposed a possible model to explain the mechanism of S1-Nal-Nal in the interaction with LPS. Firstly, S1-Nal-Nal adopts a random coil structure in aqueous solution. Then, it is attracted to LPS by the electrostatic interactions between the induced amphipathic helix and the negatively charged region of LPS. Finally, the bulky hydrophobic β-naphthylalanine (Nal) end-tags insert themselves into LPS by extra hydrophobic interactions with the lipid A region of LPS. The LPS-induced inflammation is then prohibited by the blocked lipid A region.

Based on the above-mentioned structural and functional studies of Cecropins and S1-Nal-Nal, we suspect that the binding and neutralization of LPS is not just through the sequences of Cecropins because some of the Cecropins and Cecropin analogues, although different in their sequences, can still bind to and neutralize LPS induced pro-inflammatory effects [18,40,46]. Therefore, we hypothesize that the binding and neutralization of LPS occurs through their specific structural features (i.e., amphipathic helix-AGP hinge-hydrophobic helix). Herein, we propose to extend this strategy to design AMPs with enhanced salt resistance and antiendotoxin activity. Some of the potential AMPs are listed in Table 2.

Table 2. Proposed AMPs based on amphipathic helix-AGP-hydrophobic helix.

Magainin-AGP-WR6	GIGKFLHSAKKFGKAFVGEIMNS-**AGP**-RRWWRW
KR12-AGP-Cecropin P1	KRIVQRIKDFLR-**AGP**-IAIAIQGGPR
Cecropin P1-AGP-WR6	SWLSKTAKKLENSAKKRISE-**AGP**-RRWWRW
KR12-AGP-Cecropin A	KRIVQRIKDFLR-**AGP**-AVAVVGQATQIAK
CecropinA-AGP-WR6	KWKLFKKIEKVGQNIRDGIIK-**AGP**-RRWWRW
Melittin-AGP-WR6	GIGAVLKVLTTGLPALISWIKRKRQQ-**AGP**-RRWWRW
Melittin-AGP-Cecropin P1	GIGAVLKVLTTGLPALISWIKRKRQQ-**AGP**-IAIAIQGGPR

3.8. Expression and Purification of Cecropin-like AMPs

Intein is a protein segment that can cleave itself from a whole protein sequence and ligate the remaining N-terminal and C-terminal portions (the exteins) with a peptide bond [21]. Until now, over 600 inteins with different lengths have been identified. The specific cleavage-ligation function of intein has enabled various applications, such as protein engineering, isotope labeling, biomaterials, cyclization, and protein purification [54,55]. Recently, we have created an N-terminal mutated intein that has no N-terminal cleavage activity but preserves its C-terminal cleavage activity [21]. We have shown that this mutated intein can be used as a fusion tag and can self-cleave from the target protein. Moreover, we have shown that the efficiency of the intein self-cleavage is dependent on ionic strength, pH, temperature, and reaction time [21].

In this study, we used the mutated intein as a fusion partner for the expression and purification of KR12AGPWR6. We successfully expressed His6-intein-KR12AGPWR6 in *E. coli* (Figure 2) and the recombinant His6-intein-KR12AGPWR6 was purified by Ni-NTA column (Figure 3). In order to optimize the self-cleavage efficacy of His6-intein-KR12AGPWR6, the recombinant proteins were incubated at 55 °C under various pH buffers. The optimal pH for intein self-cleavage was 10. As can be seen in Table 1, the recombinant rKR12AGPWR6 possessed strong antimicrobial activities against both Gram-positive and Gram-negative bacteria, similar to the activities of the synthesized sKR12AGPWR6. We have also shown that the expressed rKR12AGPWR6 adopts an α-helical structure in TFE which is identical to the chemical synthesized sKR12AGPWR6. In addition, we demonstrated that the ^{15}N-labeled rKR12AGPWR6 that was produced in this study may be used to understand the structural features and interactions between rKR12AGPWR6 and model membranes/microbes for the design and development of useful antimicrobial peptides.

The lower yield of KR12AGPWR6 could be due to its antimicrobial activity, which makes itself potentially fatal to the expression host. However, the yield of KR12AGPWR6 is comparable to other antimicrobial peptides expressed by using the thioredoxin fusion or the GST fusion protein systems [56]. The premature intein self-cleavage before Ni-NTA column purification is also an intrinsic problem associated with the intein system. This problem causes loss in the yield of the target peptide. On the other hand, we could easily obtain the purified product by RP-HPLC by using this intein self-cleavage system. This intein self-cleavage system also requires no auxiliary enzymes or chemicals to remove the carrier protein. Additionally, by using bioreactors, volumetric protein production (*E. coli* cell density) could be improved up to 10–34 fold using a fed-batch strategy compared to batch cultivation [57]. Therefore, this intein self-cleavage system presents a potential method to produce AMPs in *E. coli* and is beneficial to reducing the cost of production for commercial scale production.

In addition to KR12AGPWR6, we chose CecropinA-AGP-WR6 (KWKLFKKIE-KVGQNIRDGIIK-**AGP**-RRWWRW) (Table 2) to test this intein self-cleavage system. Figures S2 and S3 demonstrated that CecropinA-AGP-WR6 can be successfully obtained by using this intein self-cleavage expression and purification system. As shown in Table 3, CecropinA-AGP-WR6 also possesses strong antimicrobial activities against Gram-negative and Gram-positive bacteria.

Table 3. MIC values of recombinant CecropinA-AGP-WR6 against Gram-positive and Gram-negative bacteria.

Bacterial Strains	Minimum Inhibitory Concentration (μg/mL)
S. aureus ATCC 29213	16
E. coli ATCC 25922	8
P. aeruginosa ATCC 27853	8
A. baumannii BCRC 14B0100	2

In conclusion, we successfully expressed the cecropin-like peptides rKR12AGPWR6 and CecropinA-AGP-WR6 from *E. coli*, and purification of rKR12AGPWR6 and CecropinA-AGP-WR6 was efficiently carried out by using the optimized intein self-cleavage system. This study provided a potential strategy to produce recombinant AMPs up to a commercial scale production, and this intein self-cleavage system can be widely applied to obtain other AMPs from *E coli*.

Supplementary Materials: The following supporting information can be downloaded at: https://www.mdpi.com/article/10.3390/biomedicines10061351/s1, Figure S1: Backbone assignment of KR12AGPWR6.; Figure S2: SDS-PAGE analysis of recombinant His6-Intein-CA-AGP-WR6 expressed in *E. coli*.; Figure S3: Mass analysis of CA-AGP-WR6 from RP-HPLC.

Author Contributions: C.-L.W., Y.-H.C., H.-Y.H., K.-L.P. and Y.-Z.L. performed the experiments and analyzed the data. C.-L.W. and J.-W.C. wrote the manuscript. B.-S.Y., S.-C.S. and J.-W.C. planned the study and revised and approved the final version of the manuscript. All authors have read and agreed to the published version of the manuscript.

Funding: This work was supported by grants from the Ministry of Science and Technology of Taiwan to JWC (110-2113-M-007-021) and the National Taiwan University Hospital Hsinchu Branch to BSY.

Institutional Review Board Statement: Not applicable.

Informed Consent Statement: Not applicable.

Data Availability Statement: The data presented in this study are available on request from the corresponding author.

Acknowledgments: We thank Justin Cheng and Daniel Cheng for editing the manuscript. The authors are grateful for the NMR facility at the instrument center at National Tsing-Hua University, Taiwan.

Conflicts of Interest: The authors have no potential conflict of interest to disclose.

References

1. Sinha, S.; Harioudh, M.K.; Dewangan, R.P.; Ng, W.J.; Ghosh, J.K.; Bhattacharjya, S. Cell-selective pore forming antimicrobial peptides of the prodomain of human furin: A conserved aromatic/cationic sequence mapping, membrane disruption, and atomic-resolution structure and dynamics. *ACS Omega* **2018**, *3*, 14650–14664. [CrossRef]
2. D'Souza, A.; Wu, X.; Yeow, E.K.L.; Bhattacharjya, S. Designed Heme-Cage beta-Sheet Miniproteins. *Angew. Chem. Int. Ed. Engl.* **2017**, *56*, 5904–5908. [CrossRef]

3. Dwivedi, R.; Aggarwal, P.; Bhavesh, N.S.; Kaur, K.J. Design of therapeutically improved analogue of the antimicrobial peptide, indolicidin, using a glycosylation strategy. *Amino Acids* **2019**, *51*, 1443–1460. [CrossRef]
4. Andreu, D.; Rivas, L. Animal antimicrobial peptides: An overview. *Biopolymers* **1998**, *47*, 415–433. [CrossRef]
5. Bhattacharjya, S. NMR Structures and Interactions of Antimicrobial Peptides with Lipopolysaccharide: Connecting Structures to Functions. *Curr. Top. Med. Chem.* **2015**, *16*, 4–15. [CrossRef]
6. Mohanram, H.; Bhattacharjya, S. 'Lollipop'-shaped helical structure of a hybrid antimicrobial peptide of temporin B-lipopolysaccharide binding motif and mapping cationic residues in antibacterial activity. *Biochim. Biophys. Acta* **2016**, *1860*, 1362–1372. [CrossRef]
7. Bechinger, B.; Gorr, S.U. Antimicrobial Peptides: Mechanisms of Action and Resistance. *J. Dent. Res.* **2016**, *96*, 254–260. [CrossRef]
8. Kang, H.-K.; Kim, C.; Seo, C.H.; Park, Y. The therapeutic applications of antimicrobial peptides (AMPs): A patent review. *J. Microbiol.* **2016**, *55*, 1–12. [CrossRef]
9. Draper, L.A.; Cotter, P.D.; Hill, C.; Ross, R.P. The two peptide lantibiotic lacticin 3147 acts synergistically with polymyxin to inhibit Gram negative bacteria. *BMC Microbiol.* **2013**, *13*, 212. [CrossRef]
10. Naghmouchi, K.; Baah, J.; Hober, D.; Jouy, E.; Rubrecht, C.; Sané, F.; Drider, D. Synergistic Effect between Colistin and Bacteriocins in Controlling Gram-Negative Pathogens and Their Potential To Reduce Antibiotic Toxicity in Mammalian Epithelial Cells. *Antimicrob. Agents Chemother.* **2013**, *57*, 2719–2725. [CrossRef]
11. Etayash, H.; Alford, M.; Akhoundsadegh, N.; Drayton, M.; Straus, S.K.; Hancock, R.E.W. Multifunctional Antibiotic-Host Defense Peptide Conjugate Kills Bacteria, Eradicates Biofilms, and Modulates the Innate Immune Response. *J. Med. Chem.* **2021**, *64*, 16854–16863. [CrossRef]
12. Li, W.; O'Brien-Simpson, N.M.; Holden, J.A.; Otvos, L.; Reynolds, E.C.; Separovic, F.; Hossain, M.A.; Wade, J.D. Covalent conjugation of cationic antimicrobial peptides with a b-lactam antibiotic core. *Pept. Sci.* **2018**, *110*, e24059. [CrossRef]
13. Bhattacharjya, S. De novo Designed Lipopolysaccharide Binding Peptides: Structure Based Development of Antiendotoxic and Antimicrobial Drugs. *Curr. Med. Chem.* **2010**, *17*, 3080–3093. [CrossRef]
14. Lin, L.; Chi, J.; Yan, Y.; Luo, R.; Feng, X.; Zheng, Y.; Xian, D.; Li, X.; Quan, G.; Liu, D.; et al. Membrane-disruptive peptides/peptidomimetics-based therapeutics: Promising systems to combat bacteria and cancer in the drug-resistant era. *Acta Pharm. Sin. B* **2021**, *11*, 2609–2644. [CrossRef]
15. Guryanova, S.V.; Ovchinnikova, T.V. Immunomodulatory and Allergenic Properties of Antimicrobial Peptides. *Int. J. Mol. Sci.* **2022**, *23*, 2499. [CrossRef]
16. Wu, J.-M.; Jan, P.-S.; Yu, H.-C.; Haung, H.-Y.; Fang, H.-J.; Chang, Y.-I.; Cheng, J.-W.; Chen, H.M. Structure and function of a custom anticancer peptide, CB1a. *Peptides* **2009**, *30*, 839–848. [CrossRef]
17. Moore, A.J.; Beazley, W.D.; Bibby, M.C.; Devine, D.A. Antimicrobial activity of cecropins. *J. Antimicrob. Chemother.* **1996**, *37*, 1077–1089. [CrossRef]
18. Kim, J.K.; Lee, E.; Shin, S.; Jeong, K.W.; Lee, J.Y.; Bae, S.Y.; Kim, S.H.; Lee, J.; Kim, S.R.; Lee, D.G.; et al. Structure and Function of Papiliocin with Antimicrobial and Anti-inflammatory Activities Isolated from the Swallowtail Butterfly, Papilio xuthus. *J. Biol. Chem.* **2011**, *286*, 41296–41311. [CrossRef]
19. Lee, E.; Shin, A.; Kim, Y. Anti-inflammatory activities of cecropin a and its mechanism of action. *Arch. Insect Biochem. Physiol.* **2015**, *88*, 31–44. [CrossRef]
20. Chu, H.-L.; Chih, Y.-H.; Peng, K.-L.; Wu, C.-L.; Yu, H.-Y.; Cheng, D.; Chou, Y.-T.; Cheng, J.-W. Antimicrobial Peptides with Enhanced Salt Resistance and Antiendotoxin Properties. *Int. J. Mol. Sci.* **2020**, *21*, 6810. [CrossRef]
21. Lee, Y.Z.; Sue, S.C. Salt-sensitive intein for large-scale polypeptide production. *Methods Enzym.* **2019**, *621*, 111–130.
22. Gunasekaran, P.; Kim, E.Y.; Lee, J.; Ryu, E.K.; Shin, S.Y.; Bang, J.K. Synthesis of Fmoc-Triazine Amino Acids and Its Application in the Synthesis of Short Antibacterial Peptidomimetics. *Int. J. Mol. Sci.* **2020**, *21*, 3602. [CrossRef]
23. Andersson, L.; Blomberg, L.; Flegel, M.; Lepsa, L.; Nilsson, B.; Verlander, M. Large-scale synthesis of peptides. *Biopolymers* **2000**, *55*, 227–250. [CrossRef]
24. Li, Y. Recombinant production of antimicrobial peptides in Escherichia coli: A review. *Protein Expr. Purif.* **2011**, *80*, 260–267. [CrossRef]
25. Cheng, K.-T.; Wu, C.-L.; Yip, B.-S.; Yu, H.-Y.; Cheng, H.-T.; Chih, Y.-H.; Cheng, J.-W. High Level Expression and Purification of the Clinically Active Antimicrobial Peptide P-113 in Escherichia coli. *Molecules* **2018**, *23*, 800. [CrossRef]
26. Sang, M.; Wei, H.; Zhang, J.; Wei, Z.; Wu, X.; Chen, Y.; Zhuge, Q. Expression and characterization of the antimicrobial peptide ABP-dHC-cecropin A in the methylotrophic yeast Pichia pastoris. *Protein Expr. Purif.* **2017**, *140*, 44–51. [CrossRef]
27. Dhakal, S.; Sapkota, K.; Huang, F.; Rangachari, V. Cloning, expression and purification of the low-complexity region of RanBP9 protein. *Protein Expr. Purif.* **2020**, *172*, 105630. [CrossRef]
28. Wang, Q.; Yin, M.; Yuan, C.; Liu, X.; Hu, Z.; Zou, Z.; Wang, M. Identification of a Conserved Prophenoloxidase Activation Pathway in Cotton Bollworm Helicoverpa armigera. *Front. Immunol.* **2020**, *11*, 785. [CrossRef]
29. Gibbs, G.M.; Davidson, B.E.; Hillier, A.J. Novel Expression System for Large-Scale Production and Purification of Recombinant Class IIa Bacteriocins and Its Application to Piscicolin 126. *Appl. Environ. Microbiol.* **2004**, *70*, 3292–3297. [CrossRef]
30. Holbrook, S.R.; Cheong, C.; Tinoco, I., Jr.; Kim, S.-H. Crystal structure of an RNA double helix incorporating a track of non-Watson–Crick base pairs. *Nature* **1991**, *353*, 579–581. [CrossRef]

31. Li, Q.; Chen, A.S.; Gayen, S.; Kang, C. Expression and purification of the p75 neurotrophin receptor transmembrane domain using a ketosteroid isomerase tag. *Microb. Cell Fact.* **2012**, *11*, 1–8. [CrossRef]

32. Zhou, L.; Lai, Z.-T.; Lu, M.-K.; Gong, X.-G.; Xie, Y. Expression and Hydroxylamine Cleavage of Thymosin Alpha 1 Concatemer. *J. Biomed. Biotechnol.* **2008**, *2008*, 736060. [CrossRef]

33. Perler, F.B.; Davis, E.O.; Dean, G.E.; Gimble, F.S.; Jack, W.E.; Neff, N.; Noren, C.J.; Thorner, J.; Belfort, M. Protein splicing elements: Inteins and exteins—A definition of terms and recommended nomenclature. *Nucleic Acids Res.* **1994**, *22*, 1125–1127. [CrossRef]

34. Singleton, S.F.; A Simonette, R.; Sharma, N.C.; I Roca, A. Intein-mediated affinity-fusion purification of the Escherichia coli RecA protein. *Protein Expr. Purif.* **2002**, *26*, 476–488. [CrossRef]

35. Guo, C.; Li, Z.; Shi, Y.; Xu, M.; Wise, J.; Trommer, W.E.; Yuan, J. Intein-mediated fusion expression, high efficient refolding, and one-step purification of gelonin toxin. *Protein Expr. Purif.* **2004**, *37*, 361–367. [CrossRef]

36. Nudelman, H.; Lee, Y.Z.; Hung, Y.L.; Kolusheva, S.; Upcher, A.; Chen, Y.C.; Chen, J.Y.; Sue, S.C.; Zarivach, R. Understanding the Biomineralization Role of Magnetite-Interacting Components (MICs) From Magnetotactic Bacteria. *Front. Microbiol.* **2018**, *9*, 2480. [CrossRef]

37. Clinical and Laboratory Standards Institute (CLSI). In *Performance Standards for Antimicrobial Susceptibility Testing*; Twenty-Fifth Informational Supplement; Clinical and Laboratory Standards Institute (CLSI): Wayne, PA, USA, 2015.

38. Chen, H.M.; Wang, W.; Smith, D.; Chan, S.C. Effects of the anti-bacterial peptide cecropin B and its analogs, cecropins B-1 and B-2, on liposomes, bacteria, and cancer cells. *Biochim. Biophys. Acta* **1997**, *1336*, 171–179. [CrossRef]

39. Wu, C.; Geng, X.; Wan, S.; Hou, H.; Yu, F.; Jia, B.; Wang, L. Cecropin-P17, an analog of Cecropin B, inhibits human hepatocellular carcinoma cell HepG-2 proliferation via regulation of ROS, Caspase, Bax, and Bcl-2. *J. Pept. Sci. Off. Publ. Eur. Pept. Soc.* **2015**, *21*, 661–668. [CrossRef]

40. Fox, M.A.; Thwaite, J.E.; Ulaeto, D.O.; Atkins, T.P.; Atkins, H.S. Design and characterization of novel hybrid antimicrobial peptides based on cecropin A, LL-37 and magainin II. *Peptides* **2012**, *33*, 197–205. [CrossRef]

41. Bhunia, A.; Ramamoorthy, A.; Bhattacharjya, S. Helical hairpin structure of a potent antimicrobial peptide MSI-594 in lipopolysaccharide micelles by NMR spectroscopy. *Chemistry* **2009**, *15*, 2036–2040. [CrossRef]

42. Xiao, Y.; Herrera, A.I.; Bommineni, Y.R.; Soulages, J.L.; Prakash, O.; Zhang, G. The Central Kink Region of Fowlicidin-2, an a-Helical Host Defense Peptide, Is Critically Involved in Bacterial Killing and Endotoxin Neutralizationon. *J. Innate Immun.* **2009**, *1*, 268–280. [CrossRef]

43. Tack, B.F.; Sawai, M.V.; Kearney, W.R.; Robertson, A.D.; Sherman, M.A.; Wang, W.; Hong, T.; Boo, L.M.; Wu, H.; Waring, A.J.; et al. SMAP-29 has two LPS-binding sites and a central hinge. *JBIC J. Biol. Inorg. Chem.* **2002**, *269*, 1181–1189. [CrossRef]

44. Bhunia, A.; Domadia, P.N.; Torres, J.; Hallock, K.J.; Ramamoorthy, A.; Bhattacharjya, S. NMR structure of pardaxin, a pore-forming antimicrobial peptide in lipopolysaccharide micelles: Mechanism of outer membrane permeabilization. *J. Biol. Chem.* **2010**, *285*, 3883–3895. [CrossRef]

45. Lee, E.; Kim, J.K.; Jeon, D.; Jeong, K.W.; Shin, A.; Kim, Y. Functional Roles of Aromatic Residues and Helices of Papilioncin in its Antimicrobial and Anti-inflammatory Activities. *Sci. Rep.* **2015**, *5*, 12048. [CrossRef]

46. Pasupuleti, M.; Chalupka, A.; Mörgelin, M.; Schmidtchen, A.; Malmsten, M. Tryptophan end-tagging of antimicrobial peptides for increased potency against Pseudomonas aeruginosa. *Biochim. Biophys. Acta* **2009**, *1790*, 800–808. [CrossRef]

47. Pasupuleti, M.; Schmidtchen, A.; Chalupka, A.; Ringstad, L.; Malmsten, M. End-Tagging of Ultra-Short Antimicrobial Peptides by W/F Stretches to Facilitate Bacterial Killing. *PLoS ONE* **2009**, *4*, e5285. [CrossRef]

48. Schmidtchen, A.; Pasupuleti, M.; Mörgelin, M.; Davoudi, M.; Alenfall, J.; Chalupka, A.; Malmsten, M. Boosting Antimicrobial Peptides by Hydrophobic Oligopeptide End Tags. *J. Biol. Chem.* **2009**, *284*, 17584–17594. [CrossRef]

49. Duong, D.T.; Singh, S.; Bagheri, M.; Verma, N.K.; Schmidtchen, A.; Malmsten, M. Pronounced peptide selectivity for melanoma through tryptophan end-tagging. *Sci. Rep.* **2016**, *6*, 24952. [CrossRef]

50. Ember, J.A.; Johansen, N.L.; Hugli, T.E. Designing synthetic superagonists of C3a anaphylatoxin. *Biochemistry* **1991**, *30*, 3603–3612. [CrossRef]

51. Chu, H.-L.; Yu, H.-Y.; Yip, B.-S.; Chih, Y.-H.; Liang, C.-W.; Cheng, H.-T.; Cheng, J.-W. Boosting Salt Resistance of Short Antimicrobial Peptides. *Antimicrob. Agents Chemother.* **2013**, *57*, 4050–4052. [CrossRef]

52. Chih, Y.-H.; Lin, Y.-S.; Yip, B.-S.; Wei, H.-J.; Chu, H.-L.; Yu, H.-Y.; Cheng, H.-T.; Chou, Y.-T.; Cheng, J.-W. Ultrashort Antimicrobial Peptides with Antiendotoxin Properties. *Antimicrob. Agents Chemother.* **2015**, *59*, 5052–5056. [CrossRef]

53. Yu, H.Y.; Chen, Y.A.; Yip, B.S.; Wang, S.Y.; Wei, H.J.; Chih, Y.H.; Chen, K.H.; Cheng, J.W. Role of b-naphthylalanine end-tags in the enhancement of antiendotoxin activities: Solution structure of the antimicrobial peptide S1-Nal-Nal in complex with lipopolysaccharide. *Biochim. Biophys. Acta* **2017**, *1859*, 1114–1123. [CrossRef]

54. Li, Y. Split-inteins and their bioapplications. *Biotechnol. Lett.* **2015**, *37*, 2121–2137. [CrossRef]

55. Wood, D.W.; Camarero, J.A. Intein Applications: From Protein Purification and Labeling to Metabolic Control Methods. *J. Biol. Chem.* **2014**, *289*, 14512–14519. [CrossRef]

56. Li, Y. Carrier proteins for fusion expression of antimicrobial peptides in *Escherichia coli*. *Biotechnol. Appl. Biochem.* **2009**, *54*, 1–9. [CrossRef]

57. Ganjave, S.D.; Dodia, H.; Sunder, A.V.; Madhu, S.; Wangikar, P.P. High cell density cultivation of E. coli in shake flasks for the production of recombinant proteins. *Biotechnol. Rep.* **2021**, *33*, e00694. [CrossRef]

Article

Novel Retro-Inverso Peptide Antibiotic Efficiently Released by a Responsive Hydrogel-Based System

Angela Cesaro [1,2,3], Rosa Gaglione [4,5], Marco Chino [4], Maria De Luca [4], Rocco Di Girolamo [4], Angelina Lombardi [4], Rosanna Filosa [6,7] and Angela Arciello [4,5,*]

1 Machine Biology Group, Departments of Psychiatry and Microbiology, Institute for Biomedical Informatics, Institute for Translational Medicine and Therapeutics, Perelman School of Medicine, University of Pennsylvania, Philadelphia, PA 19104, USA; angela.cesaro@pennmedicine.upenn.edu
2 Departments of Bioengineering and Chemical and Biomolecular Engineering, School of Engineering and Applied Science, University of Pennsylvania, Philadelphia, PA 19104, USA
3 Penn Institute for Computational Science, University of Pennsylvania, Philadelphia, PA 19104, USA
4 Department of Chemical Sciences, University of Naples Federico II, Via Cintia 21, I-80126 Naples, Italy; rosa.gaglione@unina.it (R.G.); marco.chino@unina.it (M.C.); maria.deluca2@unina.it (M.D.L.); rocco.digirolamo@unina.it (R.D.G.); alombard@unina.it (A.L.)
5 Istituto Nazionale di Biostrutture e Biosistemi (INBB), I-00136 Rome, Italy
6 AMP Biotec, Research Start-Up, s.s. 7, km 256, I-82030 Apollosa, Italy; rfilosa@unisannio.it
7 Department of Sciences and Technology, University of Sannio, Via De Sanctis, I-82100 Benevento, Italy
* Correspondence: anarciel@unina.it

Abstract: Topical antimicrobial treatments are often ineffective on recalcitrant and resistant skin infections. This necessitates the design of antimicrobials that are less susceptible to resistance mechanisms, as well as the development of appropriate delivery systems. These two issues represent a great challenge for researchers in pharmaceutical and drug discovery fields. Here, we defined the therapeutic properties of a novel peptidomimetic inspired by an antimicrobial sequence encrypted in human apolipoprotein B. The peptidomimetic was found to exhibit antimicrobial and anti-biofilm properties at concentration values ranging from 2.5 to 20 μmol L^{-1}, to be biocompatible toward human skin cell lines, and to protect human keratinocytes from bacterial infections being able to induce a reduction of bacterial units by two or even four orders of magnitude with respect to untreated samples. Based on these promising results, a hyaluronic-acid-based hydrogel was devised to encapsulate and to specifically deliver the selected antimicrobial agent to the site of infection. The developed hydrogel-based system represents a promising, effective therapeutic option by combining the mechanical properties of the hyaluronic acid polymer with the anti-infective activity of the antimicrobial peptidomimetic, thus opening novel perspectives in the treatment of skin infections.

Keywords: drug design; antimicrobial peptidomimetic; hydrogel-based system; hyaluronic acid; anti-infective activity; skin infections

Citation: Cesaro, A.; Gaglione, R.; Chino, M.; De Luca, M.; Di Girolamo, R.; Lombardi, A.; Filosa, R.; Arciello, A. Novel Retro-Inverso Peptide Antibiotic Efficiently Released by a Responsive Hydrogel-Based System. *Biomedicines* **2022**, *10*, 1301. https://doi.org/10.3390/biomedicines10061301

Academic Editor: Jitka Petrlova

Received: 4 May 2022
Accepted: 29 May 2022
Published: 2 June 2022

Publisher's Note: MDPI stays neutral with regard to jurisdictional claims in published maps and institutional affiliations.

1. Introduction

The dramatic emergence of antibiotic resistance is leading to the design of novel antimicrobials that are less susceptible to resistance mechanisms than conventional antibiotics. In this context, innovative molecules such as antimicrobial peptides (AMPs) represent a promising alternative to the old generations of antibiotics since the evolution of resistance against these compounds is demonstrated to be less probable with respect to conventional antimicrobials [1–4]. Despite their numerous advantages, antimicrobial peptides' conformation often makes them extremely sensitive to endo- and exopeptidases present in biological systems [5]. A powerful tool to transform proteolytically labile peptides into compounds with improved bioactivities and pharmacokinetic profiles is represented by the development of peptidomimetics. Peptidomimetics are defined as "*compounds whose essential elements mimic a natural peptide and that retain the ability to interact with the biological target by producing the same biological effect"* [6]. Many different approaches are used to generate peptidomimetics, and the selection of the most suitable design strategy depends on the available information about the parental peptide in terms of structure, sequence, and function [6,7]. To ensure peptide proteolytic stability, we modified the primary structure of a naturally occurring antimicrobial peptide encrypted in human plasma apolipoprotein B [8–17] by designing a retro-inverso sequence composed entirely of D-enantiomeric amino acids, here named (ri)-r(P)ApoB$_S$^{Pro} [17]. In the present work, we investigate and characterize the biological activities of the designed peptidomimetic and develop a suitable topical delivery system, allowing the specific release of the selected antimicrobial in therapeutic procedures.

Several formulations, including suspensions, micro/nanoparticles, patches, and hydrogels, have been proposed to control topical drug delivery. Among different alternatives, hydrogel-based systems have been found to offer many advantages in the treatment of skin wounds and infections [18,19]. According to the definition, hydrogel-based systems are networks of natural or synthetic cross-linked polymers with a three-dimensional configuration able to imbibe high amounts of water [20]. The hydrogel's swelling ability is the result of the hydrophilic moieties in the structure, which are resistant to water dissolution thanks to the presence of cross-linkers between polymeric chains [21]. Several polymers have been chemically modified to develop synthetic hydrogels with improved properties [21]. These delivery systems are classified on the basis of their origin, physical properties, nature of swelling, method of preparation, ionic charges, rate of biodegradation, and kind of cross-linking [22]. Recently, hydrogels based on hyaluronic acid (HA) confirmed their excellent properties not only as scaffolds but also as suitable carriers of biologically active substances [20]. HA is an essential component of the natural extracellular matrix, where it plays an important role in many biological processes, including wound healing. HA is biocompatible, biodegradable, and non-immunogenic and has unique viscoelastic and rheological properties that make it an excellent polymer to build hydrogel systems with desired morphology and bioactivity [23]. Here, we developed and characterized an HA-based hydrogel system loaded with the peptidomimetic (ri)-r(P)ApoB$_S$^{Pro} with the main aim to specifically deliver this antimicrobial agent to the site of infection. We show that the combination of HA biological properties, of hydrogel system swelling and hydrating capabilities, together with the efficacy of the selected antimicrobial peptidomimetic concurs to create a promising and effective system that is able to target skin disorders and infections, thus opening interesting perspectives to the future applicability of the retro-inverso peptide in the dermatological field.

2. Materials and Methods

2.1. Materials

All the reagents were purchased from Sigma-Merck (Milan, Italy), unless differently specified. HA cross-linked with 2.5% 1,4-butanediol diglycidyl ether (BDDE) was provided by Altergon Italia s.r.l. (Morra De Sanctis, AV, Italy).

2.2. Peptide Synthesis

(ri)-r(P)ApoB$_S$^{Pro} peptide was obtained from CASLO ApS (Kongens Lyngby, Denmark).

2.3. Bacterial Strains and Growth Conditions

Four bacterial strains were used in the present study, i.e., methicillin-resistant *Staphylococcus aureus* (MRSA WKZ-2), *Staphylococcus aureus* ATCC 29213, *Escherichia coli* ATCC 25922, and *Pseudomonas aeruginosa* ATCC 27853. All bacterial strains were grown in Mueller Hinton broth (MHB; Becton Dickinson Difco, Franklin Lakes, NJ, USA) and on tryptic soy agar (TSA; Oxoid Ltd., Hampshire, UK). In all the experiments, bacteria were inoculated and grown over-night in MHB at 37 °C. After 24 h, bacteria were transferred to a fresh MHB tube and grown to mid-logarithmic phase.

2.4. Antimicrobial Activity

The antimicrobial activity of (ri)-(r)ApoB$_S^{Pro}$ peptide was assayed on methicillin-resistant *S. aureus* (MRSA WKZ-2), *S. aureus* ATCC 29213, *E. coli* ATCC 25922, and *P. aeruginosa* ATCC 27853 by using broth microdilution method. Bacteria were grown to the mid-logarithmic phase in MHB at 37 °C and then diluted to 4×10^6 CFU/mL in Difco 0.5X nutrient broth (NB; Becton-Dickenson, Franklin Lakes, NJ, USA). To perform the assay, bacterial samples were mixed 1:1 *v/v* with two-fold serial dilutions of the peptidomimetic (0–40 µmol L^{-1}) and incubated for 20 h at 37 °C. Following the incubation, each sample was diluted and plated on TSA, in order to count the number of colonies [24–26]. All the experiments were carried out as triplicates.

2.5. Anti-Biofilm Activity Assays

Anti-biofilm activity assays were performed on methicillin-resistant *S. aureus* (MRSA WKZ-2) and *E. coli* ATCC 25922. Bacterial inocula were grown overnight at 37 °C, then diluted to 4×10^9 CFU/mL in 0.5X MHB medium and mixed 1:1 *v/v* with increasing concentrations of the peptidomimetic (0–40 µmol L^{-1}). The samples were incubated at 37 °C either for 4 h, in order to test the effects on cell attachment, or for 24 h, in order to assay peptide effects on biofilm formation, as previously described [11,17]. In the case of the crystal violet assay, the bacterial biofilm was washed with phosphate buffer (PBS 1X) and then incubated with the dye (0.04%) for 20 min at room temperature. Following the incubation, samples were washed with PBS and then the dye bound to cells was dissolved in 33% acetic acid. Spectrophotometric measurements were then carried out at a wavelength of 600 nm. Confocal laser scanning microscopy (CLSM) analyses in static conditions were carried out by using Thermo Scientific™ Nunc™ Lab-Tek™ Chambered Coverglass systems (Thermo Fisher Scientific, Waltham, MA, USA). The viability of the cells within the biofilm structure was evaluated by sample staining with a LIVE/DEAD® Bacterial Viability kit (Molecular Probes Thermo Fisher Scientific, Waltham, MA, USA). Staining was performed accordingly to manufacturer's instructions. Biofilm images were collected by using a confocal laser scanning microscope (Zeiss LSM 710, Zeiss, Germany) and a 63X objective oil-immersion system. Biofilm architecture was analyzed by using the Zen Lite 2.3 software package. Each experiment was performed in triplicate, and all the images were taken under identical conditions.

2.6. Circular Dichroism (CD) Spectroscopy Analyses

CD experiments were performed by using a Jasco J-815 circular dichroism spectropolarimeter. The cell path length was 0.1 or 1 cm, depending on the peptide concentration. CD spectra were collected at 25 °C in the far-UV region (190–260 nm) at 0.2 nm intervals, with a 50 nm/min scan rate, 1.0 nm bandwidth, and a 4 s response. Either 4 or 16 accumulations were performed during titration experiments. Each spectrum was corrected by subtracting the background spectrum and reported with a fast-Fourier-transform (FFT) filter. The lyophilized peptidomimetic was dissolved in ultra-pure water (Romil, Waterbeach, Cambridge, GB, UK) at a concentration of 1600 µmol L^{-1}, determined on the basis of peptide dry weight and by the bicinchoninic acid (BCA) colorimetric assay (Thermo Fisher Scientific, Waltham, MA, USA). CD spectra of the peptidomimetic were collected in buffer alone (2.5 mM phosphate buffer, pH 7.4) or in the presence of 50% *v/v* trifluoroethanol (TFE, Sigma-Merck, Milan, Italy), lipopolysaccharide (LPS) from *E. coli* 0111:B4 strain (Sigma-Merck, Milan, Italy), or lipoteichoic acid (LTA) from *S. aureus* (Sigma-Merck, Milan, Italy). Inverse titration experiments were performed by adding known amounts of peptide in either 0.2 mg/mL LPS or LTA. After addition, samples were mixed with a magnetic stirrer for 3 min prior to CD analyses. CD spectra were corrected by subtracting every time the contribution of the compound under test at any given concentration. CD spectra deconvolutions were performed by using a Microsoft Excel-ported version of the PEPFIT program that is based on peptide-derived reference spectra, in order to estimate secondary structure contents after specular inversion of the raw data [11,27].

2.7. Preparation of HA-BDDE Hydrogel Loaded with (ri)-r(P)ApoB$_S$^{Pro} Peptidomimetic

Peptidomimetic (ri)-r(P)ApoB$_S$^{Pro} was added to hyaluronic acid hydrogel cross-linked with 2.5% 1,4-butanediol diglycidyl ether (HA-BDDE) at a ratio of 4:1 (*v/v*). A solution of peptidomimetic at a concentration of 80 or 320 µmol L^{-1} was mixed with HA-BDDE (density of 1 g mL^{-1}), properly stirred, and then lyophilized for 24 h, in order to obtain white spongy samples [28,29]. Control samples were prepared under the same experimental conditions but in the absence of the peptidomimetic. Prior to all the characterizations, dried hydrogels were rehydrated and sterilized by exposure to UV light for 20 min.

2.8. Eukaryotic Cell Culture and Biocompatibility Evaluation of (ri)-r(P)ApoB$_S$^{Pro} and of HA-BDDE Hydrogel System

Immortalized human keratinocytes (HaCaTs) and human dermal fibroblasts (HDFs) were cultured in high-glucose Dulbecco's modified Eagle's medium (DMEM) supplemented with 10% fetal bovine serum (FBS), 1% antibiotics (Pen/strep), and 1% L-glutamine and grown at 37 °C in a humidified atmosphere containing 5% CO$_2$. To evaluate peptidomimetic biocompatibility, cells were seeded into 96-well plates at a density of 3 × 10^3 cells/well in 100 µL of complete DMEM for 24 h prior to the treatment. Cells were then incubated in the presence of increasing peptidomimetic concentrations (0–20 µmol L^{-1}) for 24, 48, and 72 h. When evaluating the biocompatibility of the HA-BDDE hydrogel system, cells were seeded into 96-well plates at a density of 6 × 10^3 cells/well in 100 µL of complete DMEM for 24 h prior to the treatment. Following incubation, cells were treated with HA-BDDE hydrogel hydrated with growth medium and previously loaded with the peptidomimetic at a final concentration of 80 or 320 µmol L^{-1}. In all cases, following treatment, cell culture supernatants were replaced with 0.5 mg/mL MTT (3-(4,5-dimethylthiazol-2-yl)-2,5-diphenyltetrazolium bromide) reagent dissolved in DMEM medium without red phenol (100 µL/well). After 4 h of incubation at 37 °C, the formazan salts were solubilized in 0.01 M HCl in anhydrous isopropanol, and the absorbance of the obtained samples was measured at λ = 570 nm using an automatic plate reader spectrophotometer (Synergy™ H4 Hybrid Microplate Reader, BioTek Instruments, Inc., Winooski, VT, USA), as previously described [25]. Cell survival was expressed as the mean of the percentage values compared to control untreated cells.

2.9. Hemolytic Activity

The release of hemoglobin from human erythrocytes was used as a measure for the hemolytic activity of (ri)-r(P)ApoB$_S$^{Pro}. Briefly, human red blood cells (RBCs) were collected from EDTA anti-coagulated blood, washed three times by centrifugation at 800× *g* for 10 min, and 200-fold diluted in PBS pH 7.4. Aliquots of diluted erythrocytes (75 µL) were added to the peptide solution (0–40 µmol L^{-1}; 75 µL) in 96-well microtiter plates, and the mixtures were incubated for 1 h at room temperature. Following the incubation, the plate was centrifuged for 10 min at 1300× *g*, and 100 µL of supernatant from each well was transferred to a new 96-well plate, as previously described [25]. Absorbance values were determined at 405 nm by using an automatic plate reader (Synergy™ H4 Hybrid Microplate Reader, BioTek Instruments, Inc., Winooski, VT, USA). The percentage of hemolysis was determined by comparison with the control samples containing PBS (negative control) or 1% (*v/v*) SDS in PBS solution (positive control, complete lysis).

$$\text{Hemolysis (\%)} = \frac{(\text{Abs405 nm peptide} - \text{Abs405 nm negative control})}{(\text{Abs405 nm positive control} - \text{Abs405 nm negative control})} \times 100 \tag{1}$$

2.10. Cell Infection Assay

Immortalized human keratinocytes (HaCaTs) were seeded into 24-well plates at a density of 3 × 10^5 cells/well and allowed to attach for 24 h at 37 °C in a humidified atmosphere containing 5% CO$_2$. Following incubation, cells were washed three times with PBS and infected with *E. coli* ATCC 25922 at multiplicity of infection (MOI) of 2 in the presence or absence of (ri)-r(P)ApoB$_S$^{Pro} 10 µmol L^{-1}. To measure the number of bacteria in the wells, the keratinocytes were washed three times with PBS, and 500 µL of 1% Triton X-100 was added in order to detach and lyse the cells. Samples containing the detached keratinocytes were serially diluted, plated onto TSA agar plates, and incubated overnight at 37 °C to count the number of colonies. The number of bacteria present in the samples was analyzed over time (0–2–4 h).

2.11. Characterization of Swelling Properties of HA-BDDE Hydrogel Loaded with (ri)-r(P)ApoB$_S$^{Pro} Peptidomimetic

The hydrogel's swelling properties were characterized by measuring the gravimetric change over time as described by Jie Zhu et al. [30]. Briefly, the freeze-dried hydrogels were weighed (S_0) and then immersed into distilled water. To obtain the swelling ratio, hydrogels samples were weighed (S_1) at different time intervals until swelling equilibrium was reached. Each sample was measured in three replicates. The hydrogel's swelling ratio was calculated as follows:

$$\text{Swelling ratio } (\%) = \frac{S_1 - S_0}{S_0} \times 100 \tag{2}$$

2.12. Degradation Analyses of HA-BDDE Hydrogel Loaded with (ri)-r(P)ApoB$_S$^{Pro} Peptidomimetic

The swollen hydrogels were weighed (W_0), and degradation performances were evaluated by incubating the samples with 300 U mL^{-1} of hyaluronidase (ref. H3506, Sigma-Merck) solubilized in 0.02 M sodium phosphate, pH 7.4, containing 0.01% bovine serum albumin and 77 mM NaCl, as described by Wanxu Cao et al. [31]. The mixtures were placed in a water bath at 37 °C, and the weight loss percentage was calculated by weighing samples (W_1) at defined time intervals. Every sample was measured in three replicates. Degradation behavior was expressed as the percentage of weight loss and was calculated as follows:

$$\text{Weight change } (\%) = \frac{W_1}{W_0} \times 100 \tag{3}$$

2.13. Rheological Characterization of HA-BDDE Hydrogel Loaded with (ri)-r(P)ApoB$_S$^{Pro} Peptidomimetic

Rheological measurements were performed by using an Anton Paar MCR302 rheometer with d = 50 mm cone geometry (cone angle 1°, gap 101 µm). Briefly, 2 mL of fresh HA-BDDE hydrogel or HA-BDDE hydrogel loaded with 320 µmol L^{-1} of (ri)-r(P)ApoB$_S$^{Pro} were deposited between the cone and the plate, and the parameters (G'—storage modulus, and G''—loss modulus) were followed at 37 °C at a constant frequency of 1.59 Hz.

2.14. In Vitro Peptidomimetic Release from Hydrogel System

To determine peptidomimetic release, HA-BDDE hydrogels loaded with 80 or 320 µmol L^{-1} (ri)-r(P)ApoB$_S$^{Pro} were immersed in distilled water or saline buffer (30 mM HEPES buffer + 50 mM NaCl to reach pH 7 and 30 mM CH$_3$COOH buffer + 50 mM NaCl to reach pH 5), in order to simulate the skin ionic strength according to Traeger and co-workers [32], and then incubated at 37 °C over time. At defined time points, aliquots of 100 µL were collected and 100 µL of water were added, in order to keep a constant volume. The amounts of released (ri)-r(P)ApoB$_S$^{Pro} were measured by HPLC analyses performed by using a Shimadzu LC-20 Prominence (Shimadzu Corporation, Kyoto, Japan) mounted with a thermostated PDA detector (SPD-M20A). For this purpose, samples were loaded on a Phenomenex Aeris Peptide XB-C18 3.6 µm column (150 × 4.6 mm), and eluted with a linear gradient of ultra-pure water (A) and acetonitrile (B) (UpS solvent, Romil) from 15% to 95% of solvent B over 20 min. TFA (0.1% *v/v*) was added to both solvents. To quantify the released peptidomimetic, a five-point calibration curve was used. Each calibration standard solution was prepared by diluting the primary standard (1600 µmol L^{-1} determined on the basis of peptidomimetic dry weight) in the working buffer (ultra-pure water), in order to obtain the following concentrations: 2.5, 5, 10, 20, and 40 μmol L^{-1}. Linear regression was used to fit the peak areas as integrated from the 220 nm chromatogram ($R^2 = 0.997$). A blank sample was included in each calibration curve. Two separate analyses of peptidomimetic release were performed on a blind and randomized basis, by injecting all the samples collected at different time intervals (0, 0.5, 1, 2, 3, 24, 48, 72 h). For quality control, blank, zero, and spiked samples (5, 10, and 40 μmol L^{-1}) were injected, with the latter being found within 10% deviation of the nominal concentration.

2.15. Antimicrobial Activity of HA-BDDE Hydrogel System Loaded with the Peptidomimetic

The antimicrobial activity of HA-BDDE hydrogel loaded with (ri)-r(P)ApoB$_S$^{Pro} was assayed on methicillin-resistant *S. aureus* (MRSA WKZ-2) and *E. coli* ATCC 25922 bacterial strains. Bacteria were grown to mid-logarithmic phase in MHB at 37 °C. Afterward, cells were diluted to 4×10^6 CFU/mL in Difco 0.5X NB (Becton-Dickenson, Franklin Lakes, NJ, USA) and mixed 1:1 *v/v* with HA-BDDE hydrogel loaded with the peptidomimetic at a final concentration of 80 or 320 μmol L^{-1}. Following overnight incubation, each sample was diluted, plated on TSA, and incubated at 37 °C for 24 h, in order to count the number of colonies. All the experiments were carried out as triplicates.

2.16. Migration Assay

To examine whether the hydrogel system loaded with the peptidomimetic is able to counteract bacterial migration across surfaces, the upper chambers of a trans-well plate (Costar 3422®, Corning Corporation, Corning, NY, USA) were coated with control saline solution, HA-BDDE hydrogel or HA-BDDE hydrogel loaded with (ri)-r(P)ApoB$_S$^{Pro}, as described by Xiaojuan Li et al. [33]. Following coating, methicillin-resistant *S. aureus* (MRSA WKZ-2) and *E. coli* ATCC 25922 bacterial cells were diluted to 4×10^6 CFU/mL in Difco 0.5X NB (Becton-Dickinson, Franklin Lakes, NJ, USA), and plated into the previously coated upper chambers. The medium in the lower wells was then analyzed at defined time intervals, in order to monitor bacterial cells' migration from the upper to the lower wells. The number of migrated bacteria was quantified by diluting and plating each sample on TSA. Following an incubation of 24 h at 37 °C, the number of colonies was counted. The experiment was performed in three independent replicates.

2.17. Scanning Electron Microscopy Analyses of Bacterial Cells Treated with HA-BDDE Hydrogel System Loaded with the Peptidomimetic

To perform scanning electron microscopy (SEM) analyses, methicillin-resistant *S. aureus* (MRSA WKZ-2) and *E. coli* ATCC 25922 cells at a density of 4×10^6 CFU/mL in Difco 0.5X NB were mixed 1:1 *v/v* with HA-BDDE hydrogel loaded with the peptidomimetic at a final concentration of 80 or 320 μmol L^{-1}. Following overnight incubation at 37 °C, untreated methicillin-resistant *S. aureus* (MRSA WKZ-2) cells were centrifuged at 10,000 rpm at 4 °C, since a complete hydrolysis of hydrogel matrix due to bacteria was observed. In the case of peptide-treated methicillin-resistant *S. aureus* (MRSA WKZ-2) cells and all *E. coli* ATCC 25922 samples, mixtures were, instead, directly placed on a glass slide and then fixed in 2.5% glutaraldehyde at the end of the incubation. After 20 h incubation, the samples were washed three times in distilled water (dH$_2$O) and dehydrated with a graded ethanol series: 25% ethanol (1×10 min); 50% ethanol (1×10 min); 75% ethanol (1×10 min); 95% ethanol (1×10 min); 100% anhydrous ethanol (3×30 min). At the end of the alcoholic dehydration process, methicillin-resistant *S. aureus* (MRSA WKZ-2) samples were also deposited into the glass substrate, and all the samples were sputter-coated with a thin layer of Au-Pd (Sputter Coater Denton Vacuum Desk V) to allow their subsequent morphological characterization using a FEI Nova NanoSEM 450 at an accelerating voltage of 5 kV with an Everhart Thornley Detector (ETD) and a Through Lens Detector (TLD) at high magnification.

3. Results

3.1. In Vitro Antimicrobial and Anti-Biofilm Activity of Synthetic Retro-Inverso r(P)ApoB$_S$^{Pro}

Firstly, we assessed the antimicrobial activity of the peptidomimetic (ri)-r(P)ApoB$_S$^{Pro} against a panel of four Gram-negative and Gram-positive bacterial strains, selected among those more frequently isolated from skin and soft tissue infections [34–36]. The peptidomimetic was found to inhibit the bacterial growth at concentration values ranging from 5 to 10 µmol L^{-1} in the case of *P. aeruginosa* ATCC 27853 and *E. coli* ATCC 25922, and at concentration values ranging from 10 to 20 µmol L^{-1} in the case of *S. aureus* ATCC 29213 and methicillin-resistant *S. aureus* (MRSA WKZ-2). Based on these results, we can conclude that the peptidomimetic is endowed with significant broad-spectrum antimicrobial properties (Figure 1a). Furthermore, (ri)-r(P)ApoB$_S$^{Pro} was found to inhibit (40–50%) biofilm adhesion and formation when tested on Gram-negative *E. coli* ATCC 25922 (Figure 1b). No significant anti-biofilm effects were detected, instead, when the peptide was tested on Gram-positive methicillin-resistant *S. aureus* (MRSA WKZ-2) by performing crystal violet assays (Figure 1b). To further investigate the peptidomimetic anti-biofilm activity on *E. coli* ATCC 25922 strain, confocal laser scanning microscopy (CLSM) analyses were also performed, and it was demonstrated that (ri)-r(P)ApoB$_S$^{Pro} at a concentration of 2.5 µmol L^{-1} is able to significantly affect the biofilm formation of *E. coli* ATCC 25922 by determining a significant reduction of biofilm biovolume (Figure 1c). Such results appear to be in agreement with the findings previously obtained for natural ApoB-derived antimicrobial peptides, whose antimicrobial properties strictly depend on specific features of the bacterial strains under test [8].

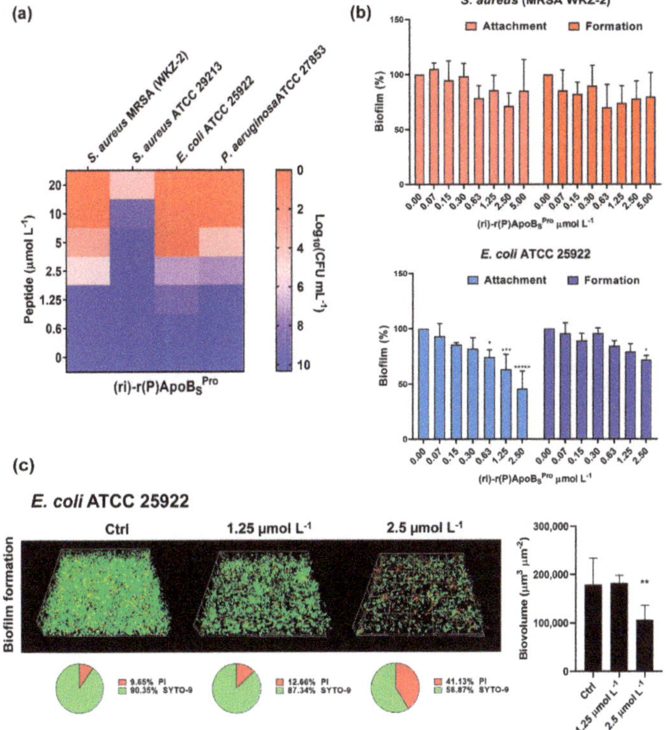

Figure 1. Antimicrobial and anti-biofilm activities of retro-inverso synthetic peptide. (**a**) Antimicrobial activity of (ri)-r(P)ApoB$_S$^{Pro} (µmol L^{-1}) against four bacterial strains; reported data refer to assays performed in triplicate and heat maps show averaged log$_{10}$(CFU mL^{-1}) values. (**b**) Anti-biofilm activity

of (ri)-r(P)ApoB$_S$^{Pro} on methicillin-resistant *S. aureus* (MRSA WKZ-2) and *E. coli* ATCC 25922 strains. The effects of increasing concentrations of (ri)-r(P)ApoB$_S$^{Pro} peptide were evaluated on cells attachment and biofilm formation. Data represent the mean (±standard deviation, SD) of at least three independent experiments, each one carried out with triplicate determinations. (c) Effects of (ri)-r(P)ApoB$_S$^{Pro} on biofilm formation in the case of *E. coli* ATCC 25922 by CLSM imaging. Two-dimensional structures of the biofilms were obtained by confocal z-stack using Zen Lite 2.3 software. Biovolume (μm^3 μm^{-2}) values were calculated by using Zen Lite 2.3 software. Significant differences were indicated as * $p < 0.05$, ** $p < 0.01$, *** $p < 0.001$, or ***** $p < 0.00001$ for treated vs. control samples, each experiment was carried out in triplicate.

3.2. Conformational Analyses of (ri)-r(P)ApoB$_S$^{Pro} Peptide by Far-UV Circular Dichroism

Far-UV CD spectroscopy was used to analyze the correlation between peptidomimetic antimicrobial activity and conformations. The retro-inverso peptide appeared largely unstructured in aqueous buffer (2.5 mM phosphate buffer, pH 7.4), as evidenced by th broad maximum centered at 199 nm in the CD spectra (Figure 2a). However, it should be highlighted that the molar residue ellipticity appeared significantly lower in absolute values with respect to that previously obtained for its (L)-amino acid variant [11] (~13 kdeg cm^2 dmol^{-1} res^{-1} vs. ~30 kdeg cm^2 dmol^{-1} res^{-1} for (ri)-r(P)ApoB$_S$^{Pro} and r(P)ApoB$_S$^{Pro}, respectively), thus suggesting the presence of a higher content of secondary structure elements in the case of the peptidomimetic. Indeed, deconvolution analyses revealed the presence of approximately 34% of β-sheet secondary structure content, whereas previously characterized r(P)ApoB$_S$^{Pro} was found to assume a fully random coil conformation [11]. When we analyzed the effects of 50% TFE, a widely adopted secondary structure inducer, (ri)-r(P)ApoB$_S$^{Pro} assumed a partial (~10%) helical conformation (Figure 2a), as evidenced by the presence of two broad minima at around 208 and 222 nm, and a maximum at <200 nm. This behavior is similar to that previously described for its (L)-amino acid variant r(P)ApoB$_S$^{Pro} peptide [11].

More interestingly, differently from its (L)-peptide counterpart [11], the retro-inverso variant's secondary structure was significantly altered upon incubation with lipopolysaccharide (LPS), the predominant glycolipid in the outer membrane of Gram-negative bacteria. When increasing amounts of peptidomimetic were incubated with 0.2 mg/mL LPS endotoxin, the maximum wavelength progressively shifted from 220 to 200 nm (Figure 2c), suggesting a major switch toward a β-strand conformation at higher LPS:peptide ratios, probably due to an interaction between the peptidomimetic and LPS. Considering that the peptidomimetic exerts a stronger direct antimicrobial activity on Gram-negative bacteria as opposed to Gram-positive bacteria, we hypothesized a less pronounced effect induced on peptide conformation by lipoteichoic acid (LTA), an anionic glycerol phosphate polymer of the cell wall of Gram-positive strains. Indeed, when increasing amounts of the peptidomimetic were incubated with 0.2 mg/mL LTA from *S. aureus*, less-pronounced variations of the CD spectrum were observed (Figure 2d). In particular, band maximum wavelength shifted only from 215 to 200 nm with a far lower effect on the molar residue ellipticity, most probably due to only minimal β-strand formation. CD spectra deconvolutions are summarized in Figure 2b. Data appear in agreement with the higher antimicrobial activity of the peptidomimetic (ri)-r(P)ApoB$_S$^{Pro} against Gram-negative bacteria compared to Gram-positive strains. Such results appear even more interesting considering the recently published observations indicating that the antimicrobial activity of two natural versions of the antimicrobial peptide identified in human ApoB depends on their interaction with specific components of bacterial surfaces, such as LPS or LTA, which induces peptides to form β-sheet-rich amyloid-like structures that are probably responsible for their antibacterial activity [15].

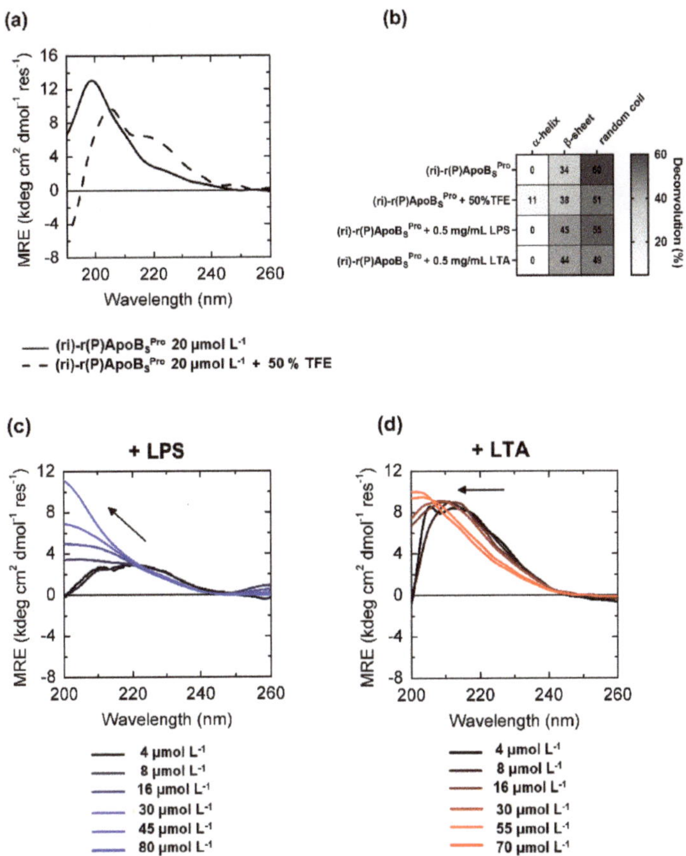

Figure 2. Conformational analyses of (ri)-r(P)ApoB$_S$^{Pro} peptide by far-UV circular dichroism. (**a**) CD spectra of peptidomimetic (ri)-r(P)ApoB$_S$^{Pro} 20 μmol L^{-1} in the absence (black line) or in the presence (black dashed line) of 50% (*v/v*) TFE. (**b**) CD spectra deconvolution percentages of (ri)-r(P)ApoB$_S$^{Pro} peptide in 2.5 mM sodium phosphate pH 7.4 buffer and in the presence of 50% TFE, 0.5 mg mL^{-1} LPS or LTA. Secondary structure percentages were calculated using CDPRO software. CD spectra at different concentrations of (ri)-r(P)ApoB$_S$^{Pro} in the presence of constant amounts (0.2 mg mL^{-1}) of either LPS (**c**) or LTA (**d**).

3.3. Biocompatibility of (ri)-r(P)ApoB$_S$^{Pro} toward Human Skin Cells

We performed additional experiments to verify the biocompatibility of (ri)-r(P)ApoB$_S$^{Pro} toward skin cell cultures and to exclude hemolytic effects that may result from incorporation of D-amino acids into peptide sequences [37]. Peptidomimetic (ri)-r(P)ApoB$_S$^{Pro} did not significantly affect cell viability, as indicated by the MTT assay results (Figure 3a). Indeed, the viability of immortalized human keratinocytes (HaCaTs) was found to be unchanged upon exposure to increasing concentrations of (ri)-r(P)ApoB$_S$^{Pro} for different time intervals (Figure 3a). Moreover, when we tested (ri)-r(P)ApoB$_S$^{Pro} on human red blood cells (RBCs), slight lytic effects were observed only at the highest peptidomimetic concentrations tested, thus indicating that the peptidomimetic is endowed with a good profile of biocompatibility (Figure 3b).

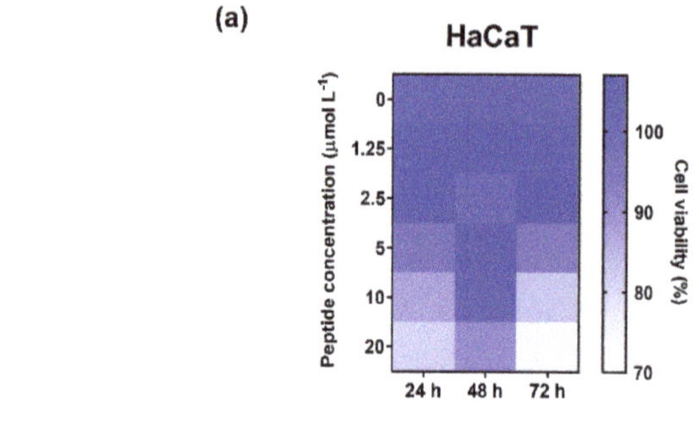
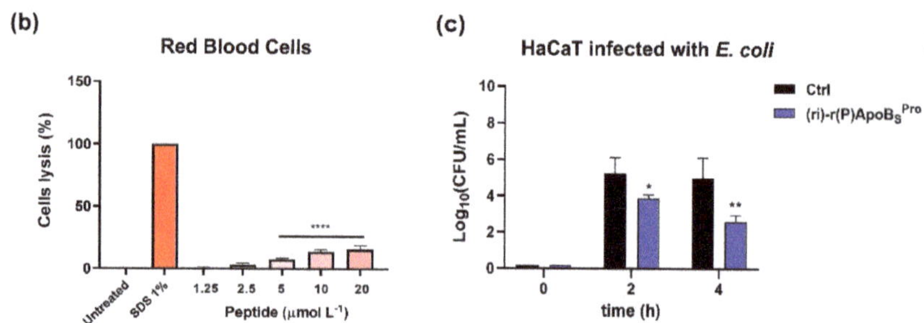

Figure 3. Effects of retro-inverso synthetic peptide on human eukaryotic cell lines. (**a**) Cytotoxic effects of increasing concentrations of (ri)-r(P)ApoB$_S$^{Pro} on HaCaT (immortalized human keratinocytes) cells. Cell viability was assessed by MTT assays and expressed as the percentage of viable cells compared to untreated cells (control). (**b**) Analysis of hemolytic effects of (ri)-r(P)ApoB$_S$^{Pro} toward human red blood cells (RBCs) after 1 h of incubation at 37 °C. Data represent the mean (±standard deviation, SD) of at least three independent experiments, each one carried out with triplicate determinations. (**c**) Effects of (ri)-r(P)ApoB$_S$^{Pro} peptide on the adhesion of *E. coli* bacterial cells to HaCaT cells. Experiments were performed three times in duplicate. Error bars represent the standard deviation of the mean. Significant differences were indicated as * $p < 0.05$, ** $p < 0.01$ or **** $p < 0.0001$ for treated vs. control samples.

Since skin and soft tissue infections occur when bacteria adhere to host cells, the ability of the peptidomimetic to prevent the adhesion of pathogens to skin cells was also evaluated. For this purpose, human keratinocytes infected with *E. coli* ATCC 25922 cells were treated with (ri)-r(P)ApoB$_S$^{Pro}. After a 4 h incubation, the synthetic retro-inverso peptide was able to reduce by 50% the number of bacterial cells infecting keratinocytes with respect to control untreated cells (Figure 3c). This property opens interesting perspectives to the future applicability of (ri)-r(P)ApoB$_S$^{Pro} in the development of therapeutic strategies to treat skin infections.

3.4. Swelling and Degradation Profiles of HA Hydrogel System Loaded with (ri)-r(P)ApoB$_S$^{Pro}

To identify a suitable system to topically deliver the peptidomimetic, a proper HA-BDDE hydrogel system loaded with (ri)-r(P)ApoB$_S$^{Pro} was prepared as described in Section 2.7 of Section 2. The ability to retain a high amount of water is one of the key properties of hydrogel-based systems. Indeed, they can swell in water under physiological conditions without dissolving. The swelling capability plays a key role in keeping the injured site moist and in controlling bleeding. Moreover, this mechanical characteristic is fundamental to allow molecules to diffuse into or to be released from the hydrogels [30]. The swelling properties of hyaluronic acid gels can be affected by pH, ionic strength, temperature, and composition of the surrounding solution [38]. We examined whether the presence of the peptidomimetic could affect swelling and degradation profiles of HA-BDDE. As shown in Figure 4a, HA-BDDE hydrogels loaded with two different concentrations of the peptidomimetic (80 or 320 μmol L^{-1}) retained a swelling behavior similar to that of control sample, thus indicating that the presence of the peptidomimetic does not affect the ability of the system to absorb water and to expand. Similarly, the degradation profile was found to be unaffected by the presence of the peptide, since the activity of hyaluronidase enzyme, which acts by cleaving the (1→4)-linkages between N-acetylglucosamine and glucuronate, was found to be unmodified in the presence of the peptidomimetic (Figure 4b). Hence, unlike covalently functionalized peptides, which significantly affect the swelling capacity of hydrogels, (ri)-r(P)ApoB$_S$^{Pro} encapsulated at high concentrations does not alter the mechanical properties of the hydrogel [39]. Altogether, these findings point to the proposed system as an effective approach to deliver the peptidomimetic to the site of infection.

3.5. Rheological Analyses of HA Hydrogel System Loaded with (ri)-r(P)ApoB$_S$^{Pro}

To analyze the rheological properties of the developed hydrogel-based systems, we evaluated the elastic modulus G′ (storage modulus) and the viscous modulus G″ (loss modulus) in amplitude, and we determined those parameters in the range of the linear viscoelastic region (LVR) for each sample (Figure 4c). The HA-BDDE hydrogel system loaded with (ri)-r(P)ApoB$_S$^{Pro} showed values of G′ and G″ (17.004 and 8.351, respectively) slightly higher than the control (13.254 and 6.326, respectively), thus indicating that the presence of the peptidomimetic within the gel might increase its viscosity by establishing weak links among the HA-filaments. However, the tan(δ) parameters of HA-BDDE hydrogel alone or loaded with the peptidomimetic are comparable (G″/G′ = 0.47 and G″/G′ = 0.49, respectively), thus demonstrating that (ri)-r(P)ApoB$_S$^{Pro} does not alter the "gel-like" structure [40] of the system (Figure 4c). This is probably due to the non-covalent bonds between the hydrogel chains and the peptide molecules that do not significantly affect the viscosity parameters as instead demonstrated for peptides that are covalently bound to the hydrogel polymers [41]. This represents an important finding, underlining that the presence of the peptidomimetic does not influence the main features that make hydrogel-based systems an ideal tool to topically deliver therapeutic agents.

3.6. Peptide Release from the Hydrogel System

Since a controlled and gradual release of the therapeutic agent is essential for the construction of a suitable drug delivery system, we performed experiments aimed at evaluating whether (ri)-r(P)ApoB$_S$^{Pro} is released over time, upon loading into the hydrogel. For this purpose, the HA-BDDE hydrogel system loaded with 80 or 320 μmol L^{-1} (ri)-r(P)ApoB$_S$^{Pro} was immersed in water or in saline buffer at pH 7 or 5, in order to simulate the pH of skin barrier in normal and pathological conditions, respectively [42]. The ionic strength of the saline buffer was designed to simulate that of sweat [32]. To evaluate peptidomimetic release, aliquots were collected at defined time intervals, in order to estimate the amount of the released peptidomimetic by HPLC analyses. Results indicate that the hydrogel was able to release 80% of the peptidomimetic amount when in contact with the saline buffer (independently from pH value) with respect to the water buffer, in which the maximum percentage of the released antimicrobial was 15% (Figure 5a). These data indicate that the selected system is able to release the antimicrobial by responding to external stimuli, such as the ionic strength, mimicking the sweat of the skin barrier. It is also important to notice that the percentage of release of the peptidomimetic always appeared to be proportional to the amount of (ri)-r(P)ApoB$_S$^{Pro} initially loaded into the hydrogel (Figure 5b). It has to be highlighted that swellable systems are generally able to release water-soluble molecules upon the entry of water molecules into the gel, swelling of the matrix, and consequent drug diffusion [43]. The behavior here observed for peptide release seems to fit a linear Fickian curve [44], since it is characterized by an initial linear increase followed by the gradual reaching of a plateau (Figure 5).

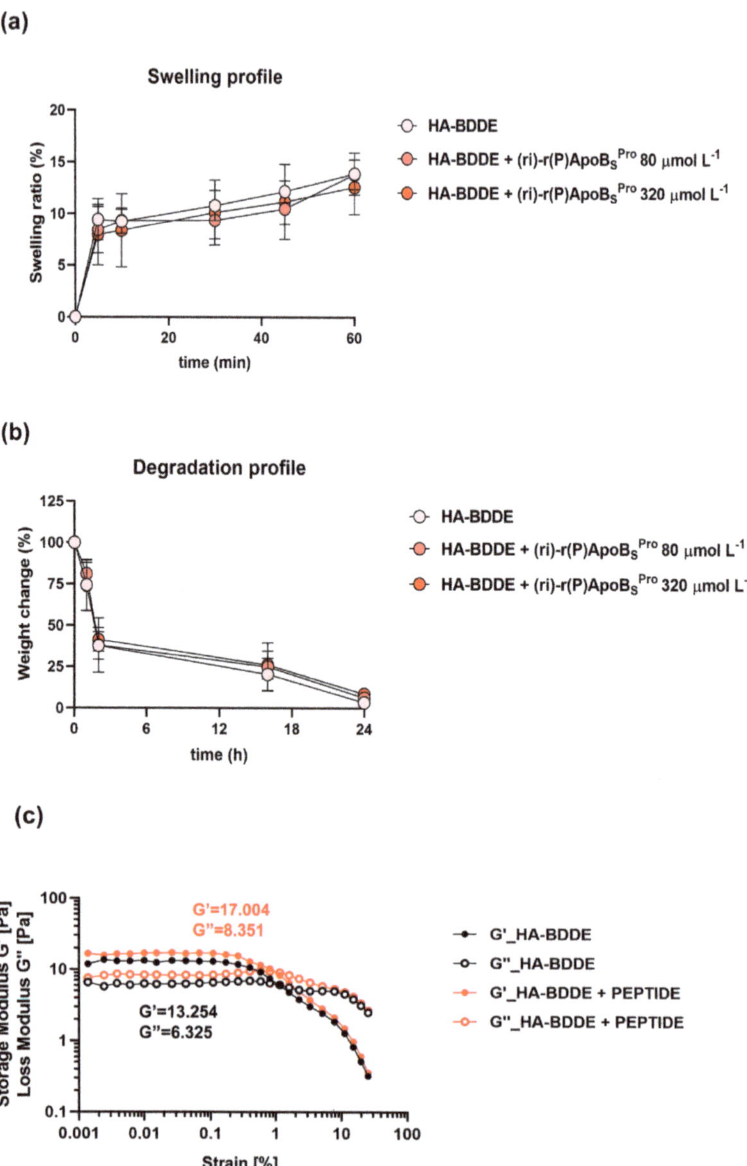

Figure 4. Physical and mechanical properties of the hydrogel-based system. Swelling (**a**) and degradation (**b**) profiles of HA-BDDE alone and loaded with two different concentrations of (ri)-r(P)ApoB$_S$^{Pro} (80 or 320 µmol L^{-1}). (**c**) Rheological characterization of HA-BDDE hydrogel alone and loaded with (ri)-r(P)ApoB$_S$^{Pro} peptidomimetic (320 µmol L^{-1}) based on the monitoring of elastic (G') and viscous moduli (G") as a function of shear stress. The values reported, 17.004 and 8.351, and 13.254 and 6.326 for G' and G", respectively, are the parameters obtained for each sample in the linear viscous range.

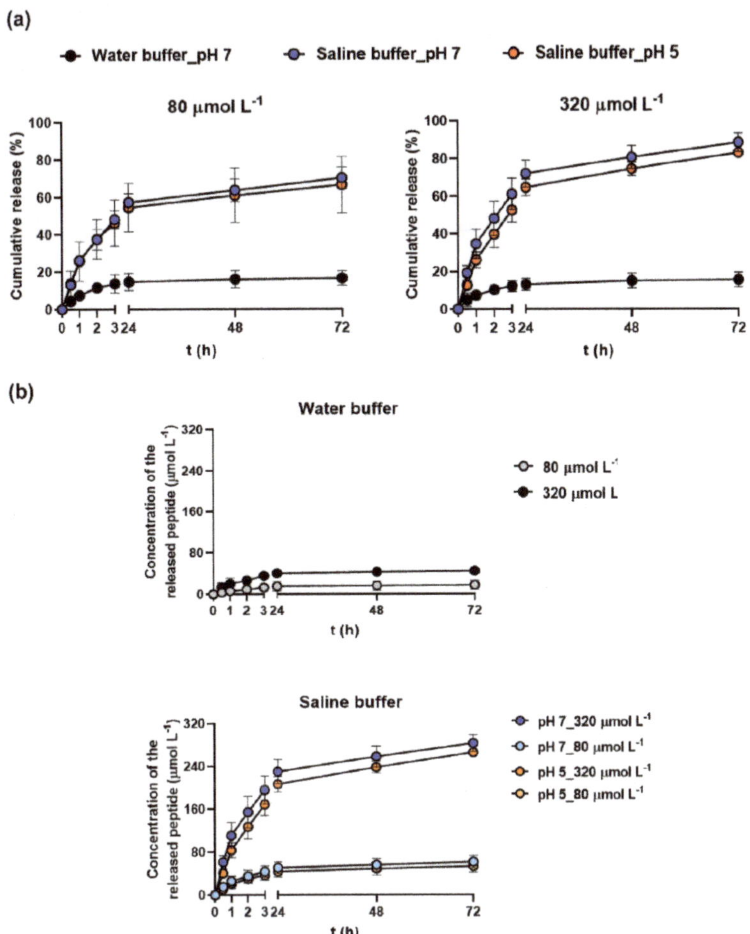

Figure 5. In vitro peptidomimetic release profile. (**a**) Evaluation of the amount of the peptidomimetic released from HA-BDDE loaded with two different (ri)-r(P)ApoB$_S$^{Pro} concentrations, i.e., 80 or 320 μmol L^{-1}. (**b**) Cumulative release of (ri)-r(P)ApoB$_S$^{Pro} expressed as percentage with respect to the total initial amount of peptidomimetic filled into gel.

3.7. Antimicrobial Properties of HA-BDDE Loaded with (ri)-r(P)ApoB$_S$^{Pro}

In order to evaluate whether the newly developed hydrogel-based system loaded with the peptidomimetic guarantees the preservation of the antimicrobial activity of (ri)-r(P)ApoB$_S$^{Pro}, we performed analyses on methicillin-resistant *S. aureus* (MRSA WKZ-2) and *E. coli* ATCC 25922, as representatives of Gram-positive and Gram-negative bacterial strains, respectively. We determined the minimal inhibitory concentration (MIC) values by counting the number of bacterial colonies obtained after an overnight incubation with HA-BDDE hydrogels loaded with (ri)-r(P)ApoB$_S$^{Pro} or unloaded. Significant antibacterial effects were detected against both bacterial strains (Figure 6a). In detail, methicillin-resistant *S. aureus* (MRSA WKZ-2) was susceptible to the hydrogel system loaded with the highest concentration of peptidomimetic (MIC$_{40}$ = 320 μmol L^{-1}) (Figure 6a). *E. coli* ATCC 25922 was even more susceptible with an MIC$_{90}$ value corresponding to 80 μmol L^{-1}, and a complete growth inhibition (MIC$_{100}$) obtained at a concentration of peptidomimetic of 320 μmol L^{-1} (Figure 6a). These observations are in agreement with previously reported data indicating that (ri)-r(P)ApoB$_S$^{Pro} exerts stronger toxic effects on Gram-negative than on Gram-positive bacterial strains (Figure 1a). To further investigate the antimicrobial properties of HA-BDDE hydrogels loaded with (ri)-r(P)ApoB$_S$^{Pro}, scanning electron microscopy (SEM) analyses were also performed. As shown in Figure 6b, when methicillin-resistant *S. aureus* (MRSA WKZ-2) cells were incubated with the HA-BDDE hydrogel system in the absence of the peptidomimetic, an almost complete degradation of the structure of the HA-hydrogel was observed upon incubation. Indeed, the activity of staphylococcal hyaluronidases has been reported to be a "spreading factor" contributing to the increase of lesion sizes in skin infections [45]. The phenomenon appeared slighter in the presence of the peptidomimetic than in to control samples. This, combined with a dose-dependent inhibition of cell viability observed in the presence of the peptidomimetic, highlights significant anti-infective properties, with the strongest effects observed at the highest peptidomimetic concentration tested (320 µmol L^{-1}) (Figure 6b). When the same experiments were performed on *E. coli* ATCC 25922 cells, similar results were obtained with the only exception that an almost complete inhibition of cell growth was already observed in the presence of the HA-BDDE hydrogel system loaded with 80 µmol L^{-1} peptidomimetic (Figure 6b). Indeed, the number of *E. coli* bacterial cells within and surrounding the gel was significantly reduced in the sample treated with the peptidomimetic than in the control. It has to be highlighted that, in the case of the *E. coli* sample, no degradation of the hydrogel matrix was observed, with a consequent spreading of cells on multiple layers and the visualization of a lower number of cells on the upper surface even in the control (Figure 6b).

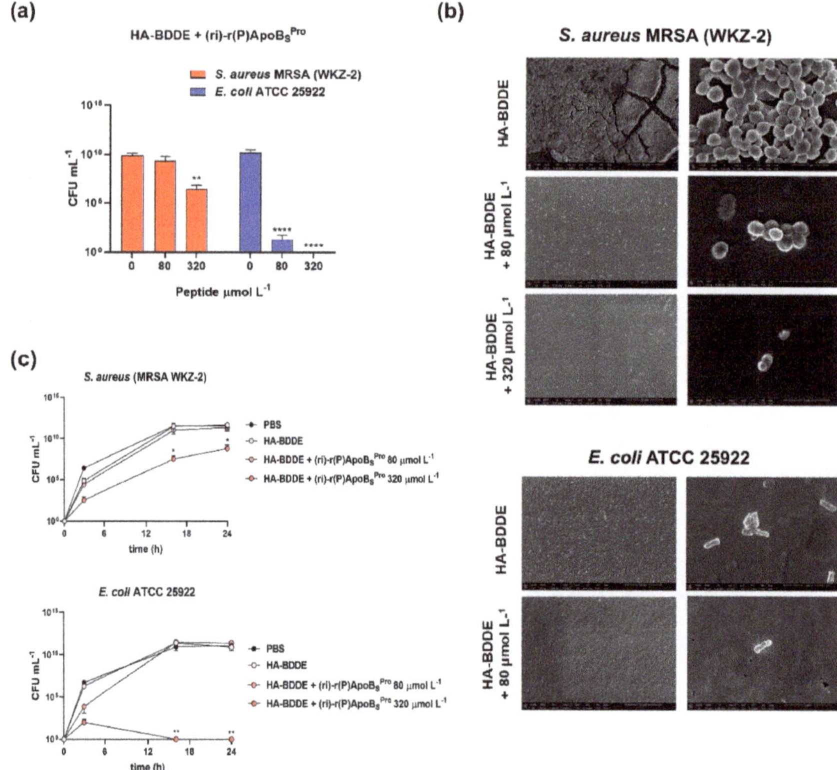

Figure 6. Antimicrobial activity of HA-BDDE hydrogel functionalized with the peptide. (a) Antimicrobial efficacy of HA-BDDE hydrogel system loaded with (ri)-r(P)ApoB$_S$^{Pro} against methicillin-resistant

S. aureus (MRSA WKZ-2) and *E. coli* ATCC 25922 bacterial strains; reported data refer to assays performed in triplicate. (**b**) Morphological analyses by SEM of methicillin-resistant *S. aureus* (MRSA WKZ-2) and *E. coli* ATCC 25922 bacterial strains treated with HA-BDDE functionalized or not with r(P)ApoB$_S$^{Pro}. (**c**) Bacterial migration across surfaces coated with PBS, unfunctionalized HA-BDDE hydrogel system, or HA-BDDE hydrogel system functionalized with (ri)-r(P)ApoB$_S$^{Pro}. Significant differences were indicated as * $p < 0.05$, ** $p < 0.0$ and **** $p < 0.0001$ for treated vs. control samples.

To investigate the antimicrobial activity of the HA-BDDE hydrogel system loaded with the peptidomimetic more deeply, we also performed experiments to evaluate whether the hydrogel system was able to counteract bacterial migration across surfaces. For this purpose, we performed trans-well migration assays by coating the upper chambers of a trans-well plate with PBS, HA-BDDE hydrogel system in the absence of the peptide, or HA-BDDE hydrogel system loaded with (ri)-r(P)ApoB$_S$^{Pro}. Upon coating, methicillin-resistant *S. aureus* (MRSA WKZ-2) and *E. coli* ATCC 25922 bacterial cells were plated into all the upper chambers. As shown in Figure 6c, only in the case of upper chambers coated with the hydrogel loaded with 320 µmol L^{-1} peptidomimetic, the migration of bacterial cells of both strains was found to be significantly attenuated.

3.8. Biocompatibility of HA-BDDE Hydrogel System Loaded with (ri)-r(P)ApoB$_S$^{Pro} on Human Skin Cell Cultures

To evaluate the biocompatibility of HA-BDDE hydrogel system loaded with (ri)-r(P)ApoB$_S$^{Pro} peptidomimetic toward human skin cell cultures, MTT assays were performed on immortalized human keratinocytes (HaCaTs) and on normal human fibroblasts (HDFs). Results indicated that HA-BDDE hydrogels loaded with 80 or 320 µmol L^{-1} (ri)-r(P)ApoB$_S$^{Pro} did not affect the viability of HaCaT and HDF cell lines (Figure 7). A significant improvement of cell viability was, instead, detected when the cells were incubated with an unfunctionalized HA-BDDE hydrogel system or with HA-BDDE hydrogel loaded with 320 µmol L^{-1} peptidomimetic (Figure 7). This can be explained by considering that, as a major component of the extracellular matrix, hyaluronic acid plays a key role in skin repair mechanisms being able to enhance the proliferation and differentiation of endothelial cells and to facilitate cell migration during the wound healing process [46,47]. The overall results highlight the biocompatibility of the HA-BDDE hydrogel system loaded with the (ri)-r(P)ApoB$_S$^{Pro} peptidomimetic and add an important aspect to the future applicability of this system.

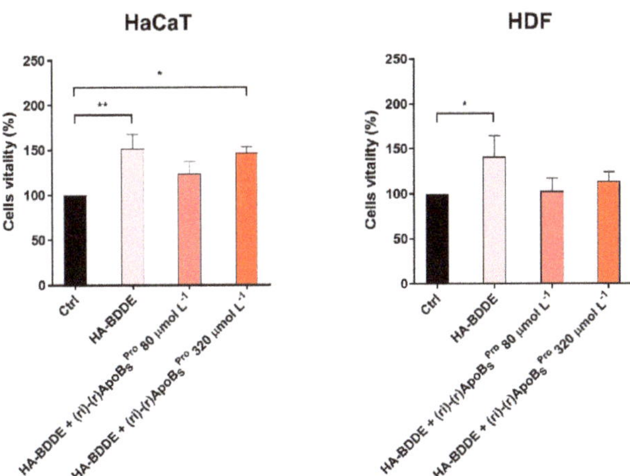

Figure 7. Effects of HA-BDDE hydrogel system functionalized with the peptidomimetic on human eukaryotic cell lines. Effects of HA-BDDE hydrogel system loaded with different concentrations of

(ri)-r(P)ApoB$_S$^{Pro} on the viability of HaCaT (immortalized human keratinocytes) and HDF (human dermal fibroblasts) cells. Cell viability was assessed by MTT assays and expressed as the percentage of viable cells compared to untreated cells (control). Experiments were performed three times. Error bars represent the standard deviation of the mean. Significant differences were indicated as * $p < 0.05$ or ** $p < 0.01$, for treated vs. control samples.

4. Discussion

The incessant spreading of drug-resistant bacteria prompts the search for novel antibiotics and for novel materials able to efficiently deliver antimicrobial agents to the site of infection. In this scenario, the discovery of novel antimicrobials less prone to induce the development of resistance phenotype is urgently needed. Antimicrobial peptidomimetics are small molecules with evident advantages over conventional antibiotics, including a lower probability of inducing resistance development [3] and improved pharmacokinetic profiles [48]. The synthetic retro-inverso peptide (ri)-r(P)ApoB$_S$^{Pro} was found to be more stable than its L-enantiomeric counterpart parental version [17] and to retain the broad-spectrum antimicrobial and anti-biofilm properties, thus being a good candidate to develop effective therapeutic strategies to treat recalcitrant infections. Importantly, the (ri)-r(P)ApoB$_S$^{Pro} peptidomimetic not only is not toxic toward eukaryotic cells, but it also exhibits a protective effect, being able to reduce the number of bacterial cells infecting keratinocytes. In order to develop a suitable system to topically deliver the peptidomimetic, in this work, a hyaluronic acid (HA)–based hydrogel system was selected and tested. HA represents an ideal candidate, since it not only contributes to viscoelasticity and lubrication of tissues but also plays an important role in physiological processes, such as inflammation, wound healing, and tissue development [49]. These properties combined with the antimicrobial features of the peptidomimetic represent an optimal starting point to develop a feasible dermatological formulation [29]. The presence of effective concentrations of the peptidomimetic in the hydrogel system was found not to affect its physiochemical properties. Even more interestingly, the bioactive peptidomimetic was found to be rapidly released by the system. The anti-infective activity of the system was confirmed by assays on both Gram-negative and Gram-positive bacterial strains, including cells characterized by resistance to conventional antibiotics. Furthermore, the HA-hydrogel system loaded with (ri)-r(P)ApoB$_S$^{Pro} was demonstrated to be able to counteract bacterial migration across surfaces, a property that opens further interesting perspectives in the treatment of dermal and epidermal infections. It has also to be highlighted that not only the HA-hydrogel system loaded with (ri)-r(P)ApoB$_S$^{Pro} is biocompatible toward human skin cells, but the presence of hyaluronic acid is also responsible for a detectable improvement of keratinocytes viability, thus suggesting putative positive effects on skin wound healing. Altogether, these findings open novel and interesting perspectives to the applicability of (ri)-r(P)ApoB$_S$^{Pro} in the treatment of skin infections and injuries. Further studies will be necessary in the future to optimize the herein developed hydrogel-based system and to test its efficacy in pre-clinical mouse models.

Author Contributions: Conceptualization, A.C. and A.A.; methodology, A.C., R.G., M.C., M.D.L., R.D.G., A.L. and R.F.; validation, A.C., R.F. and A.A.; formal analysis, A.C. and A.A.; investigation, A.C., R.G., M.C., M.D.L., R.D.G. and R.F.; data curation, A.C. and A.A.; writing—original draft preparation, A.C. and A.A.; writing—review and editing, A.C., R.G., M.C., M.D.L., R.D.G., A.L., R.F. and A.A.; supervision, A.A.; project administration, A.A.; funding acquisition, A.A. All authors have read and agreed to the published version of the manuscript.

Funding: Research reported in this publication was supported by the Programma Operativo Nazionale (PON) Ricerca e Innovazione 2014–2020 "Dottorati innovativi con caratterizzazione industriale".

Institutional Review Board Statement: Not applicable.

Informed Consent Statement: Not applicable.

Data Availability Statement: Data is contained within the article.

Conflicts of Interest: The authors declare no conflict of interest.

References

1. Fjell, C.D.; Hiss, J.A.; Hancock, R.E.W.; Schneider, G. Designing Antimicrobial Peptides: Form Follows Function. *Nat. Rev. Drug Discov.* **2011**, *11*, 37–51. [CrossRef] [PubMed]
2. Spohn, R.; Daruka, L.; Lázár, V.; Martins, A.; Vidovics, F.; Grézal, G.; Méhi, O.; Kintses, B.; Számel, M.; Jangir, P.K.; et al. Integrated Evolutionary Analysis Reveals Antimicrobial Peptides with Limited Resistance. *Nat. Commun.* **2019**, *10*, 4538. [CrossRef] [PubMed]
3. Magana, M.; Pushpanathan, M.; Santos, A.L.; Leanse, L.; Fernandez, M.; Ioannidis, A.; Giulianotti, M.A.; Apidianakis, Y.; Bradfute, S.; Ferguson, A.L.; et al. The Value of Antimicrobial Peptides in the Age of Resistance. *Lancet Infect. Dis.* **2020**, *20*, e216–e230. [CrossRef]
4. Oliva, R.; Chino, M.; Lombardi, A.; Nastri, F.; Notomista, E.; Petraccone, L.; del Vecchio, P. Similarities and Differences for Membranotropic Action of Three Unnatural Antimicrobial Peptides. *J. Pept. Sci.* **2020**, *26*, e3270. [CrossRef]
5. Pernot, M.; Vanderesse, R.; Frochot, C.; Guillemin, F.; Barberi-Heyob, M. Stability of Peptides and Therapeutic Success in Cancer. *Expert Opin. Drug Metab. Toxicol.* **2011**, *7*, 793–802. [CrossRef]
6. Lenci, E.; Trabocchi, A. Peptidomimetic Toolbox for Drug Discovery. *Chem. Soc. Rev.* **2020**, *49*, 3262–3277. [CrossRef]
7. Oliva, R.; Chino, M.; Pane, K.; Pistorio, V.; de Santis, A.; Pizzo, E.; D'Errico, G.; Pavone, V.; Lombardi, A.; del Vecchio, P.; et al. Exploring the Role of Unnatural Amino Acids in Antimicrobial Peptides. *Sci. Rep.* **2018**, *8*, 8888. [CrossRef]
8. Gaglione, R.; Cesaro, A.; Dell'Olmo, E.; Della Ventura, B.; Casillo, A.; Di Girolamo, R.; Velotta, R.; Notomista, E.; Veldhuizen, E.J.A.; Corsaro, M.M.; et al. Effects of Human Antimicrobial Cryptides Identified in Apolipoprotein B Depend on Specific Features of Bacterial Strains. *Sci. Rep.* **2019**, *9*, 6728. [CrossRef]
9. Gaglione, R.; Pane, K.; Dell'Olmo, E.; Cafaro, V.; Pizzo, E.; Olivieri, G.; Notomista, E.; Arciello, A. Cost-Effective Production of Recombinant Peptides in Escherichia Coli. *New Biotechnol.* **2019**, *51*, 39–48. [CrossRef]
10. Gaglione, R.; Cesaro, A.; Dell'Olmo, E.; Di Girolamo, R.; Tartaglione, L.; Pizzo, E.; Arciello, A. Cryptides Identified in Human Apolipoprotein B as New Weapons to Fight Antibiotic Resistance in Cystic Fibrosis Disease. *Int. J. Mol. Sci.* **2020**, *21*, 2049. [CrossRef]
11. Gaglione, R.; Dell'Olmo, E.; Bosso, A.; Chino, M.; Pane, K.; Ascione, F.; Itri, F.; Caserta, S.; Amoresano, A.; Lombardi, A.; et al. Novel Human Bioactive Peptides Identified in Apolipoprotein B: Evaluation of Their Therapeutic Potential. *Biochem. Pharmacol.* **2017**, *130*, 34–50. [CrossRef] [PubMed]
12. Gaglione, R.; Pizzo, E.; Notomista, E.; de la Fuente-Nunez, C.; Arciello, A. Host Defence Cryptides from Human Apolipoproteins: Applications in Medicinal Chemistry. *Curr. Top. Med. Chem.* **2020**, *20*, 1324–1337. [CrossRef] [PubMed]
13. Dell'Olmo, E.; Gaglione, R.; Sabbah, M.; Schibeci, M.; Cesaro, A.; di Girolamo, R.; Porta, R.; Arciello, A. Host Defense Peptides Identified in Human Apolipoprotein B as Novel Food Biopreservatives and Active Coating Components. *Food Microbiol.* **2021**, *99*, 103804. [CrossRef] [PubMed]
14. Dell'Olmo, E.; Gaglione, R.; Cesaro, A.; Cafaro, V.; Teertstra, W.R.; de Cock, H.; Notomista, E.; Haagsman, H.P.; Veldhuizen, E.J.A.; Arciello, A. Host Defence Peptides Identified in Human Apolipoprotein B as Promising Antifungal Agents. *Appl. Microbiol. Biotechnol.* **2021**, *105*, 1953–1964. [CrossRef] [PubMed]
15. Gaglione, R.; Smaldone, G.; Cesaro, A.; Rumolo, M.; de Luca, M.; di Girolamo, R.; Petraccone, L.; del Vecchio, P.; Oliva, R.; Notomista, E.; et al. Impact of a Single Point Mutation on the Antimicrobial and Fibrillogenic Properties of Cryptides from Human Apolipoprotein B. *Pharmaceuticals* **2021**, *14*, 631. [CrossRef] [PubMed]
16. Zanfardino, A.; Bosso, A.; Gallo, G.; Pistorio, V.; di Napoli, M.; Gaglione, R.; Dell'Olmo, E.; Varcamonti, M.; Notomista, E.; Arciello, A.; et al. Human Apolipoprotein E as a Reservoir of Cryptic Bioactive Peptides: The Case of ApoE 133-167. *J. Pept. Sci.* **2018**, *24*, e3095. [CrossRef]
17. Cesaro, A.; Torres, M.D.T.; Gaglione, R.; Dell'Olmo, E.; di Girolamo, R.; Bosso, A.; Pizzo, E.; Haagsman, H.P.; Veldhuizen, E.J.A.; de la Fuente-Nunez, C.; et al. Synthetic Antibiotic Derived from Sequences Encrypted in a Protein from Human Plasma. *ACS Nano* **2022**, *16*, 1880–1895. [CrossRef]
18. Bally, M.; Dendukuri, N.; Rich, B.; Nadeau, L.; Helin-Salmivaara, A.; Garbe, E.; Brophy, J.M. Risk of Acute Myocardial Infarction with NSAIDs in Real World Use: Bayesian Meta-Analysis of Individual Patient Data. *BMJ* **2017**, *357*, j1909. [CrossRef]
19. Sabbagh, F.; Kim, B.S. Recent Advances in Polymeric Transdermal Drug Delivery Systems. *J. Control. Release* **2022**, *341*, 132–146. [CrossRef] [PubMed]
20. Zasloff, M. Antimicrobial Peptides of Multicellular Organisms. *Nature* **2002**, *415*, 389–395. [CrossRef]
21. Ghasemiyeh, P.; Mohammadi-Samani, S. Hydrogels as Drug Delivery Systems; Pros and Cons. *Trends Pharm. Sci.* **2019**, *5*, 7–24.
22. Ullah, F.; Othman, M.B.H.; Javed, F.; Ahmad, Z.; Akil, H.M. Classification, Processing and Application of Hydrogels: A Review. *Mater. Sci. Eng. C* **2015**, *57*, 414–433. [CrossRef] [PubMed]
23. Xu, X.; Jha, A.K.; Harrington, D.A.; Farach-Carson, M.C.; Jia, X. Hyaluronic Acid-Based Hydrogels: From a Natural Polysaccharide to Complex Networks. *Soft Matter* **2012**, *8*, 3280. [CrossRef] [PubMed]
24. Wiegand, I.; Hilpert, K.; Hancock, R.E.W. Agar and Broth Dilution Methods to Determine the Minimal Inhibitory Concentration (MIC) of Antimicrobial Substances. *Nat. Protoc.* **2008**, *3*, 163–175. [CrossRef] [PubMed]
25. Cesaro, A.; Torres, M.; de la Fuente-Nunez, C. Methods for the Design and Characterization of Peptide Antibiotics. In *Methods in Enzymology*; Academic Press: Cambridge, MA, USA, 2022.

26. De Luca, M.; Gaglione, R.; Della Ventura, B.; Cesaro, A.; Di Girolamo, R.; Velotta, R.; Arciello, A. Loading of Polydimethylsiloxane with a Human ApoB-Derived Antimicrobial Peptide to Prevent Bacterial Infections. *Int. J. Mol. Sci.* **2022**, *23*, 5219. [CrossRef] [PubMed]

27. Monti, D.M.; Guglielmi, F.; Monti, M.; Cozzolino, F.; Torrassa, S.; Relini, A.; Pucci, P.; Arciello, A.; Piccoli, R. Effects of a Lipid Environment on the Fibrillogenic Pathway of the N-Terminal Polypeptide of Human Apolipoprotein A-I, Responsible for in Vivo Amyloid Fibril Formation. *Eur. Biophys. J.* **2010**, *39*, 1289–1299. [CrossRef]

28. Al-Sibani, M.; Al-Harrasi, A.; Neubert, R.H.H. Evaluation of In-Vitro Degradation Rate of Hyaluronic Acid-Based Hydrogel Cross-Linked with 1, 4-Butanediol Diglycidyl Ether (BDDE) Using RP-HPLC and UVeVis Spectroscopy. *J. Drug Deliv. Sci. Technol.* **2015**, *29*, 24–30. [CrossRef]

29. Gribova, V.; Boulmedais, F.; Dupret-Bories, A.; Calligaro, C.; Senger, B.; Vrana, N.E.; Lavalle, P. Polyanionic Hydrogels as Reservoirs for Polycationic Antibiotic Substitutes Providing Prolonged Antibacterial Activity. *ACS Appl. Mater. Interfaces* **2020**, *12*, 19258–19267. [CrossRef]

30. Zhu, J.; Li, F.; Wang, X.; Yu, J.; Wu, D. Hyaluronic Acid and Polyethylene Glycol Hybrid Hydrogel Encapsulating Nanogel with Hemostasis and Sustainable Antibacterial Property for Wound Healing. *ACS Appl. Mater. Interfaces* **2018**, *10*, 13304–13316. [CrossRef]

31. Cao, W.; Sui, J.; Ma, M.; Xu, Y.; Lin, W.; Chen, Y.; Man, Y.; Sun, Y.; Fan, Y.; Zhang, X. The Preparation and Biocompatible Evaluation of Injectable Dual Crosslinking Hyaluronic Acid Hydrogels as Cytoprotective Agents. *J. Mater. Chem. B* **2019**, *7*, 4413–4423. [CrossRef]

32. Traeger, N.; Shi, Q.; Dozor, A.J. Relationship between Sweat Chloride, Sodium, and Age in Clinically Obtained Samples. *J. Cyst. Fibros.* **2014**, *13*, 10–14. [CrossRef] [PubMed]

33. Li, X.; Li, A.; Feng, F.; Jiang, Q.; Sun, H.; Chai, Y.; Yang, R.; Wang, Z.; Hou, J.; Li, R. Effect of the Hyaluronic Acid-poloxamer Hydrogel on Skin-wound Healing: In Vitro and in Vivo Studies. *Anim. Models Exp. Med.* **2019**, *2*, 107–113. [CrossRef] [PubMed]

34. Petkovšek, Ž.; Eleršič, K.; Gubina, M.; Žgur-Bertok, D.; Erjavec, M.S. Virulence Potential of Escherichia Coli Isolates from Skin and Soft Tissue Infections. *J. Clin. Microbiol.* **2009**, *47*, 1811–1817. [CrossRef] [PubMed]

35. Spernovasilis, N.; Psichogiou, M.; Poulakou, G. Skin Manifestations of Pseudomonas Aeruginosa Infections. *Curr. Opin. Infect. Dis.* **2021**, *34*, 72–79. [CrossRef] [PubMed]

36. del Giudice, P. Skin Infections Caused by *Staphylococcus Aureus*. *Acta Derm.-Venereol.* **2020**, *100*, adv00110. [CrossRef]

37. Grishin, D.V.; Zhdanov, D.D.; Pokrovskaya, M.V.; Sokolov, N.N. D-Amino Acids in Nature, Agriculture and Biomedicine. *All Life* **2020**, *13*, 11–22. [CrossRef]

38. Shah, C.B.; Barnett, S.M. Swelling Behavior of Hyaluronic Acid Gels. *J. Appl. Polym. Sci.* **1992**, *45*, 293–298. [CrossRef]

39. Woerly, S.; Pinet, E.; de Robertis, L.; van Diep, D.; Bousmina, M. Spinal Cord Repair with PHPMA Hydrogel Containing RGD Peptides (NeuroGel™). *Biomaterials* **2001**, *22*, 1095–1111. [CrossRef]

40. Simões, A.; Miranda, M.; Cardoso, C.; Veiga, F.; Vitorino, C. Rheology by Design: A Regulatory Tutorial for Analytical Method Validation. *Pharmaceutics* **2020**, *12*, 820. [CrossRef]

41. Zustiak, S.P.; Durbal, R.; Leach, J.B. Influence of Cell-Adhesive Peptide Ligands on Poly(Ethylene Glycol) Hydrogel Physical, Mechanical and Transport Properties. *Acta Biomater.* **2010**, *6*, 3404–3414. [CrossRef]

42. Kuo, S.H.; Shen, C.J.; Shen, C.F.; Cheng, C.M. Role of PH Value in Clinically Relevant Diagnosis. *Diagnostics* **2020**, *10*, 107. [CrossRef] [PubMed]

43. Bao, Z.; Yu, A.; Shi, H.; Hu, Y.; Jin, B.; Lin, D.; Dai, M.; Lei, L.; Li, X.; Wang, Y. Glycol Chitosan/Oxidized Hyaluronic Acid Hydrogel Film for Topical Ocular Delivery of Dexamethasone and Levofloxacin. *Int. J. Biol. Macromol.* **2021**, *167*, 659–666. [CrossRef] [PubMed]

44. Zhao, Y.; Zhu, Z.S.; Guan, J.; Wu, S.J. Processing, mechanical properties and bio-applications of silk fibroin-based high-strength hydrogels. *Acta Biomater.* **2021**, *125*, 57–71. [CrossRef] [PubMed]

45. Tam, K.; Torres, V.J. *Staphylococcus Aureus* Secreted Toxins and Extracellular Enzymes. *Microbiol. Spectr.* **2019**, *7*, 10. [CrossRef]

46. Prosdocimi, M.; Bevilacqua, C. Exogenous Hyaluronic Acid and Wound Healing: An Updated Vision. *Panminerva Med.* **2012**, *54*, 129–135.

47. Tang, S.; Chi, K.; Xu, H.; Yong, Q.; Yang, J.; Catchmark, J.M. A Covalently Cross-Linked Hyaluronic Acid/Bacterial Cellulose Composite Hydrogel for Potential Biological Applications. *Carbohydr. Polym.* **2021**, *252*, 117123. [CrossRef]

48. Haggag, Y.A. Peptides as Drug Candidates: Limitations and Recent Development Perspectives. *Biomed. J. Sci. Tech. Res.* **2018**, *8*, 6659–6662. [CrossRef]

49. Trombino, S.; Servidio, C.; Curcio, F.; Cassano, R. Strategies for Hyaluronic Acid-Based Hydrogel Design in Drug Delivery. *Pharmaceutics* **2019**, *11*, 407. [CrossRef]

Article

Strategy to Enhance Anticancer Activity and Induced Immunogenic Cell Death of Antimicrobial Peptides by Using Non-Nature Amino Acid Substitutions

Yu-Huan Cheah [1,†], Chun-Yu Liu [1,†], Bak-Sau Yip [1,2,†], Chih-Lung Wu [1], Kuang-Li Peng [1] and Jya-Wei Cheng [1,*]

1   Department of Medical Science, Institute of Biotechnology, National Tsing Hua University, Hsinchu 300, Taiwan; s107080710@m107.nthu.edu.tw (Y.-H.C.); s109080536@m109.nthu.edu.tw (C.-Y.L.); g15004@hch.gov.tw (B.-S.Y.); s103080578@m103.nthu.edu.tw (C.-L.W.); richard850210@gapp.nthu.edu.tw (K.-L.P.)
2   Department of Neurology, National Taiwan University Hospital Hsinchu Branch, Hsinchu 300, Taiwan
*   Correspondence: jwcheng@life.nthu.edu.tw; Tel.: +886-3-5742763; Fax: +886-3-5715934
†   These authors contributed equally to this work.

Abstract: There is an urgent and imminent need to develop new agents to fight against cancer. In addition to the antimicrobial and anti-inflammatory activities, many antimicrobial peptides can bind to and lyse cancer cells. P-113, a 12-amino acid clinically active histatin-rich peptide, was found to possess anti-*Candida* activities but showed poor anticancer activity. Herein, anticancer activities and induced immunogenic cancer cell death of phenylalanine-(Phe-P-113), β-naphthylalanine-(Nal-P-113), β-diphenylalanine-(Dip-P-113), and β-(4,4′-biphenyl)alanine-(Bip-P-113) substituted P-113 were studied. Among these peptides, Nal-P-113 demonstrated the best anticancer activity and caused cancer cells to release potent danger-associated molecular patterns (DAMPs), such as reactive oxygen species (ROS), cytochrome c, ATP, and high-mobility group box 1 (HMGB1). These results could help in developing antimicrobial peptides with better anticancer activity and induced immunogenic cell death in therapeutic applications.

Keywords: antimicrobial peptides; oncolytic peptides; cancer; membrane integrity; bulky non-nature amino acid; DAMPs

Citation: Cheah, Y.-H.; Liu, C.-Y.; Yip, B.-S.; Wu, C.-L.; Peng, K.-L.; Cheng, J.-W. Strategy to Enhance Anticancer Activity and Induced Immunogenic Cell Death of Antimicrobial Peptides by Using Non-Nature Amino Acid Substitutions. *Biomedicines* 2022, 10, 1097. https://doi.org/10.3390/biomedicines10051097

Academic Editor: Jitka Petrlova

Received: 7 April 2022
Accepted: 4 May 2022
Published: 9 May 2022

Publisher's Note: MDPI stays neutral with regard to jurisdictional claims in published maps and institutional affiliations.

1. Introduction

Antimicrobial peptides (AMPs) are found in the host innate defense mechanisms and play important roles in combating microbial infections [1–4]. AMPs may work alone or in combination with antibiotics to diminish drug-resistant pathogens and improve the therapeutic effects of antibiotics [1,5–10]. Recent progress of antibiotic AMP conjugates also displays exceptional in vitro and in vivo efficacies [11,12]. Most AMPs possess two critical characteristics: a net cationicity to interact with negatively charged microbial surfaces and an amphipathic structure to incorporate into microbial cell membranes. Other than broad-spectrum antimicrobial activities, many AMPs also exert lipopolysaccharide (LPS) neutralization, as well as anticancer activities [13–15].

The histidine-rich peptide P-113 (AKRHHGYKRKFH-NH$_2$) was derived from saliva protein histatin 5 [16]. P-113 has demonstrated its efficacy in a clinical trial for HIV patients with oral candidiasis [17]. However, P-113 lost its activity in high salt conditions [18]. The antimicrobial activities of P-113 under high-salt conditions can be restored by replacing the histidine amino acids with the bulky non-natural amino acids: β-naphthylalanine (Nal), diphenylalanine (Dip), and (4,4′-biphenyl)alanine (Bip) [18,19]. In addition, these P-113 derivatives all displayed enhanced serum proteolytic stability, in vitro and in vivo LPS neutralization, and synergistic activities with antibiotics [9,18,19].

Cancer has been found to be the leading cause of death in many countries [9]. The insurgence of chemotherapeutic resistance is still a major cause of treatment failure of cancer patients [14]. Hence, the development of novel therapeutics with high specificity toward cancer cells and/or treatments stimulating the patients' own immune system to fight against cancer cells is the utmost concern in the cancer research field. Similar to the disruption of the negatively charged microbial cell membranes, some AMPs can bind to the negatively charged phosphatidylserine moieties exposed on the outer surface of cancer cell plasma membranes and cause the lysis of cancer cells [15]. In addition to direct lysis of cancer cells, the membrane-disruptive or membrane-lytic properties of AMPs may help conventional standard-of-care chemotherapeutics to fight against drug-resistant cancer cells [20]. Moreover, the release of tumor antigens and potent danger-associated molecular patterns (DAMPs) were found after treatment with these AMPs [15,21]. The release of danger signals (DAMPs) may induce complete regression and long-term protective immune responses against cancer, as previously reported [22]. However, the mode of action of the interactions between AMPs and cancer cells or the rules governing the design of AMPs with anticancer activities are still not clear.

Based on our previous studies of the interactions between microbial pathogens and P-113 and its derivatives [9,18,19], we suspected that the substitution of bulky non-natural amino acids may also affect the anticancer activity of P-113. Herein, we determined the cytotoxic effect of P-113 and its bulky non-natural amino acids substituted derivatives against five cancer cell lines. In addition, we measured the release of reactive oxygen species (ROS), high-mobility group box 1 (HMGB1), cytochrome c, and ATP after the treatment of cancer cells with P-113 derivatives.

2. Materials and Methods

2.1. Materials

All peptides were purchased from Kelowna International Scientific Inc. (Taipei, Taiwan). The identity of the peptides was checked by electrospray mass spectroscopy, and the purity (>95%) was assessed by high-performance liquid chromatography (HPLC). Sequences of P-113 and its derivatives are shown below:

P-113 (Ac-AKR **His His** GYKRKF **His**-NH$_2$)
Phe-P-113 (Ac-AKR **Phe Phe** GYKRKF **Phe**-NH$_2$)
Nal-P-113 (Ac-AKR **Nal Nal** GYKRKF **Nal**-NH$_2$)
Bip-P-113 (Ac-AKR **Bip Bip** GYKRKF **Bip**-NH$_2$)
Dip-P-113 (Ac-AKR **Dip Dip** GYKRKF **Dip**-NH$_2$)

Dulbecco's modified Eagle's medium (DMEM), Roswell Park Memorial Institute 1640 (RPMI 1640) medium, and Trypsin-EDTA were purchased from Gibco-Life Technologies (New York, NY, USA). Fetal bovine serum (FBS) was purchased from Biological industries (Beit HaEmek, Israel). Dimethyl sulfoxide (DMSO), MTT (3-(4,5-dimethylthiazol-2-yl)2.5-diphenyl tetrazolium bromide), L-Glutamine, penicillin, streptomycin, propidium iodide, and Hoechst 33342 were purchased from Sigma-Aldrich (St. Louis, MO, USA). JC-1 (5,5′,6,6′-Tetrachloro-1,1′,3,3′-tetraethyl-imidacarbocyanine iodide) was purchased from Dojindo Molecular Technologies (Rockville, MD, USA). CCCP (Carbonyl cyanide 3-chloro-phenylhydrazone) was purchased from Cayman Chemical Company (Ann Arbor, MI, USA).

2.2. Cell Lines and Cultural Conditions

The three non-small cell lung cancer cell lines, H1975, PC 9, A549, and oral squamous cell carcinoma cell line OECM-1 were cultured in RPMI 1640 medium supplemented with 10% fetal bovine serum, 2 mM L-glutamine and 1% penicillin/ streptomycin. The oral cancer cell line C9 was cultured in DMEM with 10% fetal bovine serum, 2 mM L-glutamine, and 1% penicillin/ streptomycin. All the cell lines were cultured in the humidified incubator containing 5% CO$_2$ at 37 °C.

2.3. Cell Viability Assay

The MTT assay was applied for cell viability test according to previous protocol [23]. The cells were seeded in the 96-well plate at 5000 cells/well under the environment of humidified 5% CO_2 and the temperature of 37 °C for 24 h. The medium was removed from the well before the fresh medium containing different concentrations of peptides was added and incubated for 24 h. Medium treated with PBS served as control group. All peptides were diluted serially from 100 μM to 0.78 μM in ddH_2O before adding in the medium. After 24 h of peptide treatment, fresh medium with 10% MTT solution (5 mg/mL) was incubated for 4 h. After medium was removed, 100 μL DMSO was added to dissolve the formazan crystal. Absorbance was measured at 570 nm using a microplate reader (TECAN Sunrise ELISA Reader). The cell survival was calculated from the treated cells relative to the control (100% viable cells) using the mean of three independent experiments and expressed as a 50% inhibitory concentration (IC_{50}).

The percentage of cell viability was calculated using the following formula:

$$\text{Cell viability } \% = \frac{\text{Absorbance of sample } - \text{ Blank}}{\text{Absorbance of control } - \text{ Blank}} \times 100\% \tag{1}$$

2.4. Kinetic Analysis

A total of 5000 cells were incubated with Nal-P-113, Bip-P-113, and Dip-P-113 at $2\times$ IC_{50} for 0.5, 1, 2, 3, and 4 h. After treatment, the MTT solution was further incubated for 3 h. Absorbance was measured at 570 nm on the microplate reader (TECAN Sunrise ELISA Reader) after the dissolution of the formazan crystal in 100 μL DMSO solution. The cell survival rate was calculated as aforementioned for cell viability assay.

2.5. Propidium Iodide (PI)/Hoechst 33342 Staining

To examine the membranolytic activities of peptides, the PI/Hoechst 33342 staining was performed in this study. The PI dye was used to stain damaged/dead cells, and Hoechst 33342 was used to specifically stain the nuclei of living cells [21,24]. PC 9 cells were seeded at 20,000 cells/well in a RPMI medium and allowed to adhere for 48 h. Then, the cells were replaced with serum-free RPMI medium and then treated with $2\times$ IC_{50} peptides at 37 °C for 120 min. Then, the PI and Hoechst 33342 were added at a final concentration of 1 μg/mL for 20 min. PC 9 cells were observed using the inverted fluorescent microscope Zeiss equipped with a $40\times$ oil objective lens (Carl Zeiss, Jena, Germany). The experiments were repeated three times independently.

2.6. Fluorescence Microscopy of JC-1 Staining

PC 9 cells were seeded at 40,000 cells/well in a RPMI medium and allowed to adhere for 24 h. Then, the cells were replaced with serum-free RPMI medium and treated with $2\times$ IC_{50} or 150 μM CCCP (Carbonyl cyanide 3-chloro-phenylhydrazone) at 37 °C for 120 min. The JC-1 dye was added at a final concentration of 1 μg/mL for 45 min. PC 9 cells were observed using the inverted fluorescent microscope Zeiss equipped with a $20\times$ oil objective lens (Carl Zeiss, Jena, Germany). The experiments were repeated three times independently.

2.7. Measurement of Extracellular ATP

PC 9 cells were seeded at the density of 10,000 cells/ well in the 96-well plate and allowed to adhere for 48 h. Cells were treated with Nal-P-113, Bip-P-113, and Dip-P-113 at $2\times$ IC_{50} for 0.5, 1, 1.5, and 2 h. Untreated cells were served as the control group. After being treated for different time points (0.5, 1, 1.5, and 2 h), the supernatant of cells was analyzed using the ENLITEN ATP luciferase assay kit (Promega, Madison, WI, USA). The level of extracellular ATP chemiluminescence was measured using the Wallac VICTOR 3 Multilabel plate reader (PerkinElmer, Shelton, CT, USA).

2.8. Measurement of Cellular ROS

The 2′,7′-dichlorofluorescin diacetate (DCFDA) Cellular ROS Detection Assay Kit (Abcam, Cambridge, UK, ab113851) was used to analyze the ROS release in the PC 9 cell line. DCFDA is a cell-permeable non-fluorescent probe, which diffuses into the cytoplasm and then forms a non-fluorescent moiety through deacetylation by cellular esterase. The non-fluorescent moiety could be oxidized by cellular ROS to form the fluorescent product DCF [21]. PC 9 cells were seeded at the density of 25,000 cells/ well in the 96-well black plate and allowed to adhere for 24 h. Cells were washed with a 100 µL/well of PBS to remove the RPMI 1640 medium and incubated with 100 µL of the 25 µM DCFDA solution for 45 min at 37 °C in the dark condition. An amount of 100 µL/ well of PBS buffer was washed again for the removal of 25 µM DCFDA solution. The PC 9 cells were treated with Nal-P-113, Bip-P-113, and Dip-P-113 at $2\times$ IC$_{50}$ for 2 h. The untreated cells were used as a control. The fluorescence intensity of cellular ROS was measured using the Wallac VICTOR 3 Multilabel plate reader (PerkinElmer, Shelton, CT, USA) at an excitation wavelength of 485 nm and an emission wavelength of 535 nm.

2.9. Detection of Mitochondrial Cytochrome C Release to Cytosol

PC 9 cells were seeded with 10,000 cells/well in the 96-well plate and allowed to adhere for 48 h and then treated with Nal-P-113, Bip-P-113, and Dip-P-113 at $2\times$ IC$_{50}$ for 0.5, 1, 1.5, 2, and 4 h. Untreated cells were used as a control group. Supernatants were collected after centrifugation at $16,000\times$ *g* for 10 min. Samples were analyzed using the Quantikine ELISA Rat/ Mouse Cytochrome c Immunoassay (R & D Systems, Minneapolis, MN, USA) according to the manufacturer's protocol. The optical density at the wavelength of 450 nm of each well was determined using the microplate reader (TECAN Sunrise ELISA Reader).

2.10. Detection of Extracellular HMGB1

The PC 9 cells were seeded at the density of 25,000 cells/well and left to adhere for 48 h. Cells were treated with Nal-P-113, Bip-P-113, and Dip-P-113 at $2\times$ IC$_{50}$ for 1, 2, and 4 h. After treatment of the peptides, the supernatant and lysate of the cells were collected separately for further experiments. The supernatant was collected after centrifugation at the speed of $1400\times$ *g*. The cell lysate was washed twice using PBS before the sonication of the pellet using a bio-disruptor, and the disrupted pellet was centrifuged again for 10 min under 14,000 rpm at 4 °C. The concentration of the samples was quantified using the Bradford assay. Both supernatants and cell lysate were loaded and run on the sodium dodecyl sulfate polyacrylamide gel electrophoresis (SDS-PAGE) and were then electro transferred to a polyvinylidene difluoride (PVDF) membrane (Merck-Millipore, Burlington, MA, USA) in an electro-blot system, 30 V, 400 mA, for 75 min. The membrane was incubated in blocking buffer (5% skim milk, TBST buffer) for 1 h at room temperature and washed in TBST buffer three times. The anti-HMGB1 antibody (Abcam, Cambridge, UK, ab18256) was incubated overnight at 4 °C in the dilution ratio of 1:1000. The secondary antibody of a rabbit (diluted in 1:10,000) was incubated following rinsing with PBST three times for 2 h the next day at room temperature before being developed using a luminol reagent and imaged by a detected system (ImageQuant LAS 4000 mini).

2.11. Statistical Analysis

All the statistical data represent the average of three independent experiments with standard deviation (mean $\pm$ SD) and each experiment set consisting of three replications. The evaluation of statistical analyses was calculated using one-way ANOVA and Student's *t*-tests using GraphPad Prism version 8.0 (San Diego, CA, USA). Significant differences were represented with thresholds of * $p < 0.05$; ** $p < 0.01$; *** $p < 0.001$.

3. Results

3.1. Anticancer Activities of P-113 and Its Derivatives

Five cancer cell lines, including three non-small cell lung cancer cell lines (H1975, A549, PC 9) and two oral cancer cell lines (C9, OECM-1), were used to test the cytotoxic effects of P-113 and its derivatives. H1975 is the non-small cell lung carcinoma cells, A549 is the adenocarcinomic alveolar basal epithelial cells, and PC 9 is the non-small cell lung cancer with EGFR mutation. C9 and OECM-1 cell lines are oral squamous cell carcinoma and gingival epidermoid carcinoma, respectively. P-113 and Phe-P-113 peptides have the highest survival rates (>60%) among every cell line tested, even when the peptide concentration had reached 100 μM (Figure 1). Nal-P-113, Bip-P-113, and Dip-P-113 peptides exhibited significant anticancer activities against these five cell lines (Figure 1). IC_{50} values of P-113 and its derivatives are shown in Table 1. As can be seen from Table 1, the order of the anticancer activities is Nal-P-113 > Bip-P-113 > Dip-P-113 > Phe-P-113 > P-113.

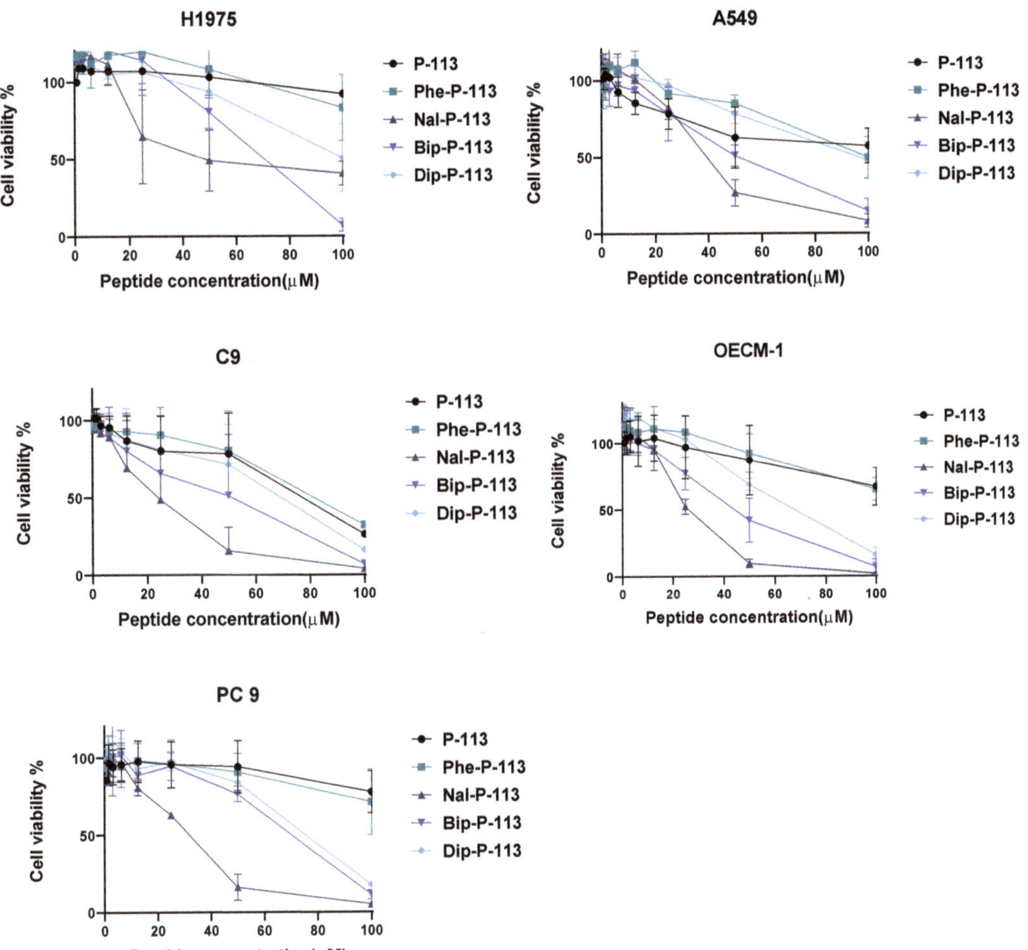

Figure 1. Anticancer activities of P-113, Phe-P-113, Nal-P-113, Bip-P-113, and Dip-P-113 against various cancer cell lines by MTT cell viability assay. Data represent mean ± SD of three independent experiments.

Table 1. IC$_{50}$ (μM) of P-113 and its derivatives.

Peptide Cell Line	P-113	Phe-P-113	Nal-P-113	Bip-P-113	Dip-P-113
H1975	110.0	106.6	58.32	63.65	100.4
A549	106.6	99.06	38.22	48.43	96.43
C9	84.27	80.49	21.37	41.97	67.89
OECM-1	140.6	119.9	25.91	42.47	62.94
PC 9	>200	172.7	28.11	64.19	71.45

The toxicity of the peptides was determined previously by measuring cell death using MTT assays against human fibroblasts (HFW) [19]. The results indicated that even at 100 μg/mL, Nal-P-113, Bip-P-113, and Dip-P-113 only caused less than 10% cell death.

3.2. Cancer Cell Killing Kinetic Analysis

The evaluation of the time killing kinetic of Nal-P-113, Bip-P-113, and Dip-P-113, which displayed excellent anticancer activity against PC 9 cell line, was employed by MTT cell viability assay. Two-fold half inhibitory concentration (2× IC$_{50}$) was chosen for three independent experiments, and the results were recorded from 30 min to 4 h. As shown in Figure 2, Nal-P-113 and Bip-P-113 started to kill cancer cells after 30 min of peptide incubation and reduced the cell survival rate to below 50%. Within 2 h, Nal-P-113 and Bip-P-113 had effectively killed more than 80% of the cells, which makes both of them the most effective peptides for anticancer activity. On the other hand, cancer cells treated with Dip-P-113 had the highest cellular survival rate at around 60% in 30 min. The overall cell survival rate of the group treated with Dip-P-113 was the highest compared to Nal-P-113 and Bip-P-113.

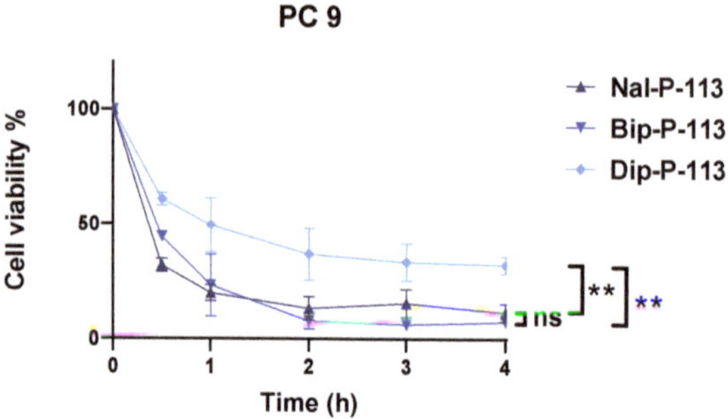

Figure 2. Time killing analysis of Nal-P-113, Bip-P-113, and Dip-P-113 against PC 9 cell line. Data represent mean ± SD of three independent experiments. ** $p < 0.01$ (Nal-P-113 vs. Dip-P-113; Bip-P-113 vs. Dip-P-113), ns = no significant differences (Nal-P-113 vs. Bip-P-113).

3.3. Loss of Plasma Membrane Integrity

The DNA binding fluorescent probe PI can only enter necrotic cells with a damaged plasma membrane [21]. In this study, PI was used to demonstrate that Nal-P-113, Bip-P-113, and Dip-P-113 can cause damage to the plasma membrane integrity of PC 9 cells and increase the amount of PI entering into PC 9 cells (Figure 3).

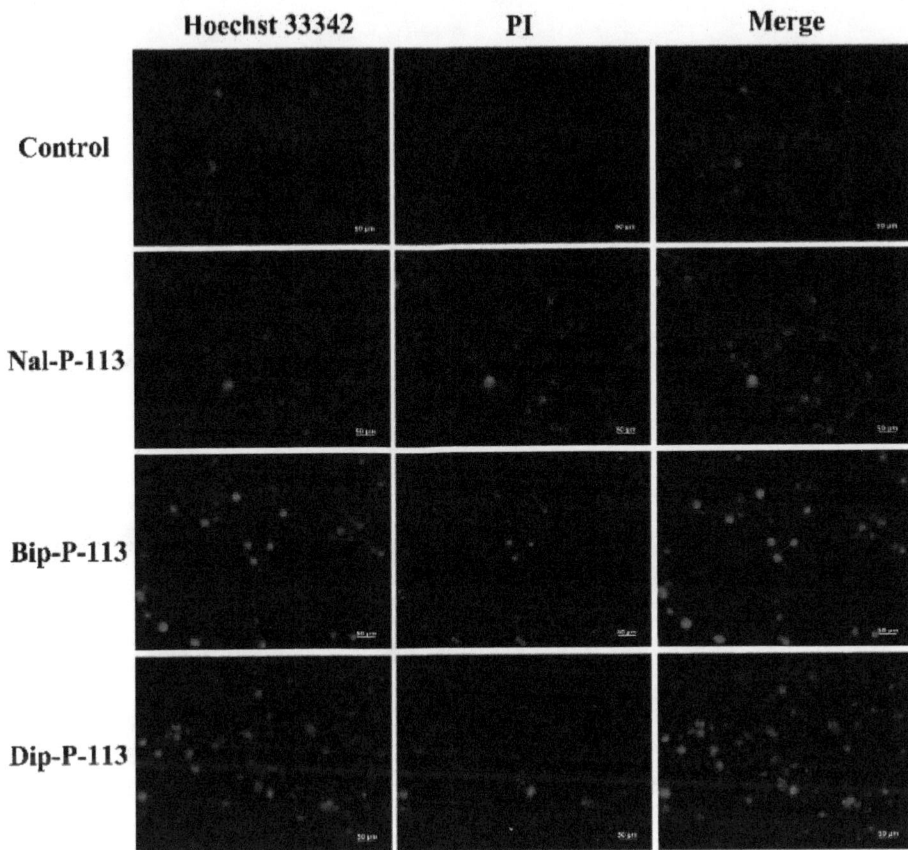

Figure 3. Fluorescence images of PC 9 cells treated with or without Nal-P-113, Bip-P-113, and Dip-P-113 at 2× IC$_{50}$ for 120 min and stained with PI (red) and nuclear dye Hoechst 33342 (blue) using fluorescence microscopy. Scale bar represents 50 μm. The experiments were repeated three times independently.

3.4. Mitochondrial Membrane Interactions

JC-1 is a cationic dye and is known to be localized exclusively in mitochondria [25]. Under sufficient mitochondrial membrane potential, JC-1 will form J-aggregates with a specific red fluorescence emission. Losing mitochondrial membrane potential (dysfunction) due to interactions with membrane-lytic peptides will lead to a decrease in JC-1 accumulation. JC-1 will then form J-monomers with a typical green fluorescence. Figure 4 shows Nal-P-113, Bip-P-113, and Dip-P-113 induced changes of the mitochondrial membrane potential in PC 9 cells, indicating that these peptides interact with the mitochondria membrane and cause dysfunction of mitochondria.

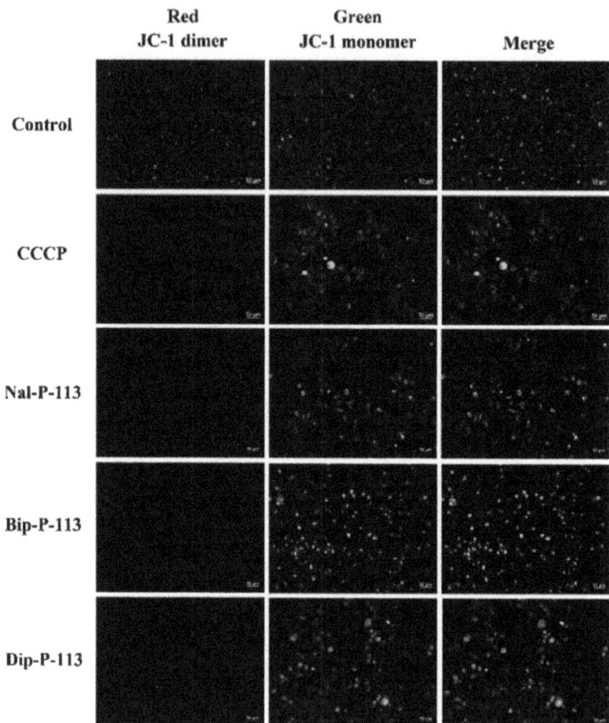

Figure 4. Fluorescence images of PC 9 cells treated with $2\times$ IC_{50} Nal-P-113, Bip-P-113, and Dip-P-113 for 120 min and stained with JC-1. Samples treated with CCCP served as a positive control. Scale bar represents 50 μm. The experiments were repeated three times independently.

3.5. Production of Reactive Oxygen Species (ROS)

Reactive oxygen species (ROS) is an important indicator in immunogenic cell death, and the level of ROS production may be caused by dysfunctional mitochondria. Figure 5 demonstrates that PC 9 cells treated with Nal-P-113, Bip-P-113, and Dip-P-113 had increasing fluorescence intensities to the level of more than 40,000. These results indicated that Nal-P-113, Bip-P-113, and Dip-P-113 may cause the damage of mitochondria and promote ROS production.

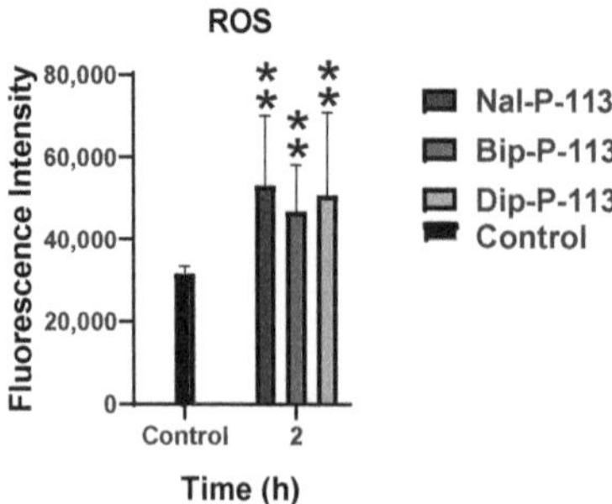

Figure 5. ROS release after the treatment with $2\times$ IC_{50} of Nal-P-113, Bip-P-113, and Dip-P-113 detected by DCFDA cellular ROS detection kit. The untreated cells were used as control. Results are presented as mean $\pm$ SD of three independent experiments, ** $p < 0.01$ compared with control.

Detection of Mitochondrial Cytochrome C release into Cytosol. The cytochrome c release was a part of the mitochondrial danger-associated molecular patterns. Therefore, the amount of cytochrome c released into the supernatant was measured to investigate the capability of Nal-P-113, Bip-P-113, and Dip-P-113 to induce the mitochondrial DAMPs. Five time points (0.5, 1, 1.5, 2, and 4 h) were used in the ELISA assay to quantify the amount of cytochrome c released, and the results were shown in Figure 6. Bip-P-113 induced a relatively high amount of cytochrome c compared to Nal-P-113 and Dip-P-113. The cytochrome c release for the Bip-P-113-treated group was more than double the amount of Nal-P-113- and Dip-P-113-treated groups. It is observed that the trend of cytochrome c release in Bip-P-113- and Nal-P-113-treated PC 9 cells was escalating from 30 min to 4 hours' time course, in contrast to the Dip-P-113 treatment group, which had a fluctuating trend. In the Bip-P-113-treated group, the amount of cytochrome c release came to a maximal value of nearly 15 ng/mL at 4 h, while the minimum value of 3 ng/mL at 30 min was observed. The highest amounts of cytochrome c detected in the supernatant for Nal-P-113 and Dip-P-113 were around 3.5 ng/mL and lower than 2 ng/mL for the least. These results suggested that these three peptides have the capability to cause cancer cells to release cytochrome c into the supernatant and thus triggering the features of immunogenic cell death.

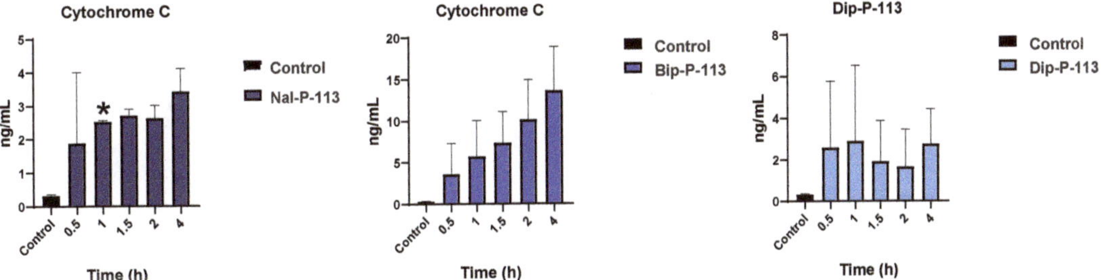

Figure 6. The cytochrome c secretion was quantified using ELISA assay. Results are presented as mean ± SD of three independent experiments. The cells were treated with Nal-P-113, Bip-P-113, and Dip-P-113 peptides at 2× IC$_{50}$ concentration for 0.5, 1, 1.5, 2, and 4 h, respectively. Untreated cells were used as a control. * $p < 0.05$ compared with control.

3.6. Extracellular ATP Release

ATP release was recognized as the iconic feature for immunogenic cell death. Following treatment with 2× IC$_{50}$ of Nal-P-113, Bip-P-113, and Dip-P-113 for 0.5, 1, 1.5, and 2 h, ATP release from PC 9 cells was measured using the luciferin-luciferase assay (Figure 7). The relative light unit (RLU) recorded the highest value for all three peptides at the treatment time of 0.5 h and declining over time. Although all three peptides reached the peak at 0.5-h, Bip-P-113 had the lowest ATP release, while Nal-P-113 and Dip-P-113 displayed similar amounts of ATP release. PC 9 cells treated with Nal-P-113 and Dip-P-113 showed greatest ATP release in the supernatant after incubating for 0.5 h, and the ATP amount detected in the supernatant decreased gradually over time.

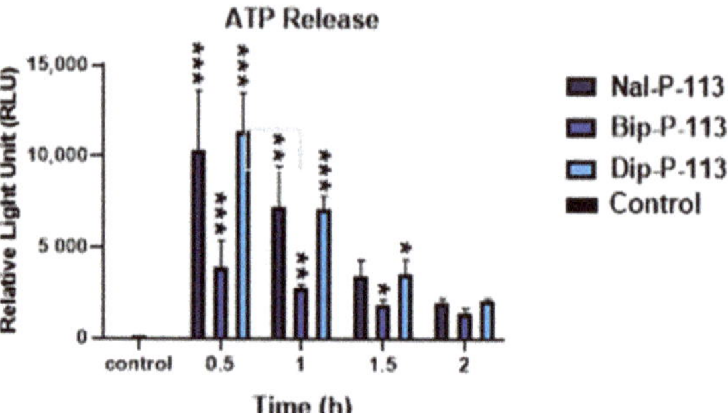

Figure 7. ATP release of PC 9 cells after treatment with Nal-P-113, Bip-P-113, and Dip-P-113. Untreated cells were served as the control group. Results are presented as mean $\pm$ SD of three independent experiments, * $p < 0.05$; ** $p < 0.01$; *** $p < 0.001$ compared with control.

3.7. Extracellular HMGB1 Release

The protein band of HMGB1 (29 kDa) for both cell lysate and supernatant was used to detect HMGB1 protein release after treatment with Nal-P-113, Bip-P-113, and Dip-P-113. No HMGB1 protein was observed in the supernatant of PC 9 cells in the 1 h treatment group (Figure 8). HMGB1 protein can clearly be seen in the supernatant of the Nal-P-113 and Dip-P-113 treatment group after 2 and 4 h (Figure 8). On the other hand, HMGB1 protein can hardly be detected in the supernatant of the Bip-P-113 treatment group after 2 and 4 h. Nevertheless, after 2 h, the cell lysate band of the Nal-P-113, Bip-P-113, and Dip-P-113 treatment groups all started to fade.

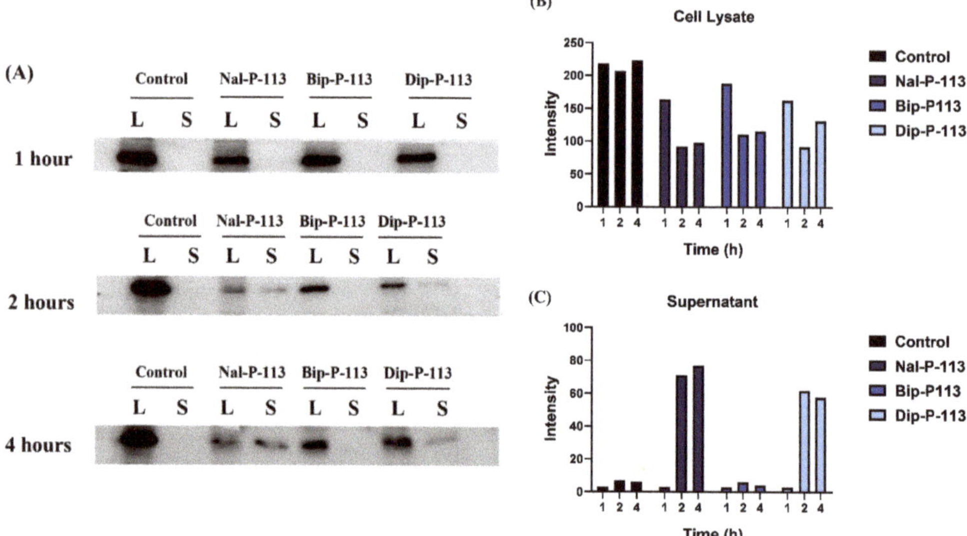

Figure 8. HMGB1 protein secreted from cell lysate to supernatant after incubation with Nal-P-113, Bip-P-113, and Dip-P-113. (**A**) The Western blot images of the PC 9 cells were treated with $2\times$ IC$_{50}$ Nal-P-113, Bip-P-113, and Dip-P-113 peptides for 1, 2, and 4 h, respectively. Untreated cells were served as a control group. L and S represented cell lysate and supernatant, respectively. The relative band intensities of HMGB1 for each group from cell lysate (**B**) and supernatant (**C**).

4. Discussion

AMPs have been proposed as promising antimicrobial and antitumor agents owing to their unique mechanism of action [14]. AMPs kill bacterial and cancer cells through the disruption or lysis of cell membranes instead of acting on a specific target, such as DNA or enzyme. Previously, we have shown that Nal-P-113, Bip-P-113, and Dip-P-113 possessed enhanced salt resistance, serum proteolytic stability, LPS neutralization, as well as synergism with antibiotics against drug-resistant bacteria [7,14,15,22]. Nal-P-113, Bip-P-113, and Dip-P-113, unlike P-113, were also found to target *C. albicans'* cell surface instead of translocating into cells [26]. In the present study, compared to the anticancer abilities of Nal-P-113, Bip-P-113, and Dip-P-113, only P-113 and Phe-P-113 demonstrated weak anticancer activities against the five cell lines studied. On the other hand, Nal-P-113, Bip-P-113, and Dip-P-113 were found to target the cell membrane of cancer cells and cause damage to plasma membrane integrity. Among these non-nature amino acids, Bip is the longest and Dip is the widest, with Nal having an intermediate length and width relative to Bip and Dip [27]. Structure–activity relationships of P-113, Phe-P-113, Nal-P-113, Dip-P-113, and Bip-P113 indicated that Bip-P-113 displayed superior antimicrobial and LPS neutralizing activities compared to Nal-P-113 and Dip-P-113 [28]. The lengthy sidechain of Bip may create extra hydrophobic interactions with the lipid A motif of LPS. However, for the five cancer cell lines studied, Nal-P-113 appeared to have the best anticancer activity and kill cancer cells rapidly. It is possible that the intermediate size and shape of Nal creates suitable hydrophobic interactions with the membrane of cancer cells.

Immunogenic cell death is the type of cell death that can induce an immune response. Cell death can be classified into two major groups: programed/regulated cell death (such as apoptosis, autophagy, and necroptosis) and accidental cell death due to non-physiological reasons (necrosis) [29]. Apoptosis normally causes cell shrinkage, plasma membrane blebbing, and formation of apoptotic bodies. Necrosis, on the other hand, causes increase in cell volume, loss of plasma membrane integrity, and leakage of cellular contents [29]. Radiotherapy and some chemotherapeutic agents can induce immunogenic cell death of cancer cells [30]. After the occurrence of immunogenic cell death of cancer cells, danger-associated molecular patterns (DAMPs), such as HSP70, ATP secretion, cytochrome c release, and HMGB1 protein, could be induced [31]. The host immune system against cancer cells can be activated by sensing the presence of DAMPs by dendritic cells [32]. Additionally, the immunogenic cell death response's efficacy is regulated by parameters such as how efficient the dendritic cells phagocytize dead cancer cells, as well as how efficiently the cytotoxic T cells target cancer cells by tumor antigen-specific killing (Figure 9). For example, PKHB1, a thrombospondin-1 peptide mimic, can induce DAMPs in breast cancer cell lines, and the supernatant of PKHB1-killed breast cancer cells can induce maturation of bone-marrow-derived DCs and elicit antitumor T cell responses [33]. Some AMPs, such as LTX-315 [21], ΔM4 [34], and TP4 [35], were shown to induce a rapid plasma membrane disruption of cancer cells and cause the release of DAMPs into the cell supernatant. In an immunocompetent mouse model, the local injection of LTX-315 into tumors resulted in complete regression and specific anticancer immune responses with long-term protective immunity [36,37]. Moreover, the results of a recent clinical trial demonstrated that LTX-315 has an acceptable safety profile and is clinically active via inducing necrosis and CD8+ T-cell infiltration into the solid tumor microenvironment [38]. An abscopal effect was also found following treatment with LTX-315 [38].

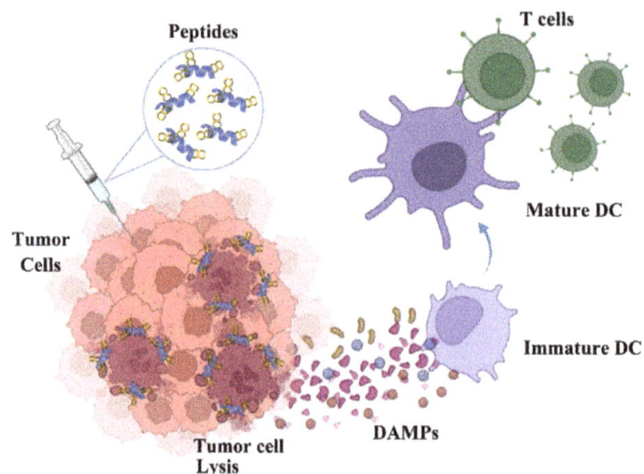

Figure 9. Proposed mechanism of anticancer action for the designed peptides. The intratumoral administration of Nal-P-113 ( ) induced tumor cellular lysis through membrane destabilization, caused the release of danger-associated molecular patterns (DAMPs), including ATP ( ), HMGB1 ( ), and ROS ( ), as well as tumor antigens into the tumor microenvironment. Thereafter, these DAMPs recruited the immature dendritic cells (DCs) and further induced the maturation of dendritic cells. The mature DCs were then priming for antigen presentation to T cells and subsequent antitumor immune responses.

In addition to anticancer activities, Nal-P-113, Bip-P-113, and Dip-P-113 also possess enhanced ability to induce immunogenic cell death. The present studies suggested that Nal-P-113 and Dip-P-113 might cause cancer cells to undergo the necrosis pathway, and HMGB1 protein can be detected as early as 2 h post-treatment. On the other hand, Bip-P-113 might depend on apoptotic cell death, as HMGB1 protein was not detected in the Western blot analysis. Undetectable HMGB1 protein in the Bip-P-113-treated group might be due to inactivation of HMGB1 protein by oxidation, and further studies should be carried out for a more concrete argument.

Reactive oxygen species (ROS) production is believed to be an important component in the intracellular danger signaling to govern immunogenic cell death [39]. ROS overproduction may help open the mitochondrial permeability transition pore and cause the leakage of mitochondrial components into the cytosol and outside of the cell. These molecules may then act as DAMPs to trigger the immune response. For example, as ROS rise, molecules such as cytochrome c may be released from mitochondria and activate the caspase cascade, which then triggers apoptosis. Excessive ROS promotes necrotic cell death. The present results also supported that Nal-P-113 and Dip-P-113 might have undergone the necrotic cell death pathway. The hypothesis that Bip-P-113 caused the cancer cells to die through the apoptotic pathway is further supported by evidence of a steady decrease in cytosolic ATP associated with a large number of dead cells [40]. ATP depletion may cause the switching from apoptotic cell death to necrotic cell death [40].

High levels of cytochrome c release were found with Bip-P-113. The mitochondria releases cytochrome c when it is damaged, and the cytochrome c then contributes to apoptotic cell death. Both Nal-P-113 and Dip-P-113 did not cause much cytochrome c release, and this might be due to the rapid switching form of cell death, from apoptosis to necrosis cell death.

In conclusion, our results demonstrated that P-113 derivatives with bulky non-nature amino acid substitutions display compelling anticancer activities against five cancer cell lines, including three non-small cell lung cancer cell lines (H1975, A549, PC 9) and two oral cancer cell lines (C9, OECM-1). Moreover, these peptides caused necrosis of cancer cells and triggered the damaged cancer cells to release potent danger-associated molecular patterns (DAMPs), such as reactive oxygen species (ROS), cytochrome c, ATP, and HMGB1. The release of DAMPs may lead to the maturation and activation of dendritic cells of the adaptive immune system (Figure 9). Among these peptides, Nal-P-113 demonstrated the best anticancer activities and induced immunogenic cell death. Our results should be useful in the development of new antimicrobial peptides and peptidomimetics with enhanced anticancer activities for potential therapeutic applications.

Author Contributions: Y.-H.C., C.-Y.L., C.-L.W. and K.-L.P. performed the experiments and analyzed the data. C.-L.W. and J.-W.C. wrote the manuscript. B.-S.Y. and J.-W.C. planned the study and revised and approved the final version of the manuscript. All authors have read and agreed to the published version of the manuscript.

Funding: This work was supported by grants from the Ministry of Science and Technology of Taiwan to J.-W.C. (110-2113-M-007-021) and National Taiwan University Hospital Hsinchu Branch to B.-S.Y.

Institutional Review Board Statement: Not applicable.

Informed Consent Statement: Not applicable.

Data Availability Statement: Not applicable.

Acknowledgments: We thank Justin Cheng and Daniel Cheng for editing the manuscript.

Conflicts of Interest: None of the authors have potential conflicts of interest to be disclosed.

Abbreviations

AMPs	antimicrobial peptides
Phe	phenylalanine
Nal	β-naphthylalanine
Dip	β-diphenylalanine
Bip	β-(4,4′-biphenyl)alanine
DAMPs	danger-associated molecular patterns
ROS	reactive oxygen species
ATP	adenosine triphosphate
HMGB1	high-mobility group box 1
HPLC	high-performance liquid chromatography
FBS	fetal bovine serum
DMSO	dimethyl sulfoxide
MTT	3-(4,5-dimethylthiazol-2-yl)2.5-diphenyl tetrazolium bromide
JC-1	5,5′,6,6′-Tetrachloro-1,1′,3,3′-tetraethyl-imidacarbocyanine iodide
CCCP	carbonyl cyanide 3-chloro-phenylhydrazone
ELISA	enzyme-linked immunosorbent assay
PI	propidium iodide

References

1. Zharkova, M.S.; Orlov, D.S.; Golubeva, O.Y.; Chakchir, O.B.; Eliseev, I.E.; Grinchuk, T.M.; Shamova, O.V. Application of Antimicrobial Peptides of the Innate Immune System in Combination with Conventional Antibiotics—A Novel Way to Combat Antibiotic Resistance? *Front. Cell. Infect. Microbiol.* **2019**, *9*, 128. [CrossRef] [PubMed]
2. Zhu, Y.; Hao, W.; Wang, X.; Ouyang, J.; Deng, X.; Yu, H.; Wang, Y. Antimicrobial peptides, conventional antibiotics, and their synergistic utility for the treatment of drug-resistant infections. *Med. Res. Rev.* **2022**. [CrossRef] [PubMed]
3. Hancock, R.E.W.; Alford, M.A.; Haney, E.F. Antibiofilm activity of host defence peptides: Complexity provides opportunities. *Nat. Rev. Microbiol.* **2021**, *19*, 786–797. [CrossRef]
4. Li, W.; Separovic, F.; O'Brien-Simpson, N.M.; Wade, J.D. Chemically modified and conjugated antimicrobial peptides against superbugs. *Chem. Soc. Rev.* **2021**, *50*, 4932–4973. [CrossRef] [PubMed]
5. Martinez, M.; Gonçalves, S.; Felício, M.R.; Maturana, P.; Santos, N.C.; Semorile, L.; Hollmann, A.; Maffía, P.C. Synergistic and antibiofilm activity of the antimicrobial peptide P5 against carbapenem-resistant Pseudomonas aeruginosa. *Biochim. Biophys. Acta* **2019**, *1861*, 1329–1337. [CrossRef] [PubMed]
6. Cassone, M.; Otvos, L.J. Synergy among antibacterial peptides and between peptides and small-molecule antibiotics. *Expert Rev. Anti Infect. Ther.* **2010**, *8*, 703–716. [CrossRef]

7. Li, J.; Fernandez-Millan P, B.E. Synergism between Host Defence Peptides and Antibiotics against Bacterial Infections. *Curr. Top. Med. Chem.* **2020**, *20*, 1238–1263. [CrossRef]

8. Sierra, J.M.; Fusté, E.; Rabanal, F.; Vinuesa, T.; Viñas, M. An overview of antimicrobial peptides and the latest advances in their development. *Expert Opin. Biol. Ther.* **2017**, *17*, 663–676. [CrossRef]

9. Wu, C.L.; Hsueh, J.Y.; Yip, B.S.; Chih, Y.H.; Peng, K.L.; Cheng, J.W. Antimicrobial Peptides Display Strong Synergy with Vancomycin against Vancomycin-Resistant E. faecium, S. aureus, and Wild-Type E. coli. *Int. J. Mol. Sci.* **2020**, *21*, 4578. [CrossRef]

10. Wu, C.L.; Peng, K.L.; Yip, B.S.; Chih, Y.H.; Cheng, J.W. Boosting Synergystic Effects of Short Antimicrobial Peptides with Conventional Antibiotics against Resistant Bacteria. *Front. Microbiol.* **2021**, *12*, 747760. [CrossRef]

11. Etayash, H.; Alford, M.; Akhoundsadegh, N.; Drayton, M.; Straus, S.K.; Hancock, R.E.W. Multifunctional Antibiotic-Host Defense Peptide Conjugate Kills Bacteria, Eradicates Biofilms, and Modulates the Innate Immune Response. *J. Med. Chem.* **2021**, *64*, 16854–16863. [CrossRef] [PubMed]

12. Li, W.; O'Brien-Simpson, N.M.; Holden, J.A.; Otvos, L.; Reynolds, E.C.; Separovic, F.; Hossain, M.A.; Wade, J.D. Covalent conjugation of cationic antimicrobial peptides with a b-lactam antibiotic core. *Peptide Sci.* **2018**, *110*, e24059. [CrossRef]

13. Bakare, O.; Gokul, A.; Wu, R.; Niekerk, L.-A.; Klein, A.; Keyster, M. Biomedical Relevance of Novel Anticancer Peptides in the Sensitive Treatment of Cancer. *Biomolecules* **2021**, *11*, 1120. [CrossRef] [PubMed]

14. Lin, L.; Chi, J.; Yan, Y.; Luo, R.; Feng, X.; Zheng, Y.; Xian, D.; Li, X.; Quan, G.; Liu, D.; et al. Membrane-disruptive peptides/peptidomimetics-based therapeutics: Promising systems to combat bacteria and cancer in the drug-resistant era. *Acta Pharm. Sin. B* **2021**, *11*, 2609–2644. [CrossRef]

15. Tornesello, A.L.; Borrelli, A.; Buonaguro, L.; Buonaguro, F.M.; Tornesello, M.L. Antimicrobial Peptides as Anticancer Agents: Functional Properties and Biological Activities. *Molecules* **2020**, *25*, 2850. [CrossRef]

16. Rothstein, D.M.; Spacciapoli, P.; Tran, L.T.; Xu, T.; Roberts, F.D.; Serra, M.D.; Buxton, D.K.; Oppenheim, F.G.; Friden, P. Anticandida Activity Is Retained in P-113, a 12-Amino-Acid Fragment of Histatin 5. *Antimicrob. Agents Chemother.* **2001**, *45*, 1367–1373. [CrossRef]

17. Helmerhorst, E.J.; Oppenheim, F.G.; Choi, L.; Cheng, J.W.; Reiner, N.E. Evaluation of a new host-derived synthetic antifungal peptide (PAC-113) in the treatment of oral candidiasis. In Proceedings of the International Meeting on Antimicrobial Chemotherapy in Clinical Practice (ACCP), Portofino, Italy, 15–17 November 2007.

18. Yu, H.-Y.; Tu, C.-H.; Yip, B.-S.; Chen, H.-L.; Cheng, H.-T.; Huang, K.-C.; Lo, H.-J.; Cheng, J.-W. Easy Strategy to Increase Salt Resistance of Antimicrobial Peptides. *Antimicrob. Agents Chemother.* **2011**, *55*, 4918–4921. [CrossRef]

19. Chih, Y.-H.; Wang, S.-Y.; Yip, B.-S.; Cheng, K.-T.; Hsu, S.-Y.; Wu, C.-L.; Yu, H.-Y.; Cheng, J.-W. Dependence on size and shape of non-nature amino acids in the enhancement of lipopolysaccharide (LPS) neutralizing activities of antimicrobial peptides. *J. Colloid Interface Sci.* **2018**, *533*, 492–502. [CrossRef]

20. Teng, Q.-X.; Luo, X.; Lei, Z.-N.; Wang, J.-Q.; Wurpel, J.; Qin, Z.; Yang, D.-H. The Multidrug Resistance-Reversing Activity of a Novel Antimicrobial Peptide. *Cancers* **2020**, *12*, 1963. [CrossRef]

21. Eike, L.-M.; Yang, N.; Rekdal, Ø.; Sveinbjørnsson, B. The oncolytic peptide LTX-315 induces cell death and DAMP release by mitochondria distortion in human melanoma cells. *Oncotarget* **2015**, *6*, 34910–34923. [CrossRef]

22. Haug, B.E.; Camilio, K.A.; Eliassen, L.T.; Stensen, W.; Svendsen, J.S.; Berg, K.; Mortensen, B.; Serin, G.; Mirjolet, J.-F.; Bichat, F.; et al. Discovery of a 9-mer Cationic Peptide (LTX-315) as a Potential First in Class Oncolytic Peptide. *J. Med. Chem.* **2016**, *59*, 2918–2927. [CrossRef] [PubMed]

23. Chu, H.-L.; Yip, B.-S.; Chen, K.-H.; Yu, H.-Y.; Chih, Y.-H.; Cheng, H.-T.; Chou, Y.-T.; Cheng, J.-W. Novel Antimicrobial Peptides with High Anticancer Activity and Selectivity. *PLoS ONE* **2015**, *10*, e0126390. [CrossRef] [PubMed]

24. Eskandari, F.; Momeni, H.R. Silymarin protects plasma membrane and acrosome integrity in sperm treated with sodium arsenite. *Int. J. Reprod. Biomed.* **2016**, *14*, 47–52. [CrossRef] [PubMed]

25. Elefantova, K.; Lakatos, B.; Kubickova, J.; Sulova, Z.; Breier, A. Detection of the Mitochondrial Membrane Potential by the Cationic Dye JC-1 in L1210 Cells with Massive Overexpression of the Plasma Membrane ABCB1 Drug Transporter. *Int. J. Mol. Sci.* **2018**, *19*, 1985. [CrossRef]

26. Cheng, K.-T.; Wu, C.-L.; Yip, B.-S.; Chih, Y.-H.; Peng, K.-L.; Hsu, S.-Y.; Yu, H.-Y.; Cheng, J.-W. The Interactions between the Antimicrobial Peptide P-113 and Living Candida albicans Cells Shed Light on Mechanisms of Antifungal Activity and Resistance. *Int. J. Mol. Sci.* **2020**, *21*, 2654. [CrossRef]

27. Haug, B.E.; Skar, M.L.; Svendsen, J.S. Bulky aromatic amino acids increase the antibacterial activity of 15-residue bovine lactoferricin derivatives. *J. Pept. Sci.* **2001**, *7*, 425–432. [CrossRef]

28. Cai, M.; Liu, J.; Gong, Y.; Krishnamoorthi, R. A pratical method for stereospecific assignments of g- and d-methylene hydrogens via estimation of vicinal 1H-1H coupling constants. *J. Magn. Reson. Ser. B* **1995**, *107*, 172–178. [CrossRef]

29. Fink, S.L.; Cookson, B.T. Apoptosis, Pyroptosis, and Necrosis: Mechanistic Description of Dead and Dying Eukaryotic Cells. *Infect. Immun.* **2005**, *73*, 1907–1916. [CrossRef]

30. Serrano-Del Valle, A.; Anel, A.; Naval, J.; Marzo, I. Immunogenic Cell Death and Immunotherapy of Multiple Myeloma. *Front Cell Dev. Biol.* **2019**, *7*, 50. [CrossRef]

31. Pitt, J.M.; Kroemer, G.; Zitvogel, L. Immunogenic and Non-Immunogenic Cell Death in the Tumor Microenvironment. *Adv. Exp. Med. Biol.* **2017**, *1036*, 65–79. [CrossRef]

32. Aaes, T.L.; Vandenabeele, P. The intrinsic immunogenic properties of cancer cell lines, immunogenic cell death, and how these influence host antitumor immune responses. *Cell Death Differ.* **2020**, *28*, 843–860. [CrossRef] [PubMed]
33. Calvillo-Rodríguez, K.M.; Mendoza-Reveles, R.; Gómez-Morales, L.; Uscanga-Palomeque, A.C.; Karoyan, P.; Martínez-Torres, A.C.; Rodríguez-Padilla, C. PKHB1, a thrombospondin-1 peptide mimic, induces anti-tumor effect through immunogenic cell death induction in breast cancer cells. *OncoImmunology* **2022**, *11*, 2054305. [CrossRef] [PubMed]
34. Santa-Gonzalez, G.A.; Patino-Gonzalez, E.; Manrique-Moreno, M. Synthetic Peptide dM4-Induced Cell Death Associated with Cytoplasmic Membrane Disruption, Mitochondrial Dysfunction and Cell Cycle Arrest in Human Melanoma Cells. *Molecules* **2020**, *25*, 5684. [CrossRef] [PubMed]
35. Su, B.-C.; Hung, G.-Y.; Tu, Y.-C.; Yeh, W.-C.; Lin, M.-C.; Chen, J.-Y. Marine Antimicrobial Peptide TP4 Exerts Anticancer Effects on Human Synovial Sarcoma Cells via Calcium Overload, Reactive Oxygen Species Production and Mitochondrial Hyperpolarization. *Mar. Drugs* **2021**, *19*, 93. [CrossRef]
36. Camilio, K.A.; Berge, G.; Ravuri, C.S.; Rekdal, Ø.; Sveinbjørnsson, B. Complete regression and systemic protective immune responses obtained in B16 melanomas after treatment with LTX-315. *Cancer Immunol. Immunother.* **2014**, *63*, 601–613. [CrossRef]
37. Nestvold, J.; Wang, M.-Y.; Camilio, K.A.; Zinöcker, S.; Tjelle, T.E.; Lindberg, A.; Haug, B.E.; Kvalheim, G.; Sveinbjørnsson, B.; Rekdal, Ø. Oncolytic peptide LTX-315 induces an immune-mediated abscopal effect in a rat sarcoma model. *OncoImmunology* **2017**, *6*, e1338236. [CrossRef]
38. Spicer, J.; Marabelle, A.; Baurain, J.-F.; Jebsen, N.L.; Jøssang, D.E.; Awada, A.; Kristeleit, R.; Loirat, D.; Lazaridis, G.; Jungels, C.; et al. Safety, Antitumor Activity, and T-Cell Responses in a Dose-Ranging Phase I Trial of the Oncolytic Peptide LTX-315 in Patients with Solid Tumors. *Clin. Cancer Res.* **2021**, *27*, 2755–2763. [CrossRef]
39. Krysko, D.V.; Garg, A.D.; Kaczmarek, A.; Krysko, O.; Agostinis, P.; Vandenabeele, P. Immunogenic cell death and DAMPs in cancer therapy. *Nat. Rev. Cancer* **2012**, *12*, 860–875. [CrossRef]
40. Zamaraeva, M.; Sabirov, R.Z.; Maeno, E.; Ando-Akatsuka, Y.; Bessonova, S.V.; Okada, Y. Cells die with increased cytosolic ATP during apoptosis: A bioluminescence study with intracellular luciferase. *Cell Death Differ.* **2005**, *12*, 1390–1397. [CrossRef]

Article

Effect and Mechanisms of Antibacterial Peptide Fraction from Mucus of *C. aspersum* against *Escherichia coli* NBIMCC 8785

Yana Topalova [1,*], Mihaela Belouhova [1], Lyudmila Velkova [2], Aleksandar Dolashki [2], Nellie Zheleva [3], Elmira Daskalova [1], Dimitar Kaynarov [2], Wolfgang Voelter [4] and Pavlina Dolashka [2,*]

[1] Faculty of Biology, Sofia University, 8 Dragan Tzankov Blvd., 1164 Sofia, Bulgaria; mihaela.kirilova@uni-sofia.bg (M.B.); baba_emi@abv.bg (E.D.)
[2] Institute of Organic Chemistry with Centre of Phytochemistry, Bulgarian Academy of Sciences, Acad. G. Bonchev Str., Bl. 9, 1113 Sofia, Bulgaria; lyudmila_velkova@abv.bg (L.V.); adolashki@yahoo.com (A.D.); mitkokaynarov@abv.bg (D.K.)
[3] Faculty of Physics, Sofia University, 5 James Bourchier Blvd., 1164 Sofia, Bulgaria; zhelevan@phys.uni-sofia.bg
[4] Institute of Biochemistry, University of Tübingen, Hoppe-Seyler-Straße 4, D-72076 Tübingen, Germany; wolfgang.voelter@uni-tuebingen.de
* Correspondence: ytopalova@sofia-uni.bg or yanatop@abv.bg (Y.T.); pavlina.dolashka@orgchm.bas.bg or pda54@abv.bg (P.D.); Tel.: +359-887193423 (P.D.)

Abstract: Peptides isolated from the mucus of *Cornu aspersum* could be prototypes for antibiotics against pathogenic bacteria. Information regarding the mechanisms, effective concentration, and methods of application is an important tool for therapeutic, financial, and ecological regulation and a holistic approach to medical treatment. A peptide fraction with MW < 10 kDa was analyzed by MALDI-TOF-TOF using Autoflex™ III. The strain *Escherichia coli* NBIMCC 8785 (18 h and 48 h culture) was used. The changes in bacterial structure and metabolic activity were investigated by SEM, fluorescent, and digital image analysis. This peptide fraction had high inhibitory effects in surface and deep inoculations of *E. coli* of 1990.00 and 136.13 $mm^2/mgPr/\mu Mol$, respectively, in the samples. Thus, it would be effective in the treatment of infections involving bacterial biofilms and homogenous cells. Various deformations of the bacteria and inhibition of its metabolism were discovered and illustrated. The data on the mechanisms of impact of the peptides permitted the formulation of an algorithm for the treatment of infections depending on the phase of their development. The decrease in the therapeutic concentrations will be more sparing to the environment and will lead to a decrease in the cost of the treatment.

Keywords: mucus of *Cornu aspersum*; peptide fraction MW < 10 kDa; *Escherichia coli* NBIMCC 8785; SEM; fluorescence and digital assays; antibacterial effect

Citation: Topalova, Y.; Belouhova, M.; Velkova, L.; Dolashki, A.; Zheleva, N.; Daskalova, E.; Kaynarov, D.; Voelter, W.; Dolashka, P. Effect and Mechanisms of Antibacterial Peptide Fraction from Mucus of *C. aspersum* against *Escherichia coli* NBIMCC 8785. *Biomedicines* **2022**, *10*, 672. https://doi.org/10.3390/biomedicines10030672

Academic Editor: Jitka Petrlova

Received: 16 February 2022
Accepted: 8 March 2022
Published: 14 March 2022

Publisher's Note: MDPI stays neutral with regard to jurisdictional claims in published maps and institutional affiliations.

1. Introduction

The discovery and testing of new therapeutic tools with a studied and clear effect on different bacterial infections is becoming increasingly topical. The requirements for using such kinds of therapeutic tools are also increasing [1–6].

The reason for this search in the first place is that the antibiotics that have been widely used previously led to the development of resistance by pathogenic microorganisms. The new tools should be such that their bio attacks on pathogens do not lead to resistance in the pathogenic microorganisms [7,8].

Two consequences related to the epidemiologic and therapeutic measures have appeared: (1) due to the wide use of antibiotics in different combinations and treatment sequences, the danger of the appearance of resistance to contemporary antibiotics and infectious pathogens that represent a serious medical risk for people and animals has been accelerated; and (2) the low biodegradability and wide distribution of antibiotics have led to their accumulation in civil wastewaters, which has led to another serious ecological risk [3,9,10].

The ecological consequences are immeasurable compared to the medical ones and from a long-term perspective. The antibiotics and their metabolic products are present in purified waters and are distributed in water basins, drinking water, plants, and animals, and through the trophic chains, are returned to humans. These pollutants in the wastewaters and in the environment combine and interact with other chemical factors. All these factors inhibit the activated sludge and microorganisms that purify the waters and lead to a lasting disturbance of the restorative processes of the resources (waters, soils, biomass, sediments, etc.) in the treatment technologies for the environment as a whole [3,4,9].

These combined interactions with human health and life and the health and future of the planet are lasting. In the future, they will impact strongly on the life of the planet [3]. The sources of the described dangers should be limited and discontinued, and intelligent solutions to the problems of the people, health, and the planet in the whole life cycle of therapeutic tools and of antibiotics should be found. The roads are three: (1) creation of new technologies for water treatment, and natural resources with included manageable detoxifying elements directed to the degradation of the toxic pollutants present and incoming in the environment; (2) replacement of the toxins in environment pollutants with their harmless alternatives; and (3) minimization of the tools used for medical and other aims that have harmful impacts on the environment from a long-term perspective. The combined consideration of the questions regarding therapeutic tools, effective concentrations, objective and personalized impacts, the life cycle, and impact on the environment and on the planet require a holistic approach and a deep understanding of the mechanisms of impact, distribution, and removal. This understanding is the basis of the concept of a less-toxic environment ensured by innovative therapeutic products and clean technologies for the sustainment of all resources [1,3,9,11,12].

The discovery of therapeutic products with an antimicrobial action and with a chemical structure close to natural compounds is a contemporary tendency and is in unison with the concept that "Nature knows best—follow it".

The World Health Organization (WHO) declared that the increase in antimicrobial resistance to conventional antibiotics in recent decades is one of the three global threats to human health worldwide, and announced the start of the "post-antibiotics era" [13,14].

Invertebrates lack the adaptive immune system that is found in vertebrate species; therefore, they rely solely upon their innate immunity, in which AMPs play a crucial role in counteracting invading pathogens [15–18]. Molluscs are a large natural sources of AMPs, in part because of their great diversity (superior only to arthropods) and their ability to adapt to almost all habitat types [17,19–23].

AMPs hold considerable potential for reducing and overcoming antibiotic resistance because they not only have an antibacterial effect, but they also increase membrane permeability, and as a consequence, their presence enhances the effect of the antibiotics [24]. To adapt to this, bacteria need to develop major changes in the structure and properties of their membranes [11]. Due to their broad spectrum of action, natural AMPs can be a model for the discovery of new antimicrobial drugs that can address the problem of multidrug resistance of pathogenic microorganisms [25]. Currently, a number of bioactive compounds from the hemolymph and mucus of snails are being studied and developed as pharmaceuticals and/or raw materials for therapeutic purposes and applications [18,19,21,22,26,27]. For example, the garden snail (*Cornu aspersum*) and the giant African snail (*Achatina fulica*) are amongst the most studied land snails [18,27]. The water gastropods also offer a variety of AMPs. Some of the biologically active molecules identified were developed into pharmaceuticals such as Prialt®, Adcetris®, and many others [28].

Several peptides from the hemolymph of the garden snails *H. lucorum* and *H. aspersa* have been reported to show a broad spectrum of antimicrobial activity against *Staphylococcus aureus*, *Staphylococcus epidermidis*, *E. coli*, *Helicobacter pylori*, and *Propionibacterium acnes* [22,26]. The mucus content of snails varies depending on the function and secretory structure of the glands and species of the gastropods [29]. Zhong et al. found the AMP "mytimacin-AF" (9700 Da) in *Achatina fulica* mucus; it exhibited antimicrobial activity against *S. aureus*, *Bacillis* spp., *Klebsiella pneumoniae*, and *Candida albicans* [27]. It is known that the antimicrobial activity of mucus from terrestrial snails is associated with the presence of AMPs, but also with some antibacterial proteins and glycoproteins. Pitt et al. showed that the mucosa of *H. aspersa* (*Cornu aspersum*) was effective against three laboratory strains of *P. aeruginosa* [30]. In addition, antibacterial proteins and glycoproteins have been reported in the mucus of various terrestrial snails (*A. fulica*, *H. aspersa*, *Cryptozona bistrialis*, *Lissachatina fulica*, and *Hemiplecta differenta*) [18,30–33].

Our previous studies described the antimicrobial activities of different fractions from the mucus of the garden snail *C. aspersum*, and characterized metabolites and antimicrobial peptides [34–37]. This study upgraded the previous data and provided additional information on new peptides and protein fractions with antibacterial activity in the mucus of *C. aspersum*. Furthermore, it shed light on the mechanism of antimicrobial action of bioactive compounds from the mucus.

One of the paths to discovery of therapeutic tools and antimicrobial agents with a natural origin that are close to the chemical structure of natural compounds is to use peptides with an antimicrobial action [34–37]. Peptides with a different molecular weight are known, including peptides and protein fractions with a studied antimicrobial action. Such peptides with antibacterial and antifungal action have been studied previously. A large part of the research was conducted using peptides and protein fractions isolated from snail slime. The present investigation contributed to the efforts to find new natural compounds that can serve as antimicrobial agents with less toxicity for humans and avoid antimicrobial resistance. It also explored the possibilities for minimization of the resources taken from nature, in accordance with some of the main features of bacterial infections.

2. Materials and Methods

2.1. Mucus Collection and Separation of Different Fractions

The mucus was collected and purified from *C. aspersum* snails grown on Bulgarian farms using a special device with low-voltage electrical stimulation and a method described in BG Utility model 2097/2015 [34–37]. After removing the coarse impurities from the obtained extract by homogenization and centrifugation, the supernatant was subjected to several cycles of filtration at 4 °C.

The obtained crude mucus extract was divided into two basic fractions by ultrafiltration, using membranes with pore sizes of 10 kDa and 20 kDa (EMD Millipore Corporation, Billerica, MA, USA; and polyethersulfone, Microdyn Nadir™ from STERLITECH Corporation, Goleta, CA, USA, respectively). The peptide fraction with an MW below 10 kDa was further divided into three peptide subfractions by using Amicon® Ultra-15 centrifugal units with 3 and 5 kDa membranes in centrifugation ($3000\times g$, 4 °C, 30 min). The used of a noninvasive technique—ultrafiltration—ensured that we obtained fractions containing intact compounds. Finally, the obtained mucus extract was divided into 5 fractions with different molecular weights:

Sample 1—fraction containing compounds with MW < 3 kDa;
Sample 2—fraction containing compounds with MW 3–5 kDa;
Sample 3—fraction containing compounds with MW 5–10 kDa;
Sample 4—fraction containing compounds with MW < 10 kDa;
Sample 5—fraction containing compounds with MW < 20 kDa.

2.2. Analysis of Peptide Fractions by Mass Spectrometric Analysis

Peptide fractions with MW < 3 kDa and with MW < 10 kDa were analyzed by MALDI-TOF-TOF mass spectrometry using an Autoflex™ III (Bruker Daltonics, Bremen, Germany) with a laser at a frequency of 200 Hz and operating at a wavelength of 355 nm. The analysis was performed after applying 1.0 µL of a mixture of 2.0 µL of the sample and 2.0 µL of a matrix solution (7 mg/mL α-cyano-4-hydroxyquinamic acid, CHCA) in 50% acetonitrile

(ACN) containing 0.1% trifluoroacetic acid (TFA), on a target plate with 192 stainless-steel wells. The mass spectrometer was calibrated with a standard mixture of angiotensin I, Glu-1-fibrinopeptide B, and ACTH (1–17); the ACTH and MS/MS spectra were obtained in reflector mode. The amino acid sequences of the peptides were identified by a MALDI-MS/MS assay using precursor ions from the MS assays.

2.3. Antibacterial Activity Testing

The experiments were performed by using a reference strain—the Gram-negative bacterial strain *E. coli* NBIMCC 8785 (NBIMCC, Sofia, Bulgaria). It could serve as a model organism for the determination of the changes in bacterial morphology, viability, and metabolic activity. These microorganisms are among the best-studied bacteria, and provide useful information that can be comparable to many other studies on the subject. They are also a member of the *Enterobacteriaceae* family and can be regarded as a model of opportunistic pathogens with high potential for antibiotic resistance.

2.4. Nutrient Media and Culture Conditions

The used nutrient media were nutrient broth (liquid medium) (Himedia Laboratories Pvt Ltd., Mumbai, India) and nutrient agar (solid media) (Himedia Laboratories Pvt Ltd., Mumbai, India). The solid media and the well-diffusion method were used for the determination of the antibacterial activity in the cultivation assays. The bacteria for the SEM and fluorescence analysis were cultivated in nutrient broth. The microbial strains were preserved as a lyophilized culture in glucose medium and peptone protector. In prior experiments, they were rehydrated and maintained at slope agar in tubes. Good laboratory practice was followed during the microbiological assays, and standard microbiological equipment and glassware were used.

2.5. Studies of Antibacterial Activities Using Different Methods

For determination of the antibacterial activity, the well-diffusion method was applied with two modifications—deep inoculation and surface cultivation. The deep inoculation included mixing the model bacteria with nutrient agar at a temperature below 40 °C. The growth of the bacteria in the agar volume was a model for infections deeper in the skin. The surface cultivation could serve as model for surface infections. It was performed by using a standardized microbial suspension (50 µL with a density of 10^9 cells/mL) spread on the top of the agar. The resulting sterile zones were calculated as mm^2/mgProtein/µMol of the sample. For details, see Dolashki et al. [34] (Figure 1).

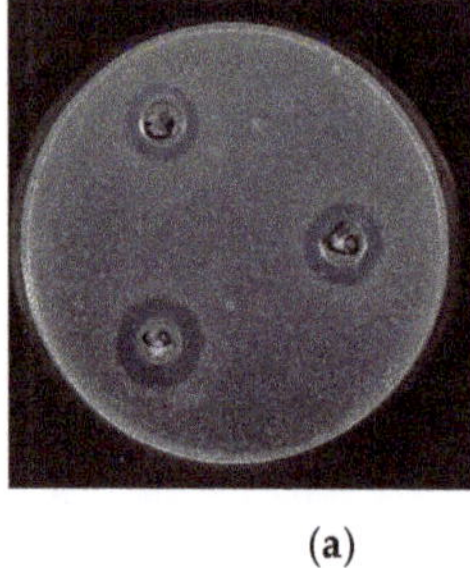 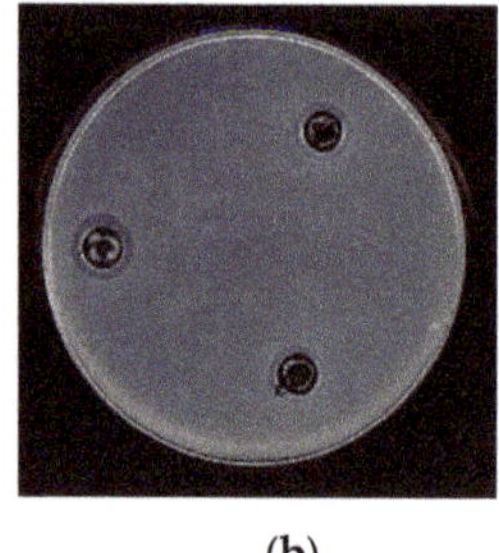

(**a**)   (**b**)

Figure 1. Images of sterile areas found in the investigation of antibacterial activity against model bacteria: antibacterial effect of peptide fraction MW < 10 kDa on *E. coli* NBIMCC 8785 at surface cultivation (**a**) and at inoculation in depth (**b**).

The incubation of bacterial cells with the peptide fractions was 1 or 6 h, depending on the variants used in the investigation.

The following variants of the inhibitory effect of the peptide fraction with MW < 10 kDa were investigated: (1) 48 h bacterial culture + 50% peptide fraction—1 h incubation; (2) 18 h bacterial culture + 50% peptide fraction—1 h incubation; and (3) 18 h bacterial culture + 1%, 5%, and 10% peptide fraction—6 h incubation.

Cultivation for the investigation of the inhibition of the activity of *E. coli* NBIMCC 8785 was carried out in the nutrient broth liquid medium. The 18 h and 48 h cultures were obtained by means of cultivation at 36 °C in 300 cm^3 flasks. The obtained microbial material was divided in 30 cm^3 flasks. Peptide fractions of 1%, 5%, 10%, or 50% *v/v* were added to every flask. The concentration of proteins in the peptide fraction was 0.5 mg/mL.

These variants corresponded to different infections treated with a peptide fraction with MW < 10 kDa in different concentrations.

The bacterial cultures were used in different phases of their development. The 18 h culture was at the late exponential phase of growth, when the microorganisms had reached their maximal activity. The 48 h culture was a model for a culture in the stationary phase, when the bacteria were closer to the ones in the real infections.

2.6. Electron Microscopic Assays

The scanning electron microscopy (SEM) was performed on samples treated with a series of ethyl alcohol in increasing concentrations, as described in Dolashki et al., 2020 [34]. The analysis was used to study the morphological changes in the bacteria at an individual level as a result of the presence of the active compounds from the *C. aspersum* mucus.

2.7. CTC/DAPI Staining and Digital Image Analysis

CTC/DAPI staining: For determination of the share of live and dead cells in the samples and for estimation of their metabolic activity, a fluorescent tetrazolium salt was used —5cyano-2,3-ditolyl tetrazolium chloride (CTC) (Merck KGaA, Darmstadt, Germany). The samples were stained for 45 min with a 5 mM concentration of the dye. For visualization of all bacteria, DAPI (4′,6-diamidino-2-phenylindole) (Merck KGaA, Darmstadt, Germany) staining was applied at a concentration of 1 μM/mL of the indole dye.

The fluorescence images were taken with a Leica DM6 B (Leica Camera AG, Wetzlar, Germany) epifluorescent microscope. The settings of the camera were identical for all pictures taken during the experiment.

Digital image analysis: The fluorescence images from the CTC/DAPI staining were subjected to digital image analysis using the software daime 2.2 (University of Vienna, Vienna, Austria). The threshold criteria for image segmentation were chosen manually. The digital analysis was made using several parameters.

Intensity of the CTC fluorescence—the intensity of the fluorescence emitted by the CTC stained cells corresponded to their level of metabolic activity. The data were obtained by extracting the mean intensity of the microorganisms on a given image. The analysis was made using three to five CTC images taken in a given sample.

Mean area of the cells—using daime, an analysis of the mean area of the cells on the fluorescent images was made. These data provided information on whether the cells had increased or decreased in volume. The latter were morphological changes that could indicate inhibition of the bacteria's development. The calculations were based on DAPI staining because the cell morphology was more visible.

Perimeter of the cells—in addition to the cells' mean area, the mean perimeter of the cells on an image was also analyzed. This could show different morphological changes in the bacterial cells, such as changes in their shape. These deformations could be linked to the antibacterial effect of the peptide fractions. The calculations were also based on the DAPI images.

Circularity—the mean circularity of the objects on the images (e.g., bacterial cells) was also analyzed with daime. The maximum value of this parameter is 1 (ideal circle). The lowering of the circularity indicated an elongation of the objects in the images. This parameter allowed the estimation of the morphological changes in the bacterial cells related to inhibition of the mechanisms of separation of the newly formed cells during division.

The share of live bacteria after the application of the antimicrobial peptides was calculated based on the number of CTC-stained cells (live cells) and DAPI (all cells, including dead ones). The calculations were performed on at least 10 fluorescent images for each sample. The share was represented as a percentage of the live cells from the total bacterial cells in the sample.

All investigations were realized in 3–6 repetitions. The data are mean values, and were statistically treated according to Student and Fisher with a guaranteed probability of 95%.

3. Results

3.1. Analysis and Physicochemical Characteristics of the Fraction

The discovery of new antimicrobial peptides (AMPs) from natural sources is of great public health importance due to the effective antimicrobial activity of AMPs and low levels of resistance. In the present study, five mucus fractions from the garden snail *C. aspersum* (four peptide fractions and one fraction containing peptides and polypeptides with MWs below 20 kDa) were used.

The extraction of crud mucus from the *C. aspersum* snails grown on Bulgarian farms was performed using a special patented technology, without injuring any snails [34–37]. After a series of purifications, the crude mucus extract, which was a multicomponent mixture of substances with different masses and properties, was divided by ultrafiltration into two basic fractions with molecular weights below 10 kDa and below 20 kDa. To characterize it in more detail, the peptide fraction with a molecular weight below 10 kDa after ultrafiltration by using Amicon® Ultra-15 with 3 and 5 kDa membranes in centrifugation (3000× g, 4 °C, 30 min) was further divided into three subfractions with MW < 3 kDa, MW of 3–5 kDa, and MW of 5–10 kDa, as well as a fraction containing peptides and polypeptides with MW < 20 kDa, as described in the Section 2.

The exact molecular weights of the peptides in the fraction with MW < 3 kDa were determined as protonated molecular ions [M + H]⁺ by the MALDI-TOF/MS analysis performed in positive ionization mode. The MS spectrum of the fraction presented in Figure 2 shows that the snail mucus contained various peptides with different masses, primarily in the region between 800–2500 Da.

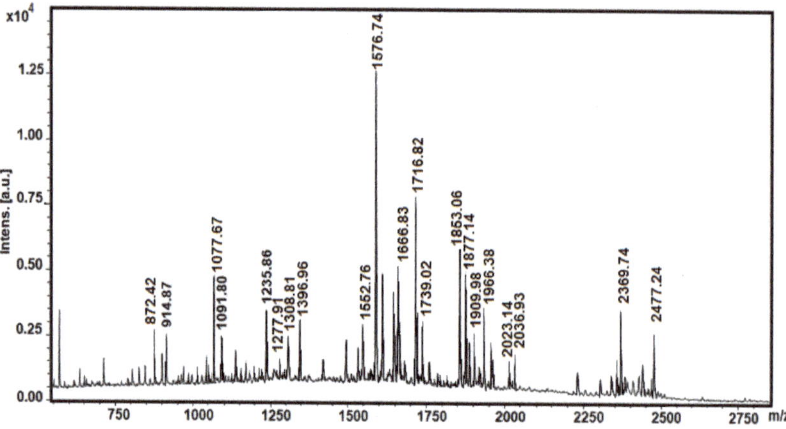

Figure 2. MALDI-MS spectrum of fraction with MW < 3 kDa. Standard peptide solution was used to calibrate the mass scale of the Autoflex™ III High-Performance MALDI-TOF and TOF/TOF Systems (Bruker Daltonics, Bremen, Germany).

The fraction with MW < 10 kDa, in addition to the peptides described above, contained peptides with higher molecular weights. They were determined by MALDI-TOF-MS analysis, and the MS spectrum was dominated by peptides represented as protonated molecular ions at m/z 3912.91 Da, 4041.15 Da, and 5197.26 Da, and various ions with lower intensities of 6194.62–7060.37 Da (Figure 3). In a previous study, the glycosylation screening by orcinol/H_2SO_4 showed that some of the mucus fractions were glycosylated [34], which confirmed the presence of glycopeptides.

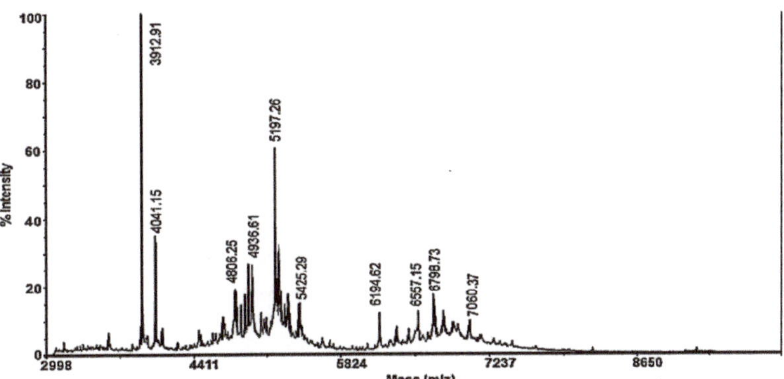

Figure 3. MALDI-MS spectrum recorded in the region of 3000–10000 m/z of the fraction with MW < 10 kDa. Standard peptide solution was used to calibrate the mass scale of the Autoflex™ III High-Performance MALDI-TOF and TOF/TOF Systems (Bruker Daltonics, Bremen, Germany).

Information on the amino acid sequences of the peptides with low molecular weights was obtained by MALDI-MS/MS analysis of the protonated molecule ions [M + H]⁺.

As shown in Figure 4, after following of fragment y- and b-ions, the amino acid sequence of a peptide, represented as a molecular ion [M + H]⁺ at m/z 1396.96 Da, was identified: EPGGGGGEGGGLLGVAL. Using the same method, the primary structures of the peptides presented in Table 1 were determined by de novo sequencing experiments (MS/MS analysis) of the protonated molecule ions [M + H]⁺.

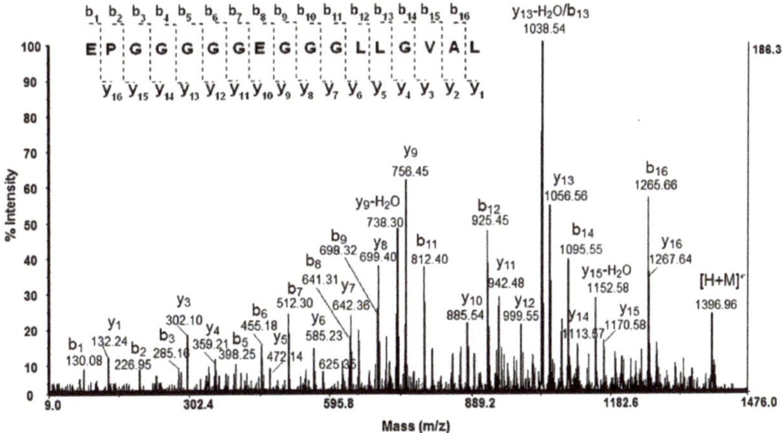

Figure 4. MALDI-MS/MS spectrum of peptide [M + H]⁺ at m/z 1396.96 Da. The sequence EPGGGGGEGGGLLGVAL was determined by de novo sequencing.

Table 1. The primary structures of the peptides presented in mucus fraction below 3 kDa, from the garden snail *C. aspersum*, identified by de novo sequencing in MALDI-MS/MS.

No.	Amino Acid Sequence of Peptides (N-C-Terminals), De Novo Sequencing.	MALDI [M + H]$^+$ (Da)	Calc. Mass (Da)	pI	GRAVY	Net Charge	Predict. Antibacterial (%)	Predict. Antiviral (%)	Predict. Antifungal (%)
1	AAGLAGAGGGGGG	872.42	871.41	5.57	0.600	0/0	58	41	32
2	DKGLGGFEA	893.51	892.43	4.37	−0.411	−2/+1	30	29	20
3	LGDLNAEFAAG	1077.67	1076.51	3.67	0.409	−2/0	27	46	30
4	AGVGAGGANPSTYVG	1277.91	1276.60	5.57	0.260	0/0	25	7.5	11
5	GAACNLEDGSCLGV	1308.81	1307.55	3.67	0.564	−2/0	58	58	53
6	EPGGGGGEGGGLLGVAL	1396.96	1395.70	3.80	0.306	−2/0	41	35	20
7	LGPLYDEMGPVGGDVG	1576.74	1574.73	3.49	0.056	−3/0	9.7	20	6.2
8	ASKGCGPGSCPPGDTVAGVG	1716.82	1715.76	5.86	0.005	−1/+1	25	12	23
9	ACSLLLGGGGVGGGKGGGGHAG	1739.02	1737.86	8.27	0.409	0/+1	83	49	67
10	ACLTPVDHFFAGMPCGGGP	1877.14	1875.81	5.08	0.542	−1/0	32	43	20
11	NGLFGGLGGGGHGGGGKGPGEGGG	1909.90	1908.88	6.75	−0.487	−1/+1	90	67	80
12	LLLLMLGGGLVGGLLGGGGKGGG	1966.24	1965.14	8.75	1.209	0/+1	92	57	76
13	PFLLGVGGLLGGSVGGGGGGGGAPL	2023.14	2022.09	5.96	0.912	0/0	69	32	38
14	GMVVKHCSAPLDSFAEFAGA	2036.93	2035.95	5.32	0.565	−2/+1	42	20	26
15 *	GLLGGGGGAGGGGLVGGLLNG	1609.94	1608.86	5.52	0.776	0/+1	90.0	53.6	65.0
16 *	MGGLLGGVNGGGKGGGGPGAP	1666.83	1665.83	8.5	0.005	0/+1	78.6	52.0	61.5

* The AASs of these peptides were also identified previously in [34].

The amino acid sequences (AASs) of the identified peptides in the fraction with MW below 3 kDa from the *C. aspersum* mucus showed the presence of various amino acid residues, but mostly glycine (G), leucine (L), valine (V), proline (P), tryptophan (W), glutamic acid (E), aspartic acid (D), phenylalanine (F), and arginine (R), which are typical for peptides with established antimicrobial activities. The physicochemical characteristics of the identified peptides (isoelectric points (pI), grand average of hydropathicity (GRAVY), and net charge) were determined using the ExPASy ProtParam tool. The analysis showed the presence of both cationic and anionic amphipathic peptide structures. Most peptide structures showed generally hydrophobic surfaces (Table 1), but only two peptides were hydrophilic (Nos. 2 and 11).

Based on the identified primary structures (Table 1), antimicrobial activity was predicted using iAMPpred software (http://cabgrid.res.in:8080/amppred) (accessed on 10 January 2022), an extensive database [38]. The iAMPpred software predicted antimicrobial peptides would be incorporated into compositional, physicochemical, and structural features.

The results showed that peptides 9, 11, 12, 13, 15, and 16 had the highest prognostic, antibacterial, and antifungal activities, while peptides 5, 11, and 12 had the highest prognostic antiviral activities.

Peptides 4, 7, and 8 had the lowest prognostic microbial activities. Some previous studies have shown [39] that if Pro residues are inserted into the sequences of α-helical AMPs, their ability to permeabilize the bacterial cytoplasmic membrane decreases substantially along with the number of Pro residues incorporated, which could explain our results. Eight peptides of the identified sequences contained one to three Pro amino acid residues in the polypeptide chain. Pro residues were incorporated in the N-terminal region of the polypeptide chain in four peptide structures, and in two peptides Pro residues were located in the C-terminal. Only one peptide contained proline residues both in the N-terminal and C-terminal polypeptide chains. Moreover, one proline residue was found in the middle of the polypeptide chain for four peptides (Nos. 4, 7, 8, and 14). Although Pro residue is commonly known as an α-helix breaker, proline residues have been found in the alpha-helical regions of many peptides and proteins, as well as AMPs. Confirmation of these data was provided by the identified structure of peptide No. 16, which contained two Pro residues located at the C-terminus, and showed high prognostic antibacterial and antifungal activities—78.6% and 61.5%, respectively.

The alignment of AASs of the mucus peptides of the garden snail *C. aspersum* presented in Table 1 was acquired with CAMPSing software (http://www.campsign.bicnirrh.res.in/blast.php) (accessed on 10 January 2022), and revealed high homology of peptides 1, 6, 9, 11, 12, 13, 15, and 16 with known AMPs such as microcin, leptoglycin, glycine-rich protein GWK from *Cucumis melo*, ctenidin-1, ctenidin-3, holotricin-3, acanthoscurrin-1 (presented in Supplementary Materials File S1), which confirmed that they belong to the AMP family. The obtained results can be considered as basic information in the study of bioactive peptides from the *C. aspersum* mucus extract and for their potential biomedical application.

3.2. The Antibacterial Activity—24 h Impact on Cells of E. coli in a Late Logarithmic Phase in a Solid Nutrient Media with Deep and Surface Inoculations of the Bacterial Material

The antibacterial effect of the peptide and protein fraction regarding *E. coli* NBIMCC 8785 was studied using surface and deep inoculations of the bacterial material. It was registered that the peptide fraction with MW < 10 kDa manifested an antibacterial effect in the deep and surface inoculations. This indicated that this reaction could be used for the treatment of infections that are based on the formation of biofilms, and as such, that are in the interior of the organs and liquids. It would be active in the studied bacterial cultures for bacterial cells immobilized on surfaces, as well as for the homogenous cells gravitating around the biofilm that reproduce the biofilm and release metabolic products of infections with toxic effects [5,40–45].

The antibacterial effects regarding *E. coli* NBIMCC 8785 in a deep and surface inoculations of different fractions of peptides and proteins isolated from snail slime is represented in Figure 5. The results showed that the peptide fraction of MW < 10 kDa had an effect on both types of inoculation. This fraction also included the peptides with an MW < 3 kDa, an MW of 3–5 kDa, and an MW of 5–10 kDa.

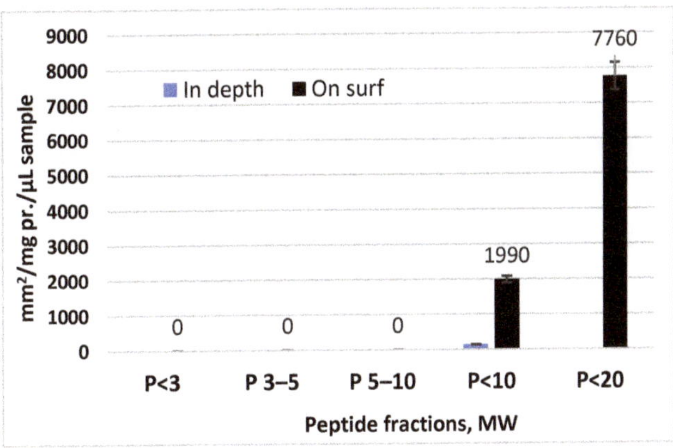

Figure 5. Antibacterial effect of the peptide and protein fractions regarding *E. coli* NBIMCC 8785 in surface and deep inoculations.

This fraction includes both peptides with MW < 3 kDa (Figure 2) and the peptides of higher molecular weights shown in the MS spectrum in Figure 3.

The activity in the surface inoculation sample reached a high value of 1990 mm²/mgPr/µMol sample, and in the deep inoculation sample, it reached 136.13 mm²/mgPr/µMol. This fraction could be applied in the treatment of biofilm of *E. coli* NBIMCC 8785, as shown by the data from the surface inoculation, but the inhibiting effect also would be distributed over the homogenous bacterial cells, which usually gravitate around the biofilm or are only on the homogenous cells [5,40–45].

The fraction with MW < 20 kDa had an inhibiting action, reaching 7760 mm^2/mgPr/μMol for *E. coli* NBIMCC 8785, but only in the surface inoculation of the bacterial material and in the nutrient media.

Our future studies will continue with this peptide fraction, aiming to demonstrate some of the mechanisms of action of the peptides in it on the bacterial cells of *E. coli*. Different experimental variants were studied, corresponding to different scenarios/algorithms for therapeutic applications. The variants chosen by us for the interaction of antimicrobial peptides and bacterial cells were consistent with the special features to be used in future therapeutic processes that will further our study of the mechanisms of practical applications of the characterized peptide fractions.

3.3. Antibacterial Activity—First Variant

The impact of the peptide fraction MW < 10 kDa at a high concentration of 50% (0.5 mg/mL protein) on a 48 h bacterial culture with an exposition of action of 1 h was determined. The experiments were conducted in a liquid medium. In terms of therapeutic impact, this meant infections were in an advanced phase, when the treated has area contained bacteria of different ages that formed a bacterial succession. Another specificity for such pathogenic cultures is that there are also metabolic molecules around them from the dying bacterial cells. For the treatment of such infections, usually high initial concentrations of the therapeutic agent are used for the fast and effective manifesting of the bactericidal and bacteriostatic effects. The specific algorithm of the impact of the bacterial cells used was: high concentration of the peptide 50% *v*/*v* (0.5 mg/mL protein) for a short time—1 h impact on a 48 h bacterial culture of *E. coli* NBIMCC 8785.

The effects were determined through the study of the changes in the morphology of the bacterial cells using a scanning electronic microscope, as shown in Figure 6.

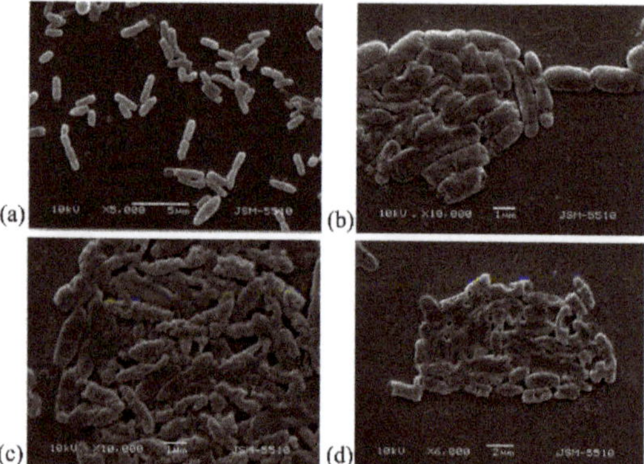

Figure 6. SEM analysis of cells of *E. coli* NBIMCC 8785 influenced with 50% peptide fraction (MW < 10 kDa): (**a,b**)—control 48 h bacterial culture at 5000× and 10,000× magnification; (**c,d**) damaged bacterial cells of *E. coli* NBIMCC 8785 after action of the peptide fraction.

In Figure 6, it can be seen that massive damage to the surface cell layers occurred and an unformed cell-free bacterial mass was created that was unable to reproduce bacteria; and at the large scale, there was a limitation of the infection [46–49].

3.4. Antibacterial Activity—Second Variant

The impact of the peptide fraction MW < 10 kDa (0.5 mg/mL protein) at a high concentration of 50% (0.5 mg/mL protein) on 18 h bacterial culture with an exposition of impact of 1 h was investigated. The experiments were conducted in a liquid medium.

In this course of experiments, we shaped the therapeutic situation in which the peptide fraction MW < 10 kDa treated infections that had arisen before 18 h. The algorithm was:

High concentration of the peptide 50% (0.5 mg/mL protein) for a short time—1 h of impact in an 18 h bacterial culture of *E. coli* NBIMCC 8785.

The analysis of the mechanisms of impact in this course of experiments was done by determining the level of damage of the metabolism of the bacterial cell through a dyeing of the *E. coli* NBIMCC 8785 cells impacted by the peptide fraction with the fluorescent tetrazolium salt 5-cyano-2,3-ditolyl tetrazolium chloride (CTC).

Figure 7 shows a bacterial culture at 18 h with a density of 1.8×10^7 cell/mL with 50% peptide fraction after dyeing with CTC. It can be clearly seen that in the control, there was a large number of metabolically active bacterial cells of *E. coli* NBIMCC 8785. After the one-hour impact on the bacterial culture with a peptide fraction MW < 10 kDa, a strong inhibiting effect was seen in the metabolism of the bacterial agent [9,50,51].

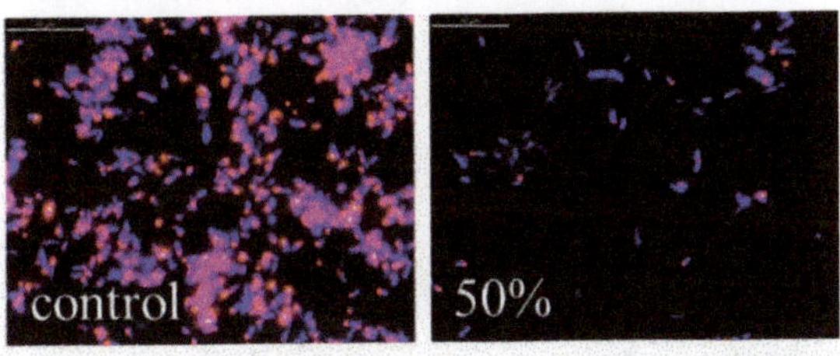

Figure 7. Fluorescence images of *E. coli* after 1 h incubation with peptide fraction with MW below 10 kDa (red—live cells; blue—dead cells). The marker corresponds to 10 μm.

The fluorescent images of the control and of the bacterial cells impacted by the peptide fraction MW < 10 kDa were digitally processed. The digital processing was performed on the following indicators: fluorescence intensity, average perimeter of the cells, roundness, and average area of the cells. The results of the changes to these parameters of the bacterial cells impacted by the peptide fraction MW < 10 kDa regarding the control are represented in Figure 8.

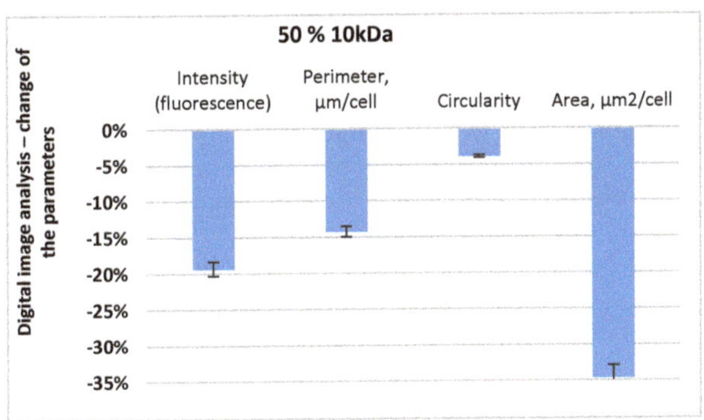

Figure 8. Fluorescence images of *E. coli* after 1 h incubation with peptide fraction with MW below 10 kDa.

The results indicated that the three parameters decreased after the impact by the peptide fraction MW < 10 kDa. This meant that the fluorescence intensity decreased by about 20%, the average perimeter decreased by about 15%, the cells' length and roundness decreased by 4%, and the average area decreased by about 35%. Using the represented fluorescent images and their digital analysis, it was found that after the 1 h impact with the peptide fraction MW < 10 kDa, the bacterial cells had inhibited functions and had changed morphology. On the other hand, using Figure 9, it was determined that at this concentration, for the 18 h culture and one hour of exposition, there was a very high level of inhibition of the bacterial cells. As a total result of the registered cell deformations and the fluorescence intensity, there was almost 100% inhibition of the structures and functions of the pathogenic bacterial cells. It was determined that the penetration of peptides with low molecular weights from the outside to the inside of bacterial cells damaged the cells' surface layers, thereby inhibiting the metabolic activity of the bacteria. Various morphological deformations of the cells were registered that were illustrated by the SEM analyses—folding of the surface cell layers, and lengthened and shortened cells [43,46–49,52].

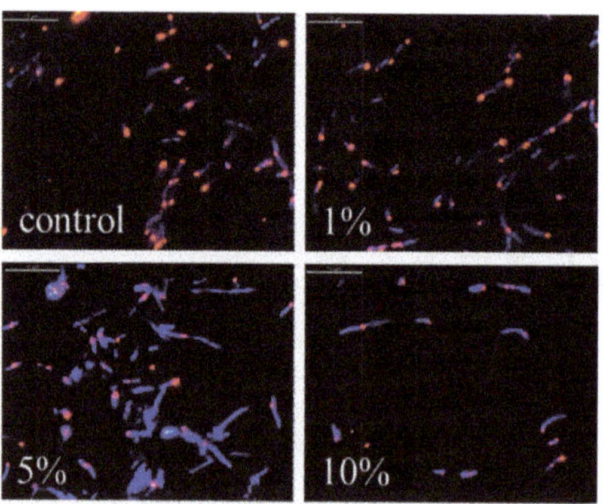

Figure 9. Fluorescence images of *E. coli* after 6 h incubation with peptide fraction with MW below 10 kDa (red—live cells; blue—dead cells). The marker corresponds to 10 µm.

3.5. Antibacterial Activity—Third Variant

Logic led us to question whether it was possible for the initial infection to have a decreased concentration of the therapeutic agent in the peptide fraction MW < 10 kDa (0.5 mg/mL protein) while, at the same time, having an increased exposition time. This algorithm for the treatment of the bacterial cells with the peptide fraction with MW < 10 kDa with the increase in the exposure time would lead to a series of positive effects: the use of a smaller quantity of peptide fractions that would be favorable regarding costs, emanation of therapeutic agents into the environment, and economization of resources at the expense of the exposition time. The studied algorithm was:

Impact of the peptide fraction MW < 10 kDa (0.5 mg/mL protein) with concentrations 1, 5, and 10% on 18 h bacterial culture with an exposition of 6 h.

The experiments were conducted in a liquid medium. Through this course of experiments, we shaped the therapeutic situation in which relatively newly formed infections (arisen before 18 h) were treated with the peptide fraction MW < 10 kDa. The aim was to choose the convenient, comparably low concentration of peptide fraction. The analysis of the mechanisms of impact in this course of experiments has been made using the level of damage of the metabolism of the bacterial cells through the dyeing of the *E. coli* NBIMCC 8785 cells impacted by the peptide fraction with the fluorescent tetrazolium salt 5-cyano-2,3-ditolyl tetrazolium chloride (CTC) and through SEM.

The fluorescent images of the obtained results are presented in Figure 9. In the presented images, it can be clearly seen that with the increase in the concentration of the peptide fraction from 1% to 10%, the number of dead inactive bacterial cells also increased. The highest effectiveness of inhibiting actions was realized using the concentration of the peptide fraction of 10%. After that, we also conducted a digital analysis of the images using the following parameters: fluorescence intensity, average perimeter of the bacterial cells, roundness, average area of the cells, and ratio of alive to dead cells.

The results shown in Figure 10 provided grounds to reach the following intermediate conclusions. In the 1% peptide fraction, only morphological changes were registered in the cells of *E. coli* NBIMCC 8785, represented by the lengthening and decrease in the average area of the cells. Regarding the impact of the 5% peptide fraction, the three morphological parameters of the cells decreased—the average perimeter and average area decreased and the cells lengthened, but the metabolic activity was not affected (See Supplementary Materials File S1).

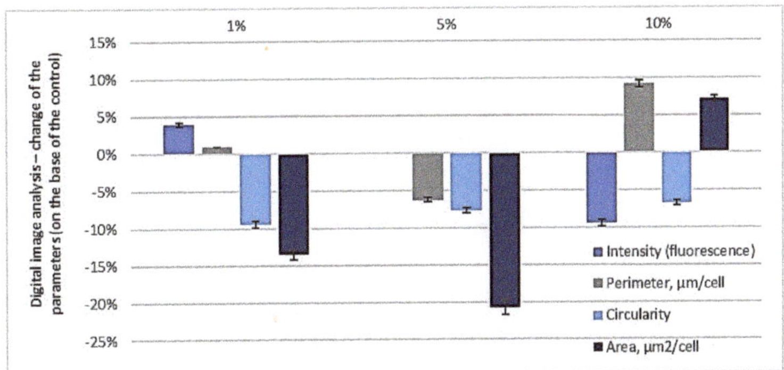

Figure 10. Changes in the fluorescence intensity, parameters of the cells, circularity, and area of *E. coli* cells compared to the control after incubation with different concentrations (1%, 5%, 10%) of the peptide fraction with MW below 10 kDa or 6 h.

In the 10% (v/v) peptide fractions with 0.5 mg/mL protein and 6 h of incubation, a high inhibiting effect of the peptide fraction regarding *E. coli* NBIMCC 8785 was registered—a strong decrease in the fluorescence intensity meant a decrease in the metabolic activity, as well as an increase in the average perimeter and the average area of the cells, and their lengthening.

After that, the mechanisms of damaging of the form and sizes of the bacterial cells in this algorithm of impact were confirmed by the SEM analyses. When comparing the effects of the damage to the cells of *E. coli* NBIMCC 8785 with the peptide fraction MW < 10 kDa at a concentration of 10% and 6 h of exposition, the following results were found. Morphologically, the bacterial cells were damaged, and the following deformations were registered: (1) strong lengthening of the cells—the lengthened cells reached three times the size of the normal cells (Figure 11a,d); (2) Cell cracking and pouring of the cells' contents into the outside cell space (Figure 11a,c); (3) shortening of the cells and pinching in the middle, folding, and concavities on the cells' surfaces (Figure 11a,b,d,f); and (4) conglutination of the cells and the obtaining of an unformed bacterial cell mass (Figure 11a,d,f).

One of the most unexpected effects of damage of bacterial cells registered by SEM was a triple increase in the size of the cells.

The strongly lengthened cells also were registered through a fluorescent technique—a dyeing with the fluorescent labelled tetrazolium salt 5-cyano-2,3-ditolyl tetrazolium chloride (CTC). The results illustrating these cells that were three times larger that were obtained after the impact of the peptide fraction MW < 10 kDa for 6 h are presented in Figure 12a,b.

Comparing the data in Figure 12a,b, we supposed that one of the effects of damage was the blocking of the division of the cells, which impeded their multiplication. Thus, the results at the cell level showed these long cell structures. This phenomenon registered and confirmed by us was accompanied by the reported effects of damage to the existing cells, leading to a fast termination of the infections that would not allow their development. In this way, the early infections were limited and terminated at a low concentration of the peptide fraction in a longer time of exposition.

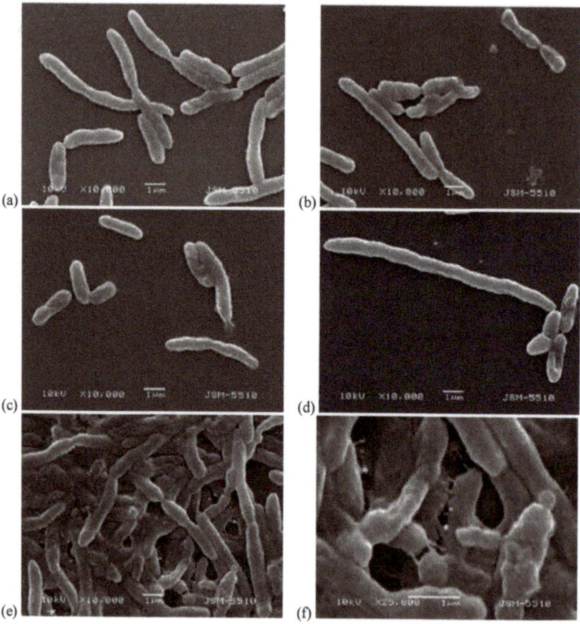

Figure 11. SEM analysis of cells of *E. coli* NBIMCC 8785 influenced with 10% peptide fraction (MW < 10 kDa): (**a**) control—18 h bacterial culture at 10,000× magnification; (**b–f**) damaged bacterial cells of *E. coli* NBIMCC 8785 after action of the peptide fraction for 6 h.

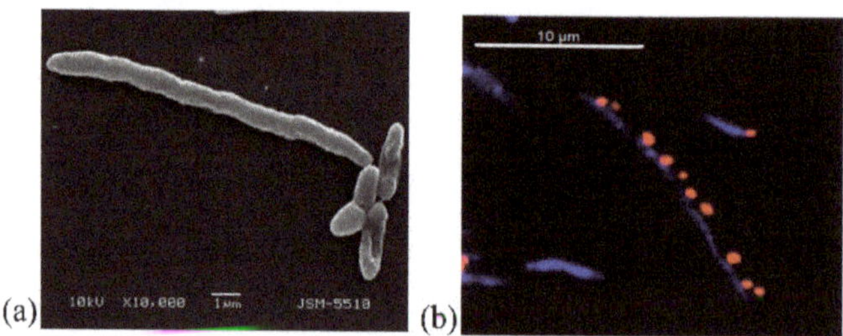

Figure 12. Illustration of long cells (results of blocking of division of cells) of *E. coli* NBIMCC 8785 after 6 h treatment with 10% peptide fraction, MW < 10 kDa: (**a**) SEM analysis at 10,000×; (**b**) fluorescence analysis, with the cells stained with DAPI and CTC.

4. Discussion

In recent years, the mucus from the garden snail *C. aspersum* has been the subject of research by our scientific team. Using tandem mass spectrometry, the primary structures of nine antimicrobial peptides with molecular weights of 1000–3000 Da in an isolated fraction with MW < 10 kDa from mucus from the garden snail *C. aspersum* with active components were identified. These peptides were rich in glycine and leucine residues, and demonstrated strong antibacterial activity against a Gram-negative bacterial strain—*E. coli* NBIMCC 878 [36]. Moreover, some peptides with MW 10–30 kDa exhibited predominantly antibacterial activity against *B. laterosporus* and *E. coli*, and another 20 kDa peptide fraction against the bacterial strain *C. perfringens* [34].

A characteristic of the structures of peptides with the highest prognostic, antibacterial, and antifungal activity was the content of high levels of glycine and leucine residues, which indicated that they belonged to a new class of antimicrobial peptides rich in Gly/Leu that demonstrated antibacterial activity mainly against Gram-negative bacteria [53]. Previous studies also showed the presence of peptides with similar amino acid sequences [34,36,37]. Furthermore, four of the nine peptides with high prognostic therapeutic values were cationic, and only one was neutral. It is known that cationic AMPs kill microbes via mechanisms that predominantly involve interactions between the peptide's positively charged residues and anionic components of the target cell's membranes. These interactions can lead to a range of effects, including membrane permeabilization, depolarization, leakage, or lysis, resulting in cell death. Some of the positively charged AMPs may penetrate into the cell to bind intracellular molecules that are crucial to the life of the cell. There are multiple models of mechanisms to explain the action of these peptides, including the toroidal pore model, the barrel-stave model, and the carpet model [54–56].

Our previously published data in Dolashki et al., 2020 showed that some peptides that were detected by an orcinol/H_2SO_4 assay were glycosylated [34]. The glycosylation of peptides can influence their antimicrobial activity and their ability to affect host immunity, target specificity, and biological stability [57–60]. Glycosylation of AMPs does not necessarily result in the generation of an efficacious peptide, and can sometimes lead to a loss in activity or functionality [61].

The observed antibacterial activity against *E. coli* NBIMCC 8785 most likely can be attributed to the peptides presented in Table 1 (Nos. 1, 11–13, 15, and 16) for which the iAMPpred software predicted high antimicrobial activity, and for which high homology with known AMPs with activity against Gram-negative bacteria was detected, such as leptoglycin, microcin, ctenidin, and acanthoscurrin (See Supplementary Materials File S2). However, based on previous research [34,62], our hypothesis was that the antibacterial activity of the mucus fraction < 10 kDa was due to a synergistic effect of cationic, anionic, and neutral peptides with MWs of 0.800–2.500 kDa and peptides with a higher MW in the range of 3–10 kDa (Figure 3); some of them were glycopeptides, as reported previously in [34]. In fact, Trapella et al., 2018, also suggested that snail mucus's potential (HelixComplex) as a therapeutic agent in wound repair were attributable to the synergistic activity of several molecules [63].

The results of our study indicated the great opportunities for the isolated and characterized peptide fractions of snail mucus to be used for treatment of infections caused by *E. coli* NBIMCC 8785. The results of the conventional study of the antibacterial activity of the fractions with MW < 3 kDa, MW 3–5 kDa, MW 5–10 kDa, MW < 10 kDa, and MW < 20 kDa indicated that the fraction of peptides MW < 10 kDa had a higher antibacterial effect in surface and deep inoculations of the bacterial material [34,36,37,57]. This fraction would be effective in the treatment of infections that form a bacterial biofilm, and it would impact on the homogenous cells that gravitate around and support the biofilm. On the other hand, this peptide fraction will be effective on the studied bacteria in surface infections and in infections situated deeper.

The peptide fraction with MW < 10 kDa would be an interesting subject of future studies of the mechanism of the antibacterial activity due to its specific actions against *E. coli* and its composition clarified from a chemical point of view. Other factors that make this fraction convenient for application are, on one hand, the lower molecular weight of the peptides means they can more easily penetrate into the bacterial cells; and on the other hand, the surplus of these therapeutic tools will degrade faster in the environment [9,10,43,50].

Depending on the type and the phase of development of the infection, different concentrations and quantities of the peptide fraction could be used. In infections in an advanced-phase 48 h bacterial culture, it was effective to apply a 50% peptide fraction. Using this concentration for a short time of 1 h, a strong inhibiting effect on the bacterial cells was registered, represented by major damage to the cells' surface layers [5,9,47–50,52,64]. In the bacterial infections in an initial phase (18 h bacterial culture), a 50% peptide fraction with one-hour impact lead to an almost 100% inhibition of the structures and functions of the *E. coli* NBIMCC 8785. The morphological parameters of the cell were changed, and the metabolic activity was inhibited at a very high level. When studying the opportunities for the achievement of the desired effect in newly arisen infections through a decrease in the concentration of the peptide fraction but with an increase in the exposition time, we found that the 10% peptide fraction led to the achievement of the desired effect—damage to the structures and functions of the bacteria [65–67]. This led to a decrease in the fluorescence as a result of the blocking of the oxide reductase apparatus of the cells, and accumulation of degenerative bacterial cells, which were shortened and round, and had highly damaged surfaces. Strongly lengthened cells were registered due to the suppression of their division and the multiplication of new bacterial cells. Similar effects were seen by other authors [43,46–49,51].

In conclusion, this paper presented 14 new antimicrobial peptides with MWs below 10 kDa isolated from the mucus of *C. aspersum* snails. Their AAS, physicochemical characteristics, and potential antibacterial activities were determined by de novo sequencing. Information on the mechanisms of antibacterial activities of peptides against *E. coli* NBIMCC was presented for the first time by determining the effective concentration and methods of administration of the active fraction with MW < 10 kDa.

Changes in the bacterial structure and metabolic activity of *E. coli* NBIMCC were determined by SEM, fluorescence, and digital image analysis.

These studies shed new light on the mechanisms of action of the peptide fraction with MW < 10 kDa, and allowed us to offer an objective algorithm for the treatment of infections, depending on the stage of development of the bacterial infection [13,68–70]. Reducing therapeutic concentrations will be more environmentally friendly and will reduce the cost of therapeutic procedures.

Supplementary Materials: The following are available online at https://www.mdpi.com/article/10.3390/biomedicines10030672/s1, File S1: Digital image analysis of the fluorescence images of *E. coli* after 6-h incubation with peptides fraction with MW below 10 kDa; File S2: An alignment of amino acid sequences of some identified peptides from *C. aspersum* mucus with known antimicrobial peptides.

Author Contributions: Conceptualization, Y.T. and P.D.; methodology, Y.T., M.B., L.V., A.D., N.Z., W.V. and P.D.; investigation, Y.T., M.B., L.V., A.D., N.Z., E.D., D.K. and D.K.; resources, Y.T. and P.D.; data curation, Y.T. and L.V.; writing—original draft preparation, Y.T., L.V., P.D. and M.B.; writing—review and editing, Y.T., L.V., P.D. and M.B.; supervision, Y.T. and P.D.; project administration, Y.T. and P.D.; funding acquisition, Y.T. and P.D. All authors have read and agreed to the published version of the manuscript.

Funding: This research was funded by the Scientific Research Fund of the Ministry of Education and Science of the Republic of Bulgaria (grant number DN 01/14 of 19 December 2016) and by Grant DO1-217/30 November 2018 and agreements DO1-323/18 December 2019 and DO1-358/17 December 2020) under the National Research Programme "Innovative Low-Toxic Bioactive Systems for Precision Medicine (BioActiveMed)" approved by DCM # 658/14 September 2018.

Data Availability Statement: The data presented in this study are available in the article.

Acknowledgments: The equipment used in the experiments was obtained under project BG05M2OP001-1.002-0019: 'Clean Technologies for Sustainable Environment—Waters, Waste, Energy for Circular Economy' funded by the Operational Program 'Science and education for smart growth', co-financed by the European Union through the European structural and investment funds.

Conflicts of Interest: The authors declare no conflict of interest.

References

1. Zaslof, M. Antimicrobial peptides of multicellular organisms. *Nature* **2002**, *415*, 389–395. [CrossRef]
2. Marr, A.K.; Gooderham, W.J.; Hancock, R.E. Antibacterial peptides for therapeutic use: Obstacles and realistic outlook. *Curr. Opin. Pharmacol.* **2006**, *6*, 468–472. [CrossRef]
3. Joo, H.S.; Fu, C.I.; Otto, M. Bacterial strategies of resistance to antimicrobial peptides. *Philos Trans. R. Soc. B Biol. Sci.* **2016**, *371*, 20150292. [CrossRef] [PubMed]
4. Wang, J.; Dou, X.; Song, J.; Lyu, Y.; Zhu, X.; Xu, L.; Li, W.; Shan, A. Antimicrobial peptides: Promising alternatives in the post feeding antibiotic era. *Med. Res. Rev.* **2019**, *39*, 831–859. [CrossRef]

5. Mandel, S.; Michaeli, J.; Nur, N.; Erbetti, I.; Zazoun, J.; Ferrari, L.; Felici, A.; Cohen-Kutner, M.; Bachnoff, N. OMN6 a novel bioengineered peptide for the treatment of multidrug resistant Gram negative bacteria. *Sci. Rep.* **2021**, *11*, 6603. [CrossRef] [PubMed]

6. Witherell, K.S.; Price, J.; Bandaranayake, A.D.; Olson, J.; Call, D.R. In vitro activity of antimicrobial peptide CDP-B11 alone and in combination with colistin against colistin-resistant and multidrug-resistant *Escherichia coli*. *Sci. Rep.* **2021**, *11*, 2151. [CrossRef] [PubMed]

7. Galdiero, S.; Falanga, A.; Berisio, R.; Grieco, P.; Morelli, G.; Galdiero, M. Antimicrobial peptides as an opportunity against bacterial diseases. *Curr. Med. Chem.* **2015**, *22*, 1665–1677. [CrossRef] [PubMed]

8. Ebbensgaard, A.; Mordhorst, H.; Overgaard, M.T.; Nielsen, C.G.; Aarestrup, F.M.; Hansen, E.B. Comparative evaluation of the antimicrobial activity of different antimicrobial peptides against a range of pathogenic bacteria. *PLoS ONE* **2015**, *10*, e0144611. [CrossRef]

9. Lima, B.; Ricci, M.; Garro, A.; Juhász, T.; Szigyártó, I.C.; Papp, Z.I.; Feresin, G.; de la Torre, J.G.; Cascales, J.L.; Fülöp, L.; et al. New short cationic antibacterial peptides. Synthesis, biological activity and mechanism of action. *Biochim. Biophys. Acta Biomembr.* **2021**, *1863*, 183665. [CrossRef]

10. Tan, P.; Fu, H.; Ma, X. Design, optimization, and nanotechnology of antimicrobial peptides: From exploration to applications. *Nano Today* **2021**, *39*, 101229. [CrossRef]

11. Peschel, A.; Sahl, H.G. The co-evolution of host cationic antimicrobial peptides and microbial resistance. *Nat. Rev. Microbiol.* **2006**, *4*, 529–536. [CrossRef] [PubMed]

12. O'Neill, J. Securing New Drugs for Future Generations: The Pipeline of Antibiotics. (The Review on Antimicrobial Resistance, 2015). Available online: https://amr-review.org/sites/default/files/SECURING%20NEW%20DRUGS%20FOR%20FUTURE%20GENERATIONS%20FINAL%20WEB_0.pdf (accessed on 10 January 2022).

13. World Health Organization. *Critically Important Antimicrobials for Human Medicine: Ranking of Antimicrobial Agents for Risk Management of Antimicrobial Resistance Due to Non-Human Use*, 5th ed.; World Health Organization: Geneva, Switzerland, 2016.

14. Totsika, M. Benefits and challenges of antivirulence antimicrobials at the dawn of the post-antibiotic era. *Curr. Med. Chem.* **2016**, *6*, 30–37. [CrossRef]

15. Jenssen, H.; Hamill, P.; Hancock, R.E. Peptide antimicrobial agents. *Clin. Microbiol Rev.* **2006**, *19*, 491–511. [CrossRef] [PubMed]

16. Zhuang, J.; Coates, C.J.; Zhu, H.; Zhu, P.; Zujian, W.; Xie, L. Identification of candidate antimicrobial peptides derived from abalone hemocyanin. *Dev. Comp. Immunol.* **2015**, *49*, 96–102. [CrossRef]

17. Li, H.; Parisi, M.G.; Parrinello, N.; Cammarata, M.; Roch, P. Molluscan antimicrobial peptides, a review from activity-based evidences to computer-assisted sequences. *Invertebr. Surviv. J.* **2011**, *8*, 85–97.

18. Pitt, S.; Graham, M.A.; Dedi, C.G.; Taylor-Harris, P.M.; Gunn, A. Antimicrobial properties of mucus from the brown garden snail *Helix aspersa*. *Br. J. Biomed. Sci.* **2015**, *72*, 174–181. [CrossRef]

19. González García, M.; Rodríguez, A.; Alba, A.; Vázquez, A.A.; Morales Vicente, F.E.; Pérez-Erviti, J.; Spellerberg, B.; Stenger, S.; Grieshober, M.; Conzelmann, C.; et al. New Antibacterial Peptides from the Freshwater Mollusk *Pomacea poeyana* (Pilsbry, 1927). *Biomolecules* **2020**, *10*, 1473. [CrossRef]

20. Wang, L.; Qiu, L.; Zhou, Z.; Song, L. Research progress on the mollusc immunity in China. *Dev. Comp. Immunol.* **2013**, *39*, 2–10. [CrossRef]

21. Dolashka, P.; Moshtanska, V.; Borisova, V.; Dolashki, A.; Stevanovic, S.; Dimanov, T.; Voelter, W. Antimicrobial proline-rich peptides from the hemolymph of marine snail *Rapana venosa*. *Peptides* **2011**, *32*, 1477. [CrossRef]

22. Dolashka, P.; Dolashki, A.; Velkova, L.; Stevanovic, S.; Molin, L.; Traldi, P.; Velikova, R.; Voelter, W. Bioactive compounds isolated from garden Snails. *J. BioSci. Biotechnol.* **2015**, 147–155.

23. Leoni, G.; De Poli, A.; Mardirossian, M.; Gambato, S.; Florian, F.; Venier, P.; Wilson, D.N.; Tossi, A.; Pallavicini, A.; Gerdol, M. Myticalins: A novel multigenic family of linear, cationic antimicrobial peptides from marine mussels (*Mytilus* spp.). *Mar. Drugs.* **2017**, *15*, 261. [CrossRef] [PubMed]

24. Cassone, M.; Otvos, L., Jr. Synergy among antibacterial peptides and between peptides and small-molecule antibiotics. *Expert Rev. Anti-Infect. Ther.* **2010**, *8*, 703–716. [CrossRef] [PubMed]

25. Park, S.-C.; Park, Y.; Hahm, K.-S. The Role of Antimicrobial Peptides in Preventing Multidrug-Resistant Bacterial Infections and Biofilm Formation. *Int. J. Mol. Sci.* **2011**, *12*, 5971–5992. [CrossRef] [PubMed]

26. Dolashka, P.; Dolashki, A.; Voelter, W.; Beeumen, J.; Stevanovic, S. Antimicrobial activity of peptides from the hemolymph of *Helix lucorum* snails. *Int. J. Curr. Microbiol. App. Sci.* **2015**, *4*, 1061–1071.

27. Zhong, J.; Wang, W.; Yang, X.; Yan, X.; Liu, R. A novel cysteine-rich antimicrobial peptide from the mucus of the snail of *Achatina fulica*. *Peptides* **2013**, *39*, 1–5. [CrossRef] [PubMed]

28. Suárez, L.; Pereira, A.; Hidalgo, W.; Uribe, N. Antibacterial, Antibiofilm and Anti-Virulence Activity of Biactive Fractions from Mucus Secretion of Giant African Snail *Achatina fulica* against *Staphylococcus aureus* Strains. *Antibiotics* **2021**, *10*, 1548. [CrossRef]

29. Smith, A.M.; Quick, T.J.; St Peter, R.L. Differences in the Composition of Adhesive and Non-Adhesive Mucus from the Limpet *Lottia limatula*. *Biol. Bull.* **1999**, *196*, 34–44. [CrossRef]

30. Pitt, S.J.; Hawthorne, J.A.; Garcia-Maya, M.; Alexandrovich, A.; Symonds, R.C.; Gunn, A. Identification and characterisation of anti-Pseudomonas aeruginosa proteins in mucus of the brown garden snail, *Cornu aspersum*. *Br. J. Biomed. Sci.* **2019**, *76*, 129–136. [CrossRef]

31. Kubota, Y.; Watanabe, Y.; Otsuka, H.; Tamiya, T.; Tsuchiya, T.; Matsumoto, J.J. Purification and characterization of an antibacterial factor from snail mucus. *Comp. Biochem. Physiol. C Comp. Pharmacol. Toxicol.* **1985**, *82*, 345–348. [CrossRef]

32. Ulagesan, S.; Kim, H.J. Antibacterial and Antifungal Activities of Proteins Extracted from Seven Different Snails. *Appl. Sci.* **2018**, *8*, 1362. [CrossRef]

33. Noothuan, N.; Apitanyasai, K.; Panha, S.; Tassanakajon, A. Snail mucus from the mantle and foot of two land snails, *Lissachatina fulica* and *Hemiplecta distincta*, exhibits different protein profile and biological activity. *BMC Res. Notes* **2021**, *14*, 138. [CrossRef] [PubMed]

34. Dolashki, A.; Velkova, L.; Daskalova, E.; Zheleva, N.; Topalova, Y.; Atanasov, V.; Voelter, W.; Dolashka, P. Antimicrobial Activities of Different Fractions from Mucus of the Garden Snail *Cornu aspersum*. *Biomedicines* **2020**, *8*, 315. [CrossRef] [PubMed]

35. Vassilev, N.G.; Simova, S.D.; Dangalov, M.; Velkova, L.; Atanasov, V.; Dolashki, A.; Dolashka, P. An ^{1}H NMR- and MS-Based Study of Metabolites Profiling of Garden Snail *Helix aspersa* Mucus. *Metabolites* **2020**, *10*, 360. [CrossRef] [PubMed]

36. Velkova, L.; Nissimova, A.; Dolashki, A.; Daskalova, E.; Dolashka, P.; Topalova, Y. Glycine-rich peptides from *C. aspersum* snail with antibacterial activity. *Bulg. Chem. Commun.* **2018**, *50*, 169–175.

37. Dolashki, A.; Nissimova, A.; Daskalova, E.; Velkova, L.; Topalova, Y.; Hristova, P.; Traldi, P.; Voelter, W.; Dolashka, P. Structure and antibacterial activity of isolated peptides from the mucus of garden snail *Cornu aspersum*. *Bulg. Chem. Commun. C* **2018**, *50*, 195–200.

38. Meher, P.; Sahu, T.; Saini, V.; Rao, A.R. Predicting antimicrobial peptides with improved accuracy by incorporating the compositional, physico-chemical and structural features into Chou's general PseAAC. *Sci. Rep.* **2017**, *7*, 42362. [CrossRef] [PubMed]

39. Álvarez-Ordóñez, A.; Begley, M.; Clifford, T.; Deasy, T.; Considine, K.; Hill, C. Structure-Activity Relationship of Synthetic Variants of the Milk-Derived Antimicrobial Peptide αs2-Casein f (183–207). *Appl. Environ. Microbiol.* **2013**, *79*, 5179–5185. [CrossRef]

40. Nickel, J.C.; Ruseska, I.; Wright, J.B.; Costerton, J.W. 1985 Tobramycin resistance of *Pseudomonas aeruginosa* cells growing as a biofilm on urinary catheter material. *Antimicrob. Agents Chemother.* **1985**, *27*, 619–624. [CrossRef]

41. Costerton, J.W.; Stewart, P.S.; Greenberg, E.P. Bacterial biofilms: A common cause of persistent infections. *Science* **1999**, *284*, 1318–1322. [CrossRef]

42. Otto, M. Bacterial evasion of antimicrobial peptides by biofilm formation. *Curr. Top. Microbiol. Immunol.* **2006**, *306*, 251–258.

43. Joo, H.S.; Otto, M. Molecular basis of in vivo biofilm formation by bacterial pathogens. *Chem. Biol.* **2012**, *19*, 1503–1513. [CrossRef]

44. Di Luca, M.; Maccari, G.; Nifosi, R. Treatment of microbial biofilms in the post-antibiotic era: Prophylactic and therapeutic use of antimicrobial peptides and their design by bioinformatics tools. *Pathog. Dis.* **2014**, *70*, 257–270. [CrossRef] [PubMed]

45. Strempel, N.; Strehmel, J.; Overhage, J. Potential application of antimicrobial peptides in the treatment of bacterial biofilm infections. *Curr. Pharm. Des.* **2015**, *21*, 67–84. [CrossRef]

46. Hurdle, J.G.; O'Neill, A.J.; Chopra, I.; Lee, R.E. Targeting bacterial membrane function: An underexploited mechanism for treating persistent infections. *Nat. Rev. Microbiol.* **2011**, *9*, 62–75. [CrossRef] [PubMed]

47. Wenzel, M.; Chiriac, A.I.; Otto, A.; Zweytick, D.; May, C.; Schumacher, C.; Gustg, R.; Albadah, H.B.; Penkovah, M.; Krämeri, U.; et al. Small cationic antimicrobial peptides delocalize peripheral membrane proteins. *Proc. Natl. Acad. Sci. USA* **2014**, *111*, E1409–E1418. [CrossRef] [PubMed]

48. Kaur, P.; Li, Y.; Cai, J.; Song, L. Selective membrane disruption mechanism of an antibacterial γ-AApeptide defned by EPR spectroscopy. *Biophys. J.* **2016**, *110*, 1789–1799. [CrossRef] [PubMed]

49. Ma, B.; Fang, C.; Lu, L.; Wang, M.; Xue, X.; Zhou, Y.; Li, M.; Hu, Y.; Luo, X.; Hou, Z. The antimicrobial peptide thanatin disrupts the bacterial outer membrane and inactivates the NDM-1 metallo-β-lactamase. *Nat. Commun.* **2019**, *10*, 3517. [CrossRef]

50. Batoni, G.; Maisetta, G.; Brancatisano, F.L.; Esin, S.; Campa, M. Use of antimicrobial peptides against microbial biofilms: Advantages and limits. *Curr. Med. Chem.* **2011**, *18*, 256–279. [CrossRef]

51. Band, V.I.; Weiss, D.S. Mechanisms of antimicrobial peptide resistance in Gram-negative bacteria. *Antibiotics* **2015**, *4*, 18–41. [CrossRef]

52. Li, Q.; Li, J.; Yu, W.; Wang, Z.; Li, J.; Feng, X.; Wang, J.; Shan, A. De novo design of a pH-triggered self-assembled β-hairpin nanopeptide with the dual biological functions for antibacterial and entrapment. *J. Nanobiotechnol.* **2021**, *19*, 183. [CrossRef]

53. Sousa, J.C.; Berto, R.F.; Gois, E.A.; Fontenele-Cardi, N.C.; Honório-Júnior, J.E.; Konno, K.; Richardson, M.; Rocha, M.F.; Camargo, A.A.; Pimenta, D.C.; et al. Leptoglycin: A new Glycine/Leucine-rich antimicrobial peptide isolated from the skin secretion of the South American frog *Leptodactylus pentadactylus* (Leptodactylidae). *Toxicon* **2009**, *54*, 23–32. [CrossRef] [PubMed]

54. Lei, J.; Sun, L.; Huang, S.; Zhu, C.; Li, P.; He, J.; Mackey, V.; Coy, D.H.; He, Q. The antimicrobial peptides and their potential clinical applications. *Am. J. Transl. Res.* **2019**, *11*, 3919–3931. [PubMed]

55. Raheem, N.; Straus, S.K. Mechanisms of Action for Antimicrobial Peptides with Antibacterial and Antibiofilm Functions. *Front. Microbiol.* **2019**, *10*, 2866. [CrossRef]

56. Chen, Y.; Guarnieri, M.T.; Vasil, A.I.; Vasil, M.L.; Mant, C.T.; Hodges, R.S. Role of peptide hydrophobicity in the mechanism of action of alpha-helical antimicrobial peptides. *Antimicrob. Agents Chemother.* **2007**, *51*, 1398–1406. [CrossRef] [PubMed]

57. Opdenakker, G.; Rudd, P.M.; Wormald, M.R.; Dwek, R.A.; Van Damme, J. Cells regulate the activities of cytokines by glycosylation. *FASEB J.* **1995**, *9*, 453–457. [CrossRef]

58. Moradi, S.V.; Hussein, W.M.; Varamini, P.; Simerska, P.; Toth, I. Glycosylation, an effective synthetic strategy to improve the bioavailability of therapeutic peptides. *Chem. Sci.* **2016**, *7*, 2492–2500. [CrossRef]

59. Lele, D.S.; Talat, S.; Kumari, S.; Srivastava, N.; Kaur, K.J. Understanding the importance of glycosylated threonine and stereospecific action of Drosocin, a Proline rich antimicrobial peptide. *Eur. J. Med. Chem.* **2015**, *92*, 637–647. [CrossRef]

60. Bednarska, N.G.; Wren, B.W.; Willcocks, S.J. The importance of the glycosylation of antimicrobial peptides: Natural and synthetic approaches. *Drug Discov. Today* **2017**, *22*, 919–926. [CrossRef]

61. Huang, C.Y.; Hsu, J.-T.; Chung, P.-H.; Cheng, W.; Jiang, Y.-N.; Ju, Y.-T. Site-specific N-glycosylation of caprine lysostaphin restricts its bacteriolytic activity toward *Staphylococcus aureus*. *Anim. Biotechnol.* **2013**, *24*, 129–147. [CrossRef]

62. Ilieva, N.; Petkov, P.; Lilkova, E.; Lazarova, T.; Dolashki, A.; Velkova, L.; Dolashka, P.; Litov, L. *In Silico Study on the Structure of Novel Natural Bioactive Peptides*; Lecture Notes in Computer Science (LNCS) 11958; Springer: Cham, Switzerland, 2020; pp. 332–339.

63. Trapella, C.; Rizzo, R.; Gallo, S.; Alogna, A.; Bortolotti, D.; Casciano, F.; Zauli, G.; Secchiero, P.; Voltan, R. Helix Complex snail mucus exhibits pro-survival, proliferative and pro-migration effects on mammalian fibroblasts. *Sci. Rep.* **2018**, *8*, 17665. [CrossRef]

64. Li, J.; Shang, L.; Lan, J.; Chou, S.; Feng, X.; Shi, B.; Wang, J.; Lyu, Y.; Shan, A. Targeted and intracellular antibacterial activity against *S. agalactiae* of the chimeric peptides based on pheromone and cell-penetrating peptides. *ACS Appl. Mater. Interfaces* **2020**, *12*, 44459–44474. [CrossRef] [PubMed]

65. Wang, Y.; Wang, M.; Shan, A.; Feng, X. Avian host defense cathelicidins: Structure, expression, biological functions, and potential therapeutic applications. *Poult. Sci.* **2020**, *99*, 34–45. [CrossRef] [PubMed]

66. Jiale, Z.; Jian, J.; Xinyi, T.; Haoji, X.; Xueqin, H.; Xiao, W. Design of a novel antimicrobial peptide 1018M targeted ppGpp to inhibit MRSA biofilm formation. *AMB Express* **2021**, *11*, 49. [CrossRef] [PubMed]

67. Mah, T.F.; O'Toole, G.A. Mechanisms of biofilm resistance to antimicrobial agents. *Trends Microbiol.* **2001**, *9*, 34–39. [CrossRef]

68. Wang, X.; Preston, J.F., III; Romeo, T. The pgaABCD locus of *Escherichia coli* promotes the synthesis of a polysaccharide adhesin required for biofilm formation. *J. Bacteriol.* **2004**, *186*, 2724–2734. [CrossRef]

69. Spinosa, M.R.; Progida, C.; Tala, A.; Cogli, L.; Alifano, P.; Bucci, C. The *Neisseria meningitidis* capsule is important for intracellular survival in human cells. *Infect. Immun.* **2007**, *75*, 3594–3603. [CrossRef] [PubMed]

70. Onime, L.A.; Oyama, L.B.; Thomas, B.J.; Gani, J.; Alexander, P.; Waddams, K.E.; Cookson, A.; Fernandez-Fuentes, N.; Creevey, C.J.; Huws, S.A. The rumen eukaryotome is a source of novel antimicrobial peptides with therapeutic potential. *BMC Microbiol.* **2021**, *21*, 105. [CrossRef] [PubMed]

Article

LL-37 and Double-Stranded RNA Synergistically Upregulate Bronchial Epithelial TLR3 Involving Enhanced Import of Double-Stranded RNA and Downstream TLR3 Signaling

Sara Bodahl, Samuel Cerps, Lena Uller and Bengt-Olof Nilsson *

Department of Experimental Medical Science, Lund University, BMC D12, SE-22184 Lund, Sweden; sara.bodahl@med.lu.se (S.B.); samuel.cerps@med.lu.se (S.C.); lena.uller@med.lu.se (L.U.)
* Correspondence: bengt-olof.nilsson@med.lu.se

Abstract: The human host defense peptide LL-37 influences double-stranded RNA signaling, but this process is not well understood. Here, we investigate synergistic actions of LL-37 and synthetic double-stranded RNA (poly I:C) on toll-like receptor 3 (TLR3) expression and signaling, and examine underlying mechanisms. In bronchial epithelial BEAS-2B cells, LL-37 potentiated poly I:C-induced TLR3 mRNA and protein expression demonstrated by qPCR and Western blot, respectively. Interestingly, these effects were associated with increased uptake of rhodamine-tagged poly I:C visualized by immunocytochemistry. The LL-37/poly I:C-induced upregulation of TLR3 mRNA expression was prevented by the endosomal acidification inhibitor chloroquine, indicating involvement of downstream TLR3 signaling. The glucocorticoid dexamethasone reduced LL-37/poly I:C-induced TLR3 expression on both mRNA and protein levels, and this effect was associated with increased IκBα protein expression, suggesting that dexamethasone acts via attenuation of NF-κB activity. We conclude that LL-37 potentiates poly I:C-induced upregulation of TLR3 through a mechanism that may involve enhanced import of poly I:C and that LL-37/poly I:C-induced TLR3 expression is associated with downstream TLR3 signaling and sensitive to inhibition of NF-κB activity.

Keywords: host defense peptide; innate immunity; NF-κB; poly I:C; toll-like receptor 3

Citation: Bodahl, S.; Cerps, S.; Uller, L.; Nilsson, B.-O. LL-37 and Double-Stranded RNA Synergistically Upregulate Bronchial Epithelial TLR3 Involving Enhanced Import of Double-Stranded RNA and Downstream TLR3 Signaling. *Biomedicines* **2022**, *10*, 492. https://doi.org/10.3390/biomedicines10020492

Academic Editor: Albrecht Piiper

Received: 21 December 2021
Accepted: 17 February 2022
Published: 19 February 2022

Publisher's Note: MDPI stays neutral with regard to jurisdictional claims in published maps and institutional affiliations.

1. Introduction

Toll-like receptors (TLRs) are important innate immune receptors that trigger defense and inflammatory responses in host cells in response to various microorganisms and pro-inflammatory agents [1]. The TLR family consists of about ten members of plasma membrane and intracellular receptors in humans, and their expression is largely dependent on cell type [2]. TLR3 is intracellularly expressed and known to recognize and respond to double-stranded (ds) RNA, usually from viral origin. The activated TLR3 receptor increases expression of transcription factors interferon regulatory factor 3 and 7, but also nuclear factor kappa B (NF-κB), leading to the production and secretion of type I interferons and pro-inflammatory cytokines [3–6].

The human host defense peptide LL-37 is part of the first line of defense against invading pathogens [7]. The peptide is mainly produced by neutrophils, but also by other types of white blood cells, skin epithelial cells and epithelial cells aligning the mucosal areas including bronchial epithelial cells [8–10]. LL-37 exerts antimicrobial activity through permeabilization of the bacterial cell wall, resulting in cell lysis, but also via neutralization of bacterial lipopolysaccharides [11,12]. Besides showing activity against microorganisms, LL-37 also modulates the immune system by acting as a chemoattractant for several types of immune cells and by promoting their expression of pro-inflammatory cytokines [13–16]. High concentrations (>1 µM) of LL-37 have been found locally in the gingival crevicular fluid collected at disease sites of patients suffering from periodontitis and in psoriatic skin lesions [17,18]. After allergen challenge, LL-37 appears on the bronchial epithelial surface as a component of plasma exudation, which is an early defense/inflammatory response in health and disease [19]. Bals et al. [20] show that LL-37 is expressed in surface epithelia in healthy human lung tissue, and moreover these authors report that the peptide shows activity against *Pseudomonas aeruginosa*, a pathogen associated with pneumonia. Thus, it is of in vivo relevance, to investigate the functional importance of LL-37 in lung epithelial cells.

Polycytidylic acid (poly I:C) is a synthetic analog to dsRNA, and a well-studied TLR3 ligand [5,21]. LL-37 has been shown to potentiate poly I:C-induced production of pro-inflammatory cytokines in human airway epithelial cells, and this effect is supposed to involve an LL-37-induced increase in the intracellular bioavailability of poly I:C [22,23]. We have recently demonstrated that LL-37 facilitates the pro-inflammatory effects of poly I:C in human coronary artery smooth muscle cells by upregulation of TLR3 expression, suggesting that LL-37 can promote poly I:C-induced inflammation also via this mechanism [24]. However, the mechanism explaining how LL-37 can cause upregulation of TLR3 expression in the presence of poly I:C has not been identified.

In the present study, we investigate the mechanisms behind LL-37/poly I:C-induced upregulation of TLR3 in the human bronchial epithelial BEAS-2B cell line. We show that LL-37 potentiates poly I:C-induced upregulation of TLR3 expression, and that this effect is associated with activation of TLR3 signaling and sensitive to pharmacological inhibition of NF-κB activity. Moreover, we demonstrate that LL-37 enhances uptake of poly I:C, suggesting that LL-37 may potentiate poly I:C-induced TLR3 expression via this mechanism.

2. Materials and Methods

2.1. Cells and Cell Culture

The human bronchial epithelial cell line BEAS-2B cells were purchased from ATCC, and cultured in RPMI 1640 medium supplemented with Glutamax, (Thermo Fisher Scientific, Waltham, MA, USA), fetal bovine serum (10%, Thermo Fisher Scientific), penicillin (50 U/mL) and streptomycin (50 μg/mL). The BEAS-2B cells were cultured under standard conditions in a cell incubator, reseeded upon reaching confluence as appropriate and counted (LUNA automated cell counter, Logos Biosystems, Anyang-si, Korea) to ensure suitable cell density. All experiments were performed in the culture medium specified above.

2.2. Real-Time RT-qPCR

Total RNA was extracted from cell lysates and purified with a miRNeasy kit (Qiagen, Venlo, Netherlands), using the QIAcube instrument (Qiagen) according to manufacturer's instructions. Concentration and quality of the RNA was determined by a NanoDrop 2000C spectrophotometer (Thermo Fisher Scientific). The RNA was analyzed in a Step-One Plus real-time cycler (Applied Biosystems, Waltham, MA, USA), using a QuantiFast SYBR green RT-PCR kit (Qiagen), according to the manufacturer's guidelines. QuantiTect primer assays (Qiagen) were used for TLR3 (Hs_TLR3_1_SG) and GAPDH (Hs_GAPDH_2_SG), with GAPDH serving as an internal control. All samples were analyzed in duplicate and gene expression was calculated using the delta CT method as previously described [25].

2.3. Western Blot

Cells were lysed in sodium dodecyl sulfate (SDS) buffer and total protein concentration in each sample determined (Bio-Rad DC protein assay, Bio-Rad, Hercules, CA, USA). Identical amounts of total protein were loaded to each lane on Criterion TGX 4–15% precast gels (Bio-Rad) and proteins separated using SDS/PAGE. The proteins were then transferred to nitrocellulose membranes (Trans-Blot Transfer system, Bio-Rad). Membranes were blocked in Tris-buffered saline (TBS, Bio-Rad) with 0.5% casein and incubated 12–15 h (4 °C) with primary TLR3 (Cell Signaling, #6961S, rabbit) or NF-κB subunits phosphorylated NF-κB p65 (Cell Signaling, #3031S, rabbit), phosphorylated NF-κB p105 (Cell signaling,

#4806S, rabbit) and IκBα (Cell Signaling, #9242S, rabbit) antibodies. The GAPDH protein (Merck Millipore, Burlington, MA, USA, #MAB374, mouse), was used in all blots as an internal control. Membranes were then incubated with horse-radish peroxidase conjugated secondary antibodies (Cell Signaling, #7074S, rabbit or #7076S, mouse) for 2 h at room temperature. TBS-Tween(T) was used to wash the membranes between each incubation step and both primary and secondary antibodies were diluted in 0.5% casein/TBS-T as appropriate. SuperSignal West Femto chemiluminescence reagent (Thermo Fisher Scientific) was used to visualize the immunoreactive bands, using a LI-COR Odyssey Fc instrument (LI-COR Biosciences, Lincoln, NE, USA).

2.4. Assessment of Cell Viability

Cell viability was assessed using the thiazolyl blue tetrazolium (MTT, Sigma-Aldrich, St Louis, MO, USA) assay. Cells were seeded in a 96-well plate and incubated with MTT solution (0.5 mg/mL, Sigma-Aldrich) for 1 h at 37 °C. The supernatants were then discarded, and the formazan product dissolved in dimethyl sulfoxide (DMSO). The plate was analyzed in a Multiscan GO Microplate Spectrophotometer (Thermo Fisher Scientific), measuring the absorbance at 540 nm.

2.5. Measurement of Poly I:C Import Using Fluorescence Imaging

Cells were grown on round glass coverslips and placed in a 24-well cell culture plate. The cells were treated with rhodamine-labeled poly I:C (InvivoGen), either alone or in combination with LL-37 for 6 and 24 h at 37 °C. The cells were fixed with 4% paraformaldehyde for 10 min and the coverslips mounted with Fluoroshield mounting media (Sigma-Aldrich) containing the nuclear marker DAPI. Phosphate buffered saline (PBS, Gibco, Waltham, MA, USA) was used to wash the cells between each incubation step. The cells were photographed and analyzed in a fluorescence microscope (Olympus, Olympus Europa, Hamburg, Germany). Quantitative assessment was performed in the Image J software. For each coverslip, average cellular fluorescence intensity was calculated by analyzing the fluorescence signal from five cells in three different areas.

2.6. Agents

LL-37 (Bachem, Bubendorf, Switzerland) and dexamethasone (Sigma-Aldrich) were dissolved in DMSO. Poly I:C (InvivoGen, San Diego, CA, USA), fluorescent rhodamine-tagged poly I:C (InvivoGen) and chloroquine (Sigma-Aldrich) were dissolved in PBS. Vehicle was added as appropriate. Cells were pre-treated with dexamethasone or chloroquine for 30 min and dexamethasone or chloroquine were then present throughout the experiment.

2.7. Statistics

Data were analyzed in GraphPad Prism9 (GraphPad Software) and presented as mean ± SEM. Each experiment was repeated 2–4 times. Each culture well was regarded to represent one biological replicate ($n = 1$), except for assessment of poly I:C import using cells grown on glass coverslips where each independent experiment was regarded to represent one biological replicate ($n = 1$). The n-values for each experiment are displayed in the figure legends. A one-way ANOVA, followed by Tukey's multiple comparison post hoc tests was used to calculate statistical significance as appropriate. p value less than 0.05 were considered statistically significant.

3. Results

3.1. Dexamethasone Reduces TLR3 Expression Induced by Combined Treatment with LL-37 and Poly I:C in BEAS-2B Cells

Initially, we investigated TLR3 mRNA expression in human bronchial epithelial BEAS-2B cells treated with poly I:C (10 µg/mL) alone or LL-37 (1 µM) and poly I:C (10 µg/mL) in combination (LL-37/poly I:C), in the presence or absence of dexamethasone (1 µM) for 6 and 24 h. We have previously demonstrated that combined treatment with LL-37 and poly I:C for 24 h triggers production of pro-inflammatory IL-6 and MCP-1 in human coronary artery smooth muscle cells, suggesting that stimulation with LL-37 and poly I:C in combinations is pro-inflammatory at this time point [24]. Hence, it is reasonable to assume that treatment with LL-37/poly I:C will affect TLR3 expression within 24 h. Both treatment with poly I:C alone and LL-37/poly I:C for 6 h increased the TLR3 transcript levels compared to control cells, though the co-treatment had a stronger effect compared to treatment with poly I:C alone (Figure 1A). Dexamethasone reduced the poly I:C-induced TLR3 mRNA expression at 6 h by 35%, whereas it reduced the LL-37/poly I:C-induced transcript levels by 60% (Figure 1A). Both treatment with poly I:C alone and LL-37/poly I:C for a longer time (24 h) increased the TLR3 mRNA levels compared to control cells (Figure 1B). Dexamethasone reduced poly I:C- and LL-37/poly I:C-induced TLR3 mRNA expression at 24 h by 45 and 60%, respectively (Figure 1B). Next, we used Western blot analysis to examine the TLR3 protein expression. Here we choose to focus on the co-treatment with LL-37 and poly I:C since this treatment caused an overall larger increase in TLR3 mRNA expression than poly I:C alone. Treatment with LL-37 (1 µM)/poly I:C (10 µg/mL) for 24 h increased the TLR3 protein expression by ten times compared to control cells (Figure 1C). Notably, dexamethasone (1 µM) reduced this effect by more than 60% (Figure 1C).

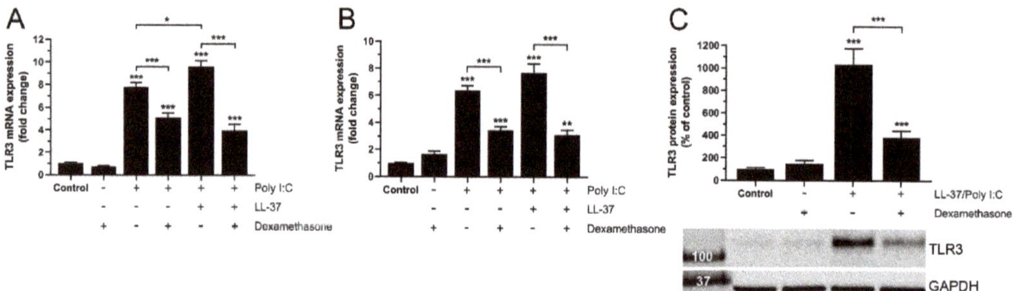

Figure 1. LL-37/poly I:C-induced stimulation of TLR3 expression is reduced by dexamethasone in BEAS-2B cells. (**A,B**) TLR3 mRNA expression was determined using quantitative real-time RT-PCR in cells treated with or without poly I:C (10 µg/mL) alone, or LL-37 (1 µM) and poly I:C (10 µg/mL) in combination in the presence or absence of dexamethasone (1 µM) for 6 h (**A**) and 24 h (**B**). (**C**) TLR3 protein expression was assessed using Western blot in cells treated with or without LL-37 (1 µM) and poly I:C (10 µg/mL) in combination (LL-37/poly I:C) in the presence or absence of dexamethasone (1 µM) for 24 h. The TLR3 immunoreactive band was observed at the expected molecular weight of 115–130 kDa and normalized to GAPDH (37 kDa), serving as internal control. (**A–C**) Data are presented as mean ± SEM, *n* = 7–8 (**A,B**) and *n* = 6 (**C**) in each group. Statistical significance was calculated using a one-way ANOVA, followed by Tukey's post hoc test. ** and *** represent *p* < 0.01 and *p* < 0.001, respectively, vs. control. For comparisons indicated by the bars, * and *** represent *p* < 0.05 and *p* < 0.001, respectively.

3.2. Dexamethasone-Induced Down-Regulation of TLR3 Expression Is Associated with Reduced NF-κB Activity in BEAS-2B Cells

To investigate if the transcription factor NF-κB is involved in the LL-37/poly I:C-induced modulation of TLR3 expression, we evaluated the protein expression of NF-κB inhibitor IκBα, NF-κB phosphorylated p65 and NF-κB phosphorylated p105 using Western blot. Cells were treated with or without LL-37 (1 µM)/poly I:C (10 µg/mL) for 24 h in the presence or absence of dexamethasone (1 µM). Both treatment with dexamethasone alone and in combination with LL-37/poly I:C increased the expression of IκBα, suggesting that dexamethasone reduces NF-κB via this mechanism (Figure 2A). The protein expression of NF-κB phosphorylated p65 was not affected by any treatment (Figure 2B). Interestingly, stimulation with LL-37/poly I:C increased the expression of NF-κB phosphorylated p105 compared to control cells, and moreover dexamethasone tended (not statistically significant) to reduce this effect (Figure 2C). Considering that stimulation with LL-37/poly I:C increases phosphorylated p105 but has no effect on either IκBα or phosphorylated p65 levels, it is reasonable to conclude that treatment with LL-37 and poly I:C in combination enhances NF-κB activity.

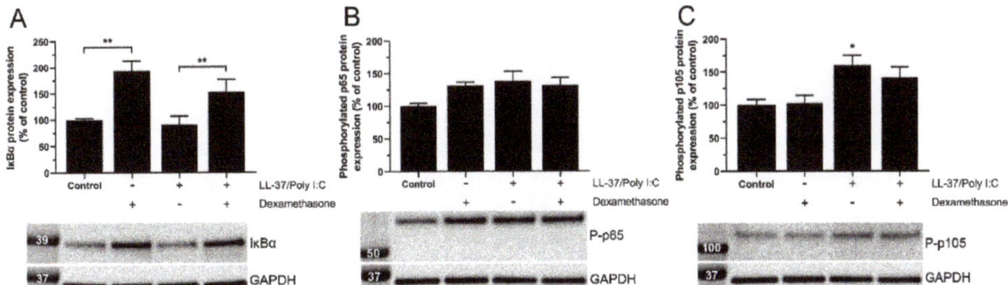

Figure 2. Dexamethasone reduces NF-κB activity in BEAS-2B cells by increasing expression of IκBα. (**A–C**) Western blot analysis of IκBα (**A**), NF-κB phosphorylated p65 (**B**) and NF-κB phosphorylated p105 (**C**) was performed in cells stimulated with or without poly I:C (10 μg/mL) and LL-37 (1 μM) in combination in the presence or absence of dexamethasone (1 μM) for 24 h. The immunoreactive bands were observed at the expected molecular weight of 39 kDa (IκBα), 65 kDa (phosphorylated p65) and 105 kDa (phosphorylated p105), and the intensity of each band was normalized to GAPDH (37 kDa), serving as internal control. Data are presented as mean $\pm$ SEM, n = 7–8 in each group. Statistical significance was calculated using a one-way ANOVA followed by Tukey's post hoc test. * represents $p < 0.05$ vs. control. For comparisons indicated by the bars, ** represents $p < 0.01$.

3.3. Upregulation of TLR3 Expression by Poly I:C and LL-37/poly I:C Involves Downstream TLR3 Signaling in BEAS-2B Cells

In the next set of experiments, we further examined LL-37-induced potentiation of poly I:C-stimulated TLR3 mRNA expression to assess its dependence on poly I:C concentration. For these experiments, we stimulated cells with or without poly I:C at low concentrations (0.2 or 2 μg/mL), LL-37 (1 μM) or the two in combination for 6 and 24 h. Treatment with poly I:C alone increased the TLR3 mRNA expression in a concentration dependent manner at both 6 and 24 h ($p < 0.01$ and $p < 0.001$ for 0.2 vs. 2 μg/mL at 6 and 24 h, respectively) (Figure 3A,B). Interestingly, co-treatment with LL-37 and poly I:C for 6 h increased the TLR3 mRNA levels by 2–3 times, compared to treatment with poly I:C alone, whereas LL-37 did not potentiate the poly I:C-induced TLR3 expression at 24 h, indicating that LL-37 has a rapid turnover (Figure 3A,B). Importantly, LL-37 alone had no effect on the mRNA expression of TLR3 (Figure 3A,B). It has previously been shown that TLR3 signaling depends on endosomal acidification [26], and to assess if LL-37/poly IC-induced upregulation of TLR3 expression is associated with activation of TLR3 signaling, we used the endosomal acidification inhibitor chloroquine. Cells were treated with or without poly I:C (0.2 μg/mL), LL-37 (1 μM) or the two in combination for 6 h, in the absence or presence of chloroquine (2 μg/mL). Treatment with chloroquine completely prevented poly I:C- and LL-37/poly I:C-induced TLR3 mRNA expression (Figure 3C). Chloroquine alone did not affect the TLR3 expression (Figure 3C).

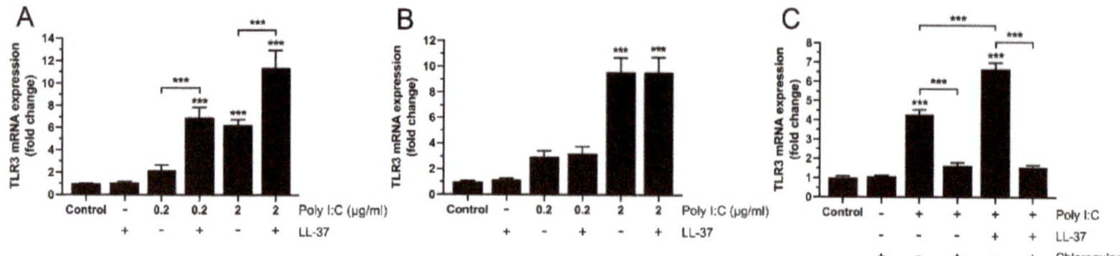

Figure 3. LL-37/poly I:C-induced stimulation of TLR3 mRNA expression is abolished by chloroquine in BEAS-2B cells. (**A,B**) TLR3 mRNA expression was analyzed using quantitative real-time RT-PCR in cells treated with or without poly I:C (0.2 or 2 μg/mL) alone, LL-37 (1 μM) alone, or the two in combination for 6 (**A**) and 24 h (**B**). (**C**) TLR3 mRNA expression was determined in cells treated with poly I:C (0.2 μg/mL), LL-37 (1 μM), or the two in combination in the presence or absence of chloroquine (2 μg/mL) for 6 h. (**A–C**) Data are presented as mean ± SEM, *n* = 8–12 (**A**), *n* = 6–8 (**B**) and *n* = 4 (**C**) in each group. Statistical significance was calculated using a one-way ANOVA, followed by Tukey's post hoc test. *** represents $p < 0.001$ vs. control or $p < 0.001$ for comparisons indicated by the bars.

3.4. High but Not Low Concentrations of LL-37 Enhance Poly I:C-Induced TLR3 Expression at 24 h, Indicating Rapid Turnover of LL-37 in BEAS-2B Cells

Since LL-37 (1 μM) potentiates poly I:C-evoked upregulation of TLR3 at 6 but not 24 h, we hypothesize that LL-37 shows rapid turnover. To test this, we examined if treatment with a high (4 μM) but not to a low (1 μM) dose of LL-37 potentiates poly I:C-induced TLR3 expression at 24 h. Interestingly, the combined stimulation with 4 μM LL-37 and 0.2 μg/mL poly I:C enhanced TLR3 mRNA expression by about five times compared to stimulation with poly I:C alone, while the lower concentration of LL-37 (1 μM) caused no potentiation of poly I:C-induced TLR3 expression (Figure 4A). Treatment with LL-37 alone (4 μM) showed no effect on mRNA levels for TLR3 (Figure 4A). Additionally, we also investigated TLR3 protein expression using Western blot analysis. The combined treatment with 4 μM LL-37 and 0.2 μg/mL poly I:C for 24 h augmented TLR3 protein expression by more than three times compared to treatment with poly I:C alone (Figure 4B). High concentrations of LL-37 can be cytotoxic for different types of human cells [27]. To exclude that LL-37 reduces cell viability at the concentrations used here (1 and 4 μM), we assessed viability of BEAS-2B cells using the MTT assay. LL-37 is considered to cause a rapid permeabilization of the plasma membrane, and this effect is associated with reduced cell viability monitored with the MTT method [27]. Treatment with 10 μM LL-37 for 4 h reduced cell viability by around 50%, whereas lower concentrations (0.1, 1 and 4 μM) had no effect (Figure 4C).

3.5. LL-37 Increases Import of Poly I:C in BEAS-2B Cells

We hypothesized that potentiation of poly I:C-induced TLR3 expression by LL-37 can be due to LL-37-stimulated import of poly I:C. To address this issue, we treated cells with rhodamine-tagged poly I:C (4 μg/mL), either alone or in combination with LL-37 (4 μM) for 6 and 24 h. A fluorescence signal (red) was observed both in cells treated with poly I:C alone and in cells co-treated with LL-37 and poly I:C, whereas no or very low fluorescence was detected in untreated control cells (Figure 5A–D). Interestingly, fluorescence was around three times stronger in cells treated with LL-37/poly I:C, compared to cells treated with poly I:C alone at both 6 and 24 h, suggesting that LL-37 increases import of poly I:C (Figure 5A–D).

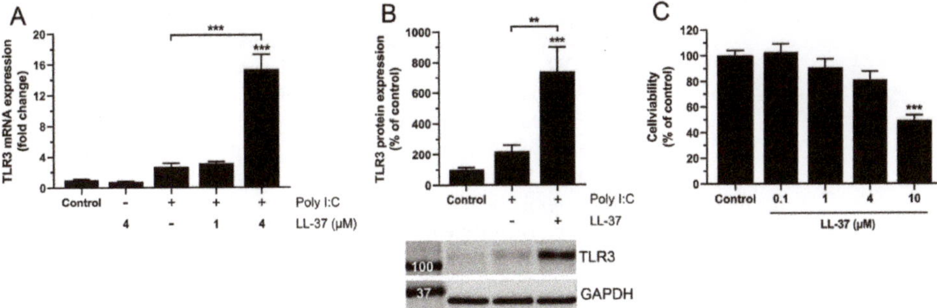

Figure 4. High but not low concentration of LL-37 stimulates poly I:C-induced upregulation of TLR3 in BEAS-2B cells. (**A**) TLR3 mRNA expression was evaluated using quantitative real-time RT-qPCR in cells treated with poly I:C (0.2 µg/mL), LL-37 (4 µM) or LL-37 (1 and 4 µM) and poly I:C (0.2 µg/mL) in combination for 24 h. (**B**) TLR3 protein expression was evaluated using Western blot in cells treated with poly I:C (0.2 µg/mL) alone or poly I:C in combination with LL-37 (4 µM) for 24 h. The TLR3 immunoreactive band was observed at the expected molecular weight of 115-130 kDa and normalized to GAPDH (37 kDa) serving as internal control. (**C**) Cell viability was assessed using the MTT assay in cells treated with LL-37 (0.1, 1, 4 and 10 µM) for 4 h. (**A-C**) Data are presented as mean ± SEM, $n = 8$ (**A,B**) and $n = 6$–8 (**C**) in each group. Statistical significance was calculated using a one-way ANOVA, followed by Tukey's post hoc test. *** represents $p < 0.001$ vs. control. For comparisons indicated by the bars, ** and *** represent $p < 0.01$ and $p < 0.001$, respectively.

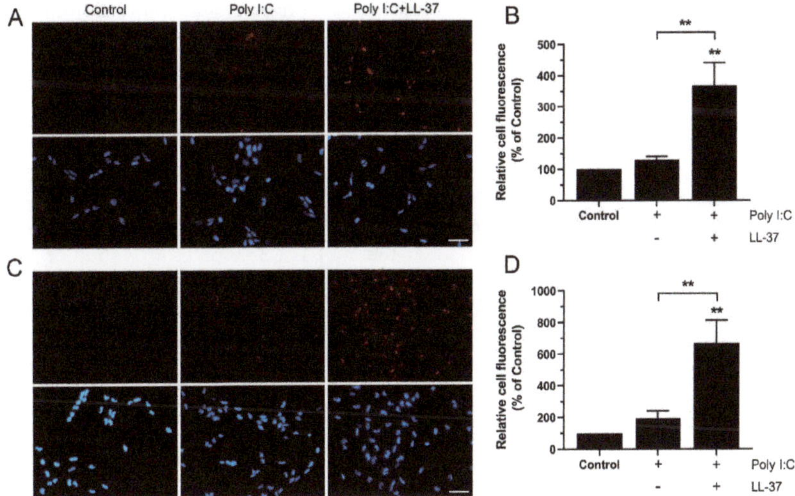

Figure 5. LL-37 triggers import of poly I:C in BEAS-2B cells. (**A–D**) Cells were treated with fluorescent rhodamine-tagged poly I:C (4 µg/mL) alone or in combination with LL-37 (4 µM) for 6 (**A,B**) and 24 h (**C,D**). The intracellular fluorescence signal (red) and the nuclei staining with DAPI (blue) were analyzed and photographed using a fluorescence microscope equipped with a digital camera. The bars in panel **A** and **C** represent 40 µm. The fluorescence intensity of five cells was measured in three different areas on each coverslip, and an average cellular fluorescence intensity was calculated for each experiment. Data are presented as mean ± SEM, $n = 4$ in each group representing the number of independent experiments. Statistical significance was calculated using a one-way ANOVA, followed by Tukey's post hoc test. ** represents $p < 0.01$ vs. control. For comparisons indicated by the bars, ** represents $p < 0.01$.

4. Discussion

In the present study, we demonstrate on both transcript and protein levels that the host defense peptide LL-37 potentiates poly I:C-induced upregulation of TLR3 expression in human bronchial BEAS-2B epithelial cells, and that this effect is associated with LL-37-evoked increase of cellular uptake of poly I:C. LL-37-induced stimulation of poly I:C import is observed already at 6 h of co-treatment with LL-37 and poly I:C, suggesting that LL-37 triggers uptake of poly I:C through a rapid process. Previously, Singh et al. [23] have reported that LL-37 promotes poly I:C-induced stimulation of pro-inflammatory cytokine production via a mechanism involving endosomal acidification, leading to increased intracellular bioavailability of poly I:C in the same cell type as used by us here, i.e., BEAS-2B. Hence, it seems that LL-37 facilitates poly I:C signaling through different mechanisms involving both enhanced uptake as demonstrated in the present study and increased processing of endosomal poly I:C. LL-37 has been shown to be internalized by human macrophages and osteoblasts via endocytosis, but inhibition of the endocytic pathways does not completely prevent import of LL-37, suggesting that also other mechanisms besides endocytosis are involved [28,29]. It is well-recognized that LL-37 forms pores in the plasma membrane, and it is plausible that LL-37/poly I:C complexes can use the LL-37 self-made pores to cross plasma membranes [27,30].

Here we show that treatment with 1 µM LL-37 for 6 h stimulates poly I:C-induced upregulation of TLR3 expression, whereas this effect is absent at 24 h, indicating a rapid turnover of the peptide. Interestingly, the peptide can also elevate poly I:C-induced upregulation of TLR3 at 24 h if the cells are treated with a higher concentration (4 µM) of LL-37, suggesting 4 µM LL-37 is necessary to maintain a high enough concentration of LL-37 over the whole 24 h period to potentiate the effect of poly I:C. Indeed, LL-37 has been reported to have a short half-life (~1 h) in cells [23]. In high concentrations (>4 µM), LL-37 may show cytotoxicity by promoting apoptosis in human host cells [27,31]. Importantly, we demonstrate that LL-37 does not reduce BEAS-2B cell viability in the concentrations (1 or 4 µM) used in the present study, arguing that our present results are not influenced by LL-37-induced cytotoxicity.

In the present study, we demonstrate that chloroquine, an inhibitor of endosomal acidification, attenuates poly I:C- and LL-37/poly I:C-induced upregulation of TLR3, suggesting that poly I:C and LL-37/poly I:C driven stimulation of TLR3 expression involves downstream TLR3 signaling. Stimulation of TLR3 expression by LL-37/poly I:C correlates with increased levels of NF-κB phosphorylated p105, suggesting that LL-37/poly I:C-induced signaling downstream of TLR3 involves activation of NF-κB. We show that the NF-κB inhibitor dexamethasone [32], reduces LL-37/poly I:C-evoked upregulation of both TLR3 mRNA expression and protein production and tend to reduce (not statistically significant) LL-37/poly I:C-induced enhancement of NF-κB phosphorylated p105. Importantly, our data show that dexamethasone strongly elevates IκBα levels. Taken together these data provide evidence that LL-37/poly I:C-induced upregulation of TLR3 involves activation of NF-κB. Interestingly, dexamethasone reduces mortality in critically ill COVID-19 patients [33]. The life-threatening cytokine storm induced by coronavirus seems to involve NF-κB activation, and inhibition of NF-κB activity by dexamethasone is an important therapeutic strategy in seriously ill COVID-19 patients [34,35]. The present data support the possibility that dexamethasone antagonizes the cytokine storm in COVID-19 patients in part by reducing expression of the virus signaling associated receptor TLR3.

Respiratory viral infection of human airways involves plasma exudation that, rather than local cells, likely is responsible for appearance of LL-37 in airway surface liquids. Hence, there is opportunity for interactions between viral dsRNA and LL-37. This interaction may be particularly pronounced in bronchial asthma where plasma exudation is exaggerated at viral infection, the latter being the most common cause of asthma exacerbations. The possibility arises that LL-37/dsRNA synergy is involved in TLR3-dependent overproduction of epithelial cytokines of importance in exacerbations asthma, Hence, it is warranted to further explore the possibility that the present synergy between dsRNA and LL-37 is partly involved in epithelium-driven worsening of asthma.

In summary, we conclude that LL-37 potentiates poly I:C-induced upregulation of TLR3 through a mechanism that may involve enhanced import of poly I:C in bronchial epithelial BEAS-2B cells. Furthermore, we demonstrate that LL-37 and poly I:C, acting in synergy, increases TLR3 expression through downstream TLR3 signaling and that this process is sensitive to inhibition of NF-κB activity.

Author Contributions: Conceptualization, S.B., S.C., L.U. and B.-O.N.; Methodology, S.B., S.C., L.U. and B.-O.N.; Validation, S.B., S.C., L.U. and B.-O.N.; Formal Analysis, S.B., S.C., L.U. and B.-O.N.; Investigation, S.B., S.C., L.U. and B.-O.N.; Software, S.B.; Resources, L.U. and B.-O.N.; Data Curation, S.B.; Writing—Original Draft Preparation, S.B. and B.-O.N.; Writing—Review and Editing, S.B., S.C., L.U. and B.-O.N.; Visualization, S.B.; Supervision, L.U. and B.-O.N.; Project Administration, S.B. and B.-O.N.; Funding Acquisition, S.B., L.U. and B.-O.N. All authors have read and agreed to the published version of the manuscript.

Funding: This study was supported by the Alfred Österlund Foundation (to B.-O.N.), the Royal Physiographic Society (to S.B.), the Research Funds for Oral Health Related Research by Region Skåne (grant number 732821 to B.-O.N.), the Swedish Heart and Lung Foundation (grant number 20180207 to L.U.) and the Swedish Medical Research Council (grant number 2020-00922 to L.U.).

Informed Consent Statement: Not applicable.

Data Availability Statement: The datasets generated during and/or analyzed during the current study are available from the corresponding author on request.

Conflicts of Interest: The authors declare no conflict of interest.

References

1. Medzhitov, R. Toll-like receptors and innate immunity. *Nat. Rev. Immunol.* **2001**, *1*, 135–145. [CrossRef] [PubMed]
2. Lancaster, G.I.; Khan, Q.; Drysdale, P.; Wallace, F.; Jeukendrup, A.E.; Drayson, M.T.; Gleeson, M. The physiological regulation of toll-like receptor expression and function in humans. *J. Physiol.* **2005**, *563*, 945–955. [CrossRef] [PubMed]
3. Kariko, K.; Ni, H.; Capodici, J.; Lamphier, M.; Weissman, D. mRNA is an endogenous ligand for toll-like receptor 3. *J. Biol. Chem.* **2003**, *279*, 12542–12550. [CrossRef] [PubMed]
4. Kawai, T.; Akira, S. Antiviral signaling through pattern recognition receptor. *J. Biochem.* **2007**, *141*, 137–145. [CrossRef]
5. Alexopoulou, L.; Holt, A.C.; Medzhitov, R.; Flavell, R.A. Recognition of double-stranded RNA and activation of NF-κB by toll-like receptor 3. *Nature* **2001**, *413*, 732–738. [CrossRef]
6. Oshimumi, H.; Matsumoto, M.; Funami, K.; Akazawa, T.; Seya, T. TICAM-1, an adaptor molecule that participates in toll-like receptor 3-mediated interferon-β induction. *Nat. Immunol.* **2003**, *4*, 161–167. [CrossRef]
7. Mansour, S.C.; Pena, O.M.; Hancock, R.E. Host defense peptides: Front-line immunomodulators. *Trends Immunol.* **2014**, *35*, 443–450. [CrossRef]
8. Sørensen, O.E.; Follin, P.; Johnsen, A.H.; Calafat, J.; Tjabringa, G.S.; Hiemstra, P.S.; Borregaard, N. Human cathelicidin, hCAP-18, is processed to the antimicrobial peptide LL-37 by extracellular cleavage with proteinase 3. *Blood* **2001**, *97*, 3951–3959. [CrossRef]
9. Yamasaki, K.; Schauber, J.; Coda, A.; Lin, H.; Dorschner, R.A.; Schechter, N.M.; Bonnart, C.; Descargues, P.; Hovnanian, A.; Gallo, R.L. Kallikrein-mediated proteolysis regulates the antimicrobial effects of cathelicidins in skin. *FASEB* **2006**, *20*, 2068–2080. [CrossRef]
10. Turner, J.; Cho, Y.; Dinh, N.-N.; Waring, A.J.; Lehrer, R.I. Activities of LL-37, a cathelin-associated antimicrobial peptide of human neutrophils. *Antimicrob Agents Chemother* **1998**, *42*, 2206–2214. [CrossRef]
11. Xhindoli, D.; Pacor, S.; Benincasa, M.; Scocchi, M.; Gennaro, R.; Tossi, A. The human cathelicidin LL-37—A pore-forming antibacterial peptide and host-cell modulator. *Biochim. Biophys. Acta (BBA)* **2016**, *1858*, 546–566. [CrossRef] [PubMed]
12. Larrick, J.W.; Hirata, M.; Balint, R.F.; Lee, J.; Zhong, J.; Wright, S.C. Human CAP18: A novel antimicrobial lipopolysaccharide-binding protein. *Infect. Immun.* **1995**, *63*, 1291–1297. [CrossRef] [PubMed]
13. Agier, J.; Efenberger, M.; Brzezinska-Blaszczyk, E. Cathelicidin impact on inflammatory cells. *Cent. Eur. J. Immunol.* **2015**, *40*, 225–235. [CrossRef] [PubMed]
14. Cederlund, A.; Gudmundsson, G.H.; Agerberth, B. Antimicrobial peptides important in innate immunity. *FEBS J.* **2011**, *278*, 3942–3951. [CrossRef] [PubMed]
15. Nijnik, A.; Pistolic, J.; Filewood, N.C.J.; Hancock, R.E.W. Signaling pathways mediating chemokine induction in keratinocytes by cathelicidin LL-37 and flagellin. *J. Innate Immun.* **2012**, *4*, 377–386. [CrossRef] [PubMed]

16. Yang, D.; Chen, Q.; Schmidt, A.P.; Anderson, G.M.; Wang, J.M.; Wooters, J.; Oppenheim, J.J.; Chertov, O. LL-37, the neutrophil granule– and epithelial cell–derived cathelicidin, utilizes formyl peptide receptor–like 1 (FPRL1) as a receptor to chemoattract human peripheral blood neutrophils, monocytes, and T cells. *J. Exp. Med.* **2000**, *2000*, 1069–1074. [CrossRef] [PubMed]

17. Türkoglu, O.; Emingil, G.; Eren, G.; Atmaca, H.; Kutukculer, N.; Atilla, G. Gingival crevicular fluid and serum hCAP18/LL-37 levels in generalized aggressive periodontitis. *Clin. Oral Investig.* **2017**, *21*, 763–769. [CrossRef]

18. Morizane, S.; Gallo, R.L. Antimicrobial peptides in pathogenesis of psoriasis. *J. Dermatol.* **2012**, *39*, 225–230. [CrossRef]

19. Liu, M.C.; Xiao, H.Q.; Brown, A.J.; Ritter, C.S.; Schroeder, J. Association of vitamin D and antimicrobial peptide production during late-phase allergic responses in the lung. *Clin. Exp. Allergy* **2012**, *42*, 383–391. [CrossRef]

20. Bals, R.; Wang, X.; Zasloff, M.; Wilson, J.M. The peptide antibiotic LL-37/hCAP-18 is expressed in epithelia of the human lung where it has broad antimicrobial activity at the airway surface. *Proc. Natl. Acad. Sci. USA* **1998**, *95*, 9541–9546. [CrossRef]

21. Kumar, A.; Zhang, J.; Yu, F.-S.X. Toll-like receptor 3 agonist poly(I:C)-induced antiviral response in human corneal epithelial cells. *Immunology* **2005**, *117*, 11–21. [CrossRef] [PubMed]

22. Lai, Y.; Adhikarakunnathu, S.; Bhardwaj, K.; Ranjith-Kumar, C.T.; Wen, Y.; Jordan, J.L.; Wu, L.H.; Dragnea, B.; Mateo, L.S.; Kao, C.C. LL37 and cationic peptides enhance TLR3 signaling by viral double-stranded RNAs. *PLoS ONE* **2011**, *6*, e26632. [CrossRef] [PubMed]

23. Singh, D.; Vaughan, R.; Kao, C.C. LL-37 peptide enhancement of signal transduction by toll-like receptor 3 is regulated by pH. *J. Biol. Chem.* **2014**, *289*, 27614–27624. [CrossRef] [PubMed]

24. Dahl, S.; Cerps, S.; Rippe, C.; Swärd, K.; Uller, L.; Svensson, D.; Nilsson, B.O. Human host defense peptide LL-37 facilitates double-stranded RNA pro-inflammatory signaling through up-regulation of TLR3 expression in vascular smooth muscle cells. *Inflamm. Res.* **2020**, *69*, 579–588. [CrossRef]

25. Pfaffl, M.W. A new mathematical model for relative quantification in real-time RT-PCR. *Nucleic Acids Res.* **2001**, *29*, e45. [CrossRef] [PubMed]

26. Gleeson, P.A. The role of endosomes in innate and adaptive immunity. *Semin. Cell Dev. Biol.* **2014**, *31*, 64–72. [CrossRef] [PubMed]

27. Svensson, D.; Wilk, L.; Morgelin, M.; Herwald, H.; Nilsson, B.O. LL-37-induced host cell cytotoxicity depends on cellular expression of the globular C1q receptor (p33). *Biochem. J.* **2016**, *4731*, 87–98. [CrossRef]

28. Tang, X.; Basavarajappa, D.; Haeggström, J.Z.; Wan, M. P2X7 receptor regulates internalization of antimicrobial peptide LL-37 by human macrophages that promotes intracellular pathogen clearance. *J. Immunol.* **2015**, *195*, 1191–1201. [CrossRef]

29. Anders, E.; Dahl, S.; Svensson, D.; Nilsson, B.O. LL-37-induced human osteoblast cytotoxicity and permeability occurs independently of cellular LL-37 uptake through clathrin-mediated endocytosis. *BBRC* **2018**, *501*, 280–285. [CrossRef]

30. Burton, M.F.; Steel, P.G. The chemistry and biology of LL-37. *Nat. Prod. Rep.* **2009**, *26*, 1572–1584. [CrossRef]

31. Aarbiou, J.; Tjabringa, G.S.; Verhoosel, R.M.; Ninaber, D.K.; White, S.R.; Peltenburg, L.T.C.; Rabe, K.F.; Hiemstra, P.S. Mechanisms of cell death induced by the neutrophil antimicrobial peptides a-defensins and LL-37. *Inflamm. Res.* **2006**, *55*, 119–127. [CrossRef]

32. Clark, A.R. Anti-inflammatory functions of glucocorticoid-induced genes. *Mol. Cell Endocrinol.* **2007**, *275*, 79–97. [CrossRef] [PubMed]

33. Horby, P.; Lim, W.S.; Emberson, J.R.; Mafham, M.; Bell, J.L.; Linsell, L.; Staplin, N.; Brightling, C.; Ustianowski, A.; Elmahi, E.; et al. Dexamethasone in Hospitalized Patients with COVID-19. *N. Engl. J. Med.* **2021**, *384*, 693–704. [CrossRef] [PubMed]

34. Morris, G.; Bortolasci, C.C.; Puri, B.K.; Marx, W.; O'Neil, A.; Athan, E.; Walder, K.; Berk, M.; Olive, L.; Carvalho, A.F.; et al. The cytokine storms of COVID-19, H1N1 influenza, CRS and MAS compared. Can one sized treatment fit all? *Cytokine* **2021**, *144*, 155593. [CrossRef] [PubMed]

35. Attiq, A.; Yao, L.J.; Afzal, S.; Khan, M.A. The triumvirate of NF-κB, inflammation and cytokine storm in COVID-19. *Int. Immunopharmacol.* **2021**, *101*, 108255. [CrossRef] [PubMed]

 *biomedicines*

Article

Host Defense Peptides LL-37 and Lactoferrin Trigger ET Release from Blood-Derived Circulating Monocytes

Frederic V. Schwäbe, Lotta Happonen, Sofie Ekestubbe and Ariane Neumann *

Department of Clinical Sciences, Division of Infection Medicine, Lund University, SE-22184 Lund, Sweden; f.schwaebe@gmail.com (F.V.S.); lotta.happonen@med.lu.se (L.H.); sofie.ekestubbe@skane.se (S.E.)
* Correspondence: ariane.neumann@med.lu.se; Tel.: +46-462226807

Abstract: Neutrophils are commonly regarded as the first line of immune response during infection or in tissue injury-induced inflammation. The rapid influx of these cells results in the release of host defense proteins (HDPs) or formation of neutrophil extracellular traps (NETs). As a second wave during inflammation or infection, circulating monocytes arrive at the site. Earlier studies showed that HDPs LL-37 and Lactoferrin (LTF) activate monocytes while neutrophil elastase facilitates the formation of extracellular traps (ETs) in monocytes. However, the knowledge about the impact of HDPs on monocytes remains sparse. In the present study, we investigated the effect of LL-37 and LTF on blood-derived CD14$^+$ monocytes. Both HDPs triggered a significant release of TNFα, nucleosomes, and monocyte ETs. Microscopic analysis indicated that ET formation by LL-37 depends on storage-operated calcium entry (SOCE), mitogen-activated protein kinase (MAPK), and ERK1/2, whereas the LTF-mediated ET release is not affected by any of the here used inhibitors. Quantitative proteomics mass spectrometry analysis of the neutrophil granular content (NGC) revealed a high abundance of Lactoferrin. The stimulation of CD14$^+$ monocytes with NGC resulted in a significant secretion of TNFα and nucleosomes, and the formation of monocyte ETs. The findings of this study provide new insight into the complex interaction of HDPs, neutrophils, and monocytes during inflammation.

Keywords: host defense peptides; monocytes; neutrophils; neutrophil–monocyte interaction; extracellular traps

Citation: Schwäbe, F.V.; Happonen, L.; Ekestubbe, S.; Neumann, A. Host Defense Peptides LL-37 and Lactoferrin Trigger ET Release from Blood-Derived Circulating Monocytes. *Biomedicines* **2022**, *10*, 469. https://doi.org/10.3390/biomedicines10020469

Academic Editor: Jean A. Boutin

Received: 31 December 2021
Accepted: 14 February 2022
Published: 17 February 2022

Publisher's Note: MDPI stays neutral with regard to jurisdictional claims in published maps and institutional affiliations.

1. Introduction

In response to bacterial invasion or tissue injuries, neutrophils undergo the process of degranulation, releasing a broad range of proteins from distinct granules [1]. Since neutrophils are often seen as initiators of the immune response, it may be hypothesized that their deployed molecules orchestrate subsequent defense mechanisms. Inflammation driven by neutrophils is an underlying factor in various clinical conditions. While usually being considered short-lived cells, neutrophils display an unnormal prolonged life span during chronic inflammation [2].

Upon stimulation by neutrophil or endothelial secretion products, e.g., due to a microbial breach [3], circulating monocytes arrive as a second wave of immune cells. Due to their plasticity, monocyte behavior is majorly distinct to the tissue environment [4]. They account for around 10% of all leukocytes [5]. Monocytes are traditionally divided into three sup-populations, dependent on their expression of the surface markers CD14 and CD16 [5]. Based on this, classical monocytes (CD14$^+$CD16$^-$) are associated with phagocytosis and immune responses and intermediate monocytes (CD14$^+$CD16$^+$) are connected to cytokine secretion and antigen presentation while non-classical monocytes (CD14^{lo}CD16$^+$) play a role in complement and adhesion [6]. Changes in the monocyte subsets have been linked to bacterial and viral infections, auto-immune disorders, or chronic inflammation [5]. In addition, different subsets of monocytes are distinct in the secretion of TNFα and IL-6 [7]. While classical monocytes were found to be the most efficient producers of cytokines, non-classical monocytes released the lowest cytokine levels [7]. The stimulus together with the released cytokines might, in turn, give an indication towards which type of monocyte is responding and thus if the cell response is directed towards diminishing or perpetuating an inflammation. Intermediate monocytes are often characterized by their ability to produce proinflammatory cytokines and reactive oxygen species (ROS) upon toll-like receptor (TLR) stimulation [6]. Monocytes have long been solely recognized as macrophage precursor cells, and their role in host defense and immunity, however, remains mostly unclear [3].

A correlation between neutrophils and monocyte recruitment to sites of inflammation has been proposed by Janardhan et al. [8]. Infiltration of monocytes mediated by neutrophils has additionally been shown in viral-induced encephalitis [9]. Furthermore, neutrophil-associated host defense peptides LL-37 and LTF have been identified to mobilize inflammatory monocytes [10,11]. LTF triggers the recruitment of macrophages, upregulation of surface markers, and secretion of proinflammatory cytokines from peripheral blood or monocyte-derived dendritic cells [11]. LL-37 can also enhance chemokine expression and, together with another HDP, heparin-binding protein (HBP), plays a direct role in chemoattraction of monocytes [12]. Degranulation of LL-37 and HBP by neutrophils triggers the polarization of macrophages into M1 proinflammatory phenotype [10,13] and adhesion of monocytes to the endothelial tissue [14,15]. In addition, LL-37 induces the formation of neutrophil extracellular traps (NETs; [16]), suggesting that neutrophils may activate their own kind. Stimulation of neutrophils and monocytes leading to ET formation can be acquired by various (bio)chemical inducers, such as phorbol-12-myristate-13-acetate (PMA; [17]) or lipopolysaccharide (LPS) [18], and with a broad range of bacteria, viruses, and parasites [17,19–23]. Recently, it has been demonstrated that host-derived molecules, such as neutrophil elastase and histones, can trigger ET formation in monocytes [6]. In this study, we analyzed the effect of neutrophil-associated HDPs LL-37 and LTF on monocyte extracellular trap formation (ETosis) to shed light on the complex relationship between neutrophils and monocytes.

2. Materials and Methods

2.1. Reagents

Phorbol 12-myristate 13-acetate (PMA), 2-Aminoethyl diphenylborinate (2-APB), U0126 ethanolate, SB203580, and Cytochalasin D were all purchased from Merck KGaA (Darmstadt, Germany). LL-37 was purchased from Schafer-N (Copenhagen, Denmark). LTF was purchased from Merck KGaA (Darmstadt, Germany).

2.2. Isolating CD14$^+$ Monocytes from the PBMC Fraction

CD14$^+$ monocytes were isolated by density gradient centrifugation and magnetic bead separation. Here, 10 mL of leukocyte concentrate (pooled, project identification code 2021:10, 21.05.2021, Clinical Immunology and Transfusion Medicine, SUS Lund) were diluted 1:1 with 0.9% sodium chloride (NaCl). In total, 20 mL of the mixture were layered onto 20 mL of Lymphoprep™(Thermo Fisher Scientific, Waltham, MA, USA) and centrifuged for 20 min at $700 \times g$ without brake. Erythrocytes were lysed with H_2O for 15 s. Purified PBMCs were resuspended in Magnetic-Activated Cell Sorting (MACS) buffer and CD14 microbeads (both Miltenyi Biotec, Bergisch Gladbach, Germany) were added. Cells were then sorted with an LS column using an MACS separator (both Miltenyi Biotec, Bergisch Gladbach, Germany). After the separation, cells were resuspended in RPMI 1640 (Thermo Fisher Scientific, Waltham, MA, USA), and cell numbers were adjusted to the respective experiments.

2.3. TNFα and Nucleosome Release

For the analysis of the nucleosome release, 1×10^6/mL CD14$^+$ cells were incubated with NGC, PMA, or peptides for 3 h. Cells were then spun down for 5 min at $400 \times g$, and supernatants were collected and stored at $-80\ ^\circ$C. Cell Death Detection ELISAPLUS (Roche Holding, Basel, Switzerland) was used for the analysis of nucleosome release according to the manufacturer's recommendations. TNFα and IL-6 secretion was analyzed using an ELISA kit from Thermo according to the manufacturer. Absorbance for both assays was measured at 450 nm.

2.4. Oxidative Burst

The production of reactive oxygen species (ROS) was applied to analyze neutrophil activation. In total, 1×10^6/mL monocytes were seeded out in triplicates for each sample. Cells were labeled by adding 100 µL of 2,7 di-chlorofluorescein diacetate (DCF-DA; Thermo Fisher Scientific, Waltham, MA, USA), at a final concentration of 50 µg/mL, and incubating for 20 min at RT. Unlabeled cells served as a control for autofluorescence. After incubation, the plate was centrifuged for 5 min at $370 \times g$ and RT. The supernatant was removed and 100 µL of stimuli were added to the cells. Fluorescence was measured at an excitation wavelength of 485 nm and emission wavelength of 530 nm using a Victor 3 microplate reader (PerkinElmer Inc., Waltham, MA, USA) at t_{0h} and t_{3h}.

2.5. Monocyte ET Induction and Immunofluorescence Microscopy

To observe and quantify monocyte extracellular trap (MoET) formation, immunofluorescence microscopy was used, as it was described for neutrophils [24]. Cover slips were placed in 48-well plates and coated with poly-L-lysine (Thermo Fisher Scientific, Waltham, MA, USA) for 20 min. In total, 5×10^5 cells were seeded with 100 µL per well. Then, 100 µL of stimuli were added and the plate was incubated at 37 °C for 3 h. For the inhibition experiments, cells were pre-incubated with the respective inhibitors for 30 min, and then stimuli were added for an additional 3 h of incubation. Samples were fixed with 2% PFA, permeabilized with 1X PBS + 0.5% Triton X-100 for 1 min, and blocked with 1X PBS + 0.05% Tween 20 + 2% BSA for 20 min at RT. Anti-DNA/Histone H1 Antibody (Merck KGaA, Darmstadt, Germany) diluted 1:5000 in blocking solution was added and the plate was incubated for 1 h at RT. Goat anti-Mouse IgG (H + L) Cross-Adsorbed Secondary Antibody, Alexa Fluor 488 (Thermo Fisher Scientific, Waltham, MA, USA) diluted 1:1000 in blocking solution was added. The plate was incubated shielded from light for 1 h at RT. Cells were embedded in ProLong™ Gold Antifade Mountant with DAPI (Thermo Fisher Scientific, Waltham, MA, USA). Samples were visualized using an Eclipse Ti-E inverted microscope system (Nikon, Tokyo, Japan). In total, 4 images were taken per stimulus with a "Plan Apo 40X DIC M N2" objective with a numeric aperture (NA) of 0.9. Images were analyzed using Fiji software version 2.1.0/1.53c [1]. Cells were marked either negative or positive for MoETosis and the percentage of MoET release was calculated. One data point in the graph represents one microscopy image analyzed.

2.6. Collection and Analysis of Neutrophil-derived Granulation Content (NGC)

PMNs were isolated as described previously [24]. Briefly, freshly drawn blood from healthy donors was collected in citrate tubes (ethical permit 2008/657 Lund University). Blood was layered on PolymorphPrep (Thermo Fisher Scientific, Waltham, MA, USA), and density gradient centrifugation was performed. Erythrocytes were lysed with endotoxin-free water and cells were finally resuspended in RPMI 1640 (Thermo Fisher Scientific, Waltham, MA, USA). After 4 h of incubation at 37 °C, cells were then mechanically activated by thermal shock. For this, samples were quickly switched between 45 °C and 0 °C to 45 °C, vortexed, and then spun for 20 min at $22,000 \times g$ to collect proteins. The samples were pooled from several donors and stored at -20 °C in aliquots until further usage. The overall protein content was determined using Pierce 660 nm assay (Thermo Fisher Scientific, Waltham, MA, USA). For SDS-PAGE, 2 µg were mixed (1:1) with reducing loading buffer (Thermo Fisher Scientific, Waltham, MA, USA), denatured at 95 °C for 10 min, and then loaded on a 10–20% Novex Tricine pre-cast gel (Thermo Fisher Scientific, Waltham, MA, USA). Electrophoresis was performed at 120 V for 90 min. The gel was stained using Coomassie Brilliant blue (Thermo Fisher Scientific, Waltham, MA, USA), and images were obtained using a Gel Doc Imager (Bio-Rad Laboratories Inc., Hercules, CA, USA).

2.7. Sample Preparation for Mass Spectrometry

For in-solution digestion, 50 µL of the NGC at a protein concentration of 70 µg/mL were denatured with 8 M urea, 100 mM ammonium bicarbonate, 5 mM tris(2-carboxyethyl)-phosphine (TCEP) in a final volume of 100 µL at 37 °C, 800 rpm for 60 min in triplicates. The disulfide bonds were reduced with a final concentration of 10 mM iodoacetamide at 22 °C, for 30 min in the dark. The urea was diluted to a concentration of below 1.5 M with the addition of 250 µL of 100 mM ammonium bicarbonate, and the proteins digested with 2 µL of 0.5 µg/µL of sequencing-grade trypsin (Promega, Fitchburg, WI, USA) at 37 °C, 800 rpm for 18 h. The digested samples were acidified with 10% formic acid to a pH of approximately 3.0 and cleaned for mass spectrometry using silica C18 reverse phase MacroSpin columns (Thermo Fisher Scientific, Waltham, MA, USA) according to the manufacturer's instructions. For in gel digestion, 2 µg of protein were mixed (1:1) with reducing loading buffer (Thermo Fisher Scientific, Waltham, MA, USA), denatured at 95 °C for 10 min, and then loaded on a 10–20% Novex Tricine pre-cast gel (Thermo Fisher Scientific, Waltham, MA, USA). Electrophoresis was performed at 120 V for 90 min. The gel was stained using Coomassie Brilliant blue (Thermo Fisher Scientific, Waltham, MA, USA), and images were obtained using a Gel Doc Imager (Bio-Rad Laboratories, Hercules, MA, USA). The most prominent bands were excised and prepared for mass spectrometry as described [25]. The peptide concentration of all samples was measured on a Nanodrop (Thermo Fisher Scientific, Waltham, MA, USA) and approximately 150 ng of peptides were analyzed by mass spectrometry.

2.8. Liquid Chromatography Tandem Mass Spectrometry (LC-MS/MS)

All peptides were analyzed on an Orbitrap Eclipse mass spectrometer connected to an ultra-high-performance liquid chromatography Dionex Ultra300 system (both Thermo Fisher Scientific, Waltham, MA, USA). The peptides were loaded and concentrated on an Acclaim PepMap 100 C18 precolumn (75 µm × 2 cm) and then separated on an Acclaim PepMap RSLC column (75 µm × 25 cm, nanoViper, C18, 2 µm, 100 Å) (both columns Thermo Fisher Scientific, Waltham, MA, USA), at a column temperature of 45 °C and a maximum pressure of 900 bar. A linear gradient of 3% to 38% of 80% acetonitrile in aqueous 0.1% formic acid was run for 90 min. In total, 1 full MS scan (resolution 120,000; mass range of 350–1400 m/z) was followed by MS/MS scans (resolution 15,000) of the 20 most abundant ion signals. The precursor ions were isolated with a 1.6 m/z isolation window and fragmented using higher-energy collisional-induced dissociation (HCD) at a normalized collision energy of 30. The dynamic exclusion was set to 45 s.

2.9. Data Analysis

Acquired MS raw spectra were analyzed using Proteome Discoverer 2.5 (Thermo Fisher Scientific, Waltham, MA, USA) against an in-house compiled dataset containing the reviewed Homo sapiens proteome, UniProt ID UP000005640. Fully tryptic digestion was used allowing 2 missed cleavages. Carbamidomethylation (C) was set to static and protein N-terminal acetylation and oxidation (M) to variable modifications. Mass tolerance for precursor ions was set to 10 ppm, and for fragment ions to 0.02 Da. The protein false discovery rate (FDR) was set to 1%. Proteins quantified by 2 or more unique peptides were considered as relevant and used in R [2] for total ion current (TIC) normalization of the data. The MS non-TIC normalized raw data for the NGC in-solution digest is presented in Table S1, and that of the SDS-PAGE fractionated gel bands in Table S2. The mass spectrometry data were deposited to the ProteomeXchange [26] consortium via the MassIVE partner repository (https://massive.ucsd.edu/; accessed on 2 February 2022) with the dataset identifier PXD031375, and are presented in Tables S1 and S2.

2.10. Statistical Analysis

Data were analyzed by using GraphPad Prism v7.0 (GraphPad Software, San Diego, CA, USA). Differences between 2 groups were analyzed by using a paired, 1-tailed

Student *t* test or 1-way ANOVA with the Bonferroni post hoc test. Significance is indicated as * $p \leq 0.05$, ** $p \leq 0.01$, *** $p \leq 0.001$, and **** $p \leq 0.0001$. ns indicates no statistical significance.

3. Results

3.1. LL-37 Triggers Secretion of TNFα, Nucleosomes, and ETs from CD14+ Monocytes

Neutrophils and monocytes share a complex relationship, orchestrating each other in addition to other immune cells [27]. For the first set of experiments, we thus sought to analyze the effect of LL-37 and LTF on monocytes regarding ROS production, TNF and IL-6 release, and ET formation. Here, we incubated CD14+ monocytes with LL-37 and LTF at concentrations described in the literature [28,29]. PMA served as a positive control for the activation of monocytes [17] (Figure 1a–c). To exclude the cytotoxic effects of the stimulation, we measured the LDH release (Figure S1). This analysis showed no detrimental effect on the cells mediated by the used agents. The interaction of peptides with monocytes has been reported to affect cytokine secretion [10,30]. As seen in Figure 1a, the incubation with PMA resulted in a significant secretion of TNFα within 3 h. While the stimulation with LL-37 or LTF was not as efficient compared to PMA, a significant increase in TNFα release was still detected after treatment with LL-37 or LTF treatment (Figure 1a). Analysis of IL-6 secretion revealed that LTF had a significant impact (Figure 1b), whereas no response was detected when cells were treated with PMA or LL-37. Next, we sought to analyze the effect of PMA and the peptides regarding monocyte ET formation. Various studies have reported that PMA is a potent inducer of neutrophil extracellular traps [18,31] and recently also for human CD14+ monocytes [17]. Correlating with the findings of Granger and colleagues, in our experimental set up, CD14+ monocytes responded to PMA with the secretion of DNA-histone complexes/nucleosomes (Figure 1c). The formation of neutrophil ET in response to 5 μM of LL-37 was reported earlier [16]. Here, we found that 5 μM LL-37 also significantly induced the release of nucleosomes from monocytes within 3 h (Figure 1c). In addition, 75 nM LTF showed significant nucleosome release from CD14+ monocytes with only low standard error of the mean (Figure 1c).

For neutrophils, ROS-dependent and -independent mechanisms of ET formation have been described. PMA or live *Staphylococcus aureus* induce the assembly and activation of NADPH oxidase [32] while depletion of cholesterol using cyclodextrin led to the release of NET fibers, independently of NADPH oxidase [33]. PMA, an activator of NADPH oxidases, has been shown to induce oxidative burst in PBMC-derived monocytes and macrophages [34]. Here, we found PMA significantly triggered an oxidative burst in primary monocytes within 3 h (Figure 1d). Analyzing the production of ROS in response to LL-37 and LTF revealed no detectable increase but rather a quenching of fluorescence in the presence of the proteins (Figure 1d).

DNA-intercalating dyes, such as SytoxGreen™, as a method for ET quantification are commonly used in the literature [35]. However, we have previously shown that LL-37 quenches the fluorescence of the DNA-intercalating dye PicoGreen™, suggesting a competitive binding of LL-37 to the DNA [36]. We therefore decided to additionally analyze the interaction of monocytes with LL-37 and LTF by immunofluorescence microscopy. ETs were visualized with an antibody against DNA-histone complexes. The representative images shown in Figure 2a reveal that monocytes treated with all three stimuli significantly released extracellular DNA-histone fibers.

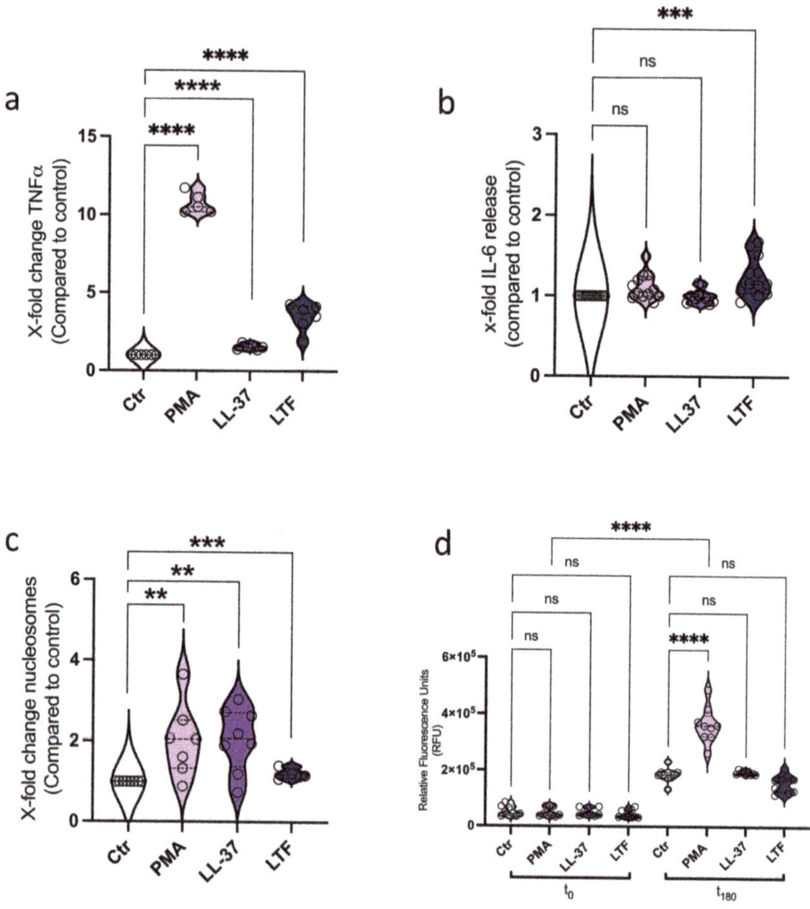

Figure 1. HDPs LL-37 and LTF activate CD14$^+$ monocytes. (**a**). Analysis of TNFα secretion in response to 50 nM PMA, 5 μM LL-37, and 75 nM LTF. (**b**). Release of IL-6 after PMA, LL-37, and LTF treatment. (**c**). Nucleosome release/NET formation triggered by PMA, LL-37, and LTF. (**d**). ROS production in response to PMA, LL-37, and LTF measured at t0 and t180. All data represent mean +/− SEM of 3–4 independent experiments. ns = not significant, ** $p \leq 0.01$ *** $p \leq 0.001$, **** $p \leq 0.0001$.

Quantitative analysis of the images further confirmed that this phenotype is significant (Figure 2b). The highest percentage of released ETs was detected when cells were stimulated with LL-37 (Figure 2b). Various studies have demonstrated that neutrophil ETs from different animal species are released in an ERK1/2-, MAPK-, or store-operated calcium signaling (SOCE) -dependent fashion [37–40]. This variety of mechanisms led us to further investigate the pathways involved in monocyte ET formation. We therefore pre-incubated the monocytes with cytochalasin D (inhibitor of actin rearrangement), 2-ABP (inhibitor of SOCE), SB203580 (inhibitor of MAPK; [41]), and U0126 (inhibitor of ERK1/2). As seen in Figure 3a,b, ET formation mediated by PMA can be completely diminished by blocking of the ERK1/2 pathway. LL-37-mediated ET release was affected by pre-incubation with inhibitors of the intracellular calcium storage (SOCE), MAPK, and ERK pathway (Figure 3a,c). Previously, it was reported that LTF blocks ET release in neutrophils, thus acting as an intrinsic inhibitor of NET formation in circulation [42]. In contrast to these earlier findings, we observed that LTF significantly facilitated the release of ETs from monocytes (Figure 3a,d). Further analysis indicated that none of the used pathway inhibitors (ERK,

MAPK, SOCE, actin rearrangement) significantly affected the LTF-mediated ET formation (Figure 3d). These data suggest that different stimuli activate ET formation via different pathways, as previously reported by Kenny and colleagues [43].

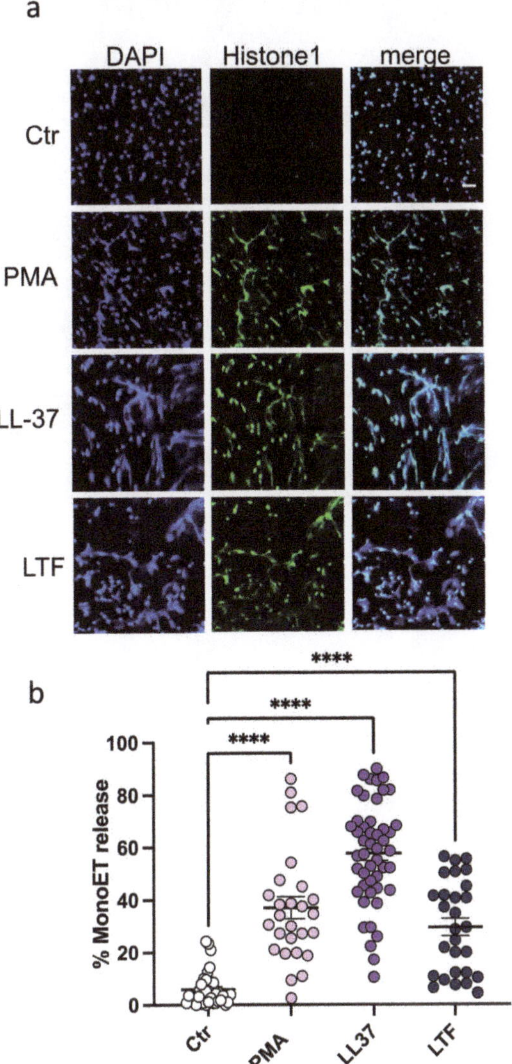

Figure 2. ETs released from CD14+ monocytes in response to LL-37 and LTF. (**a**). Cells were stimulated with different stimuli for 3 h and stained for extracellular DNA (blue) and Histone 1 (green). Scale bar is 50 μm. (**b**). Quantitative analysis of microscopical images shown in (**a**). Every data point represents the relative number of cells undergoing MoET formation in a microscopic image. All data represent mean +/− SEM of 6–11 independent experiments. **** $p \leq 0.0001$.

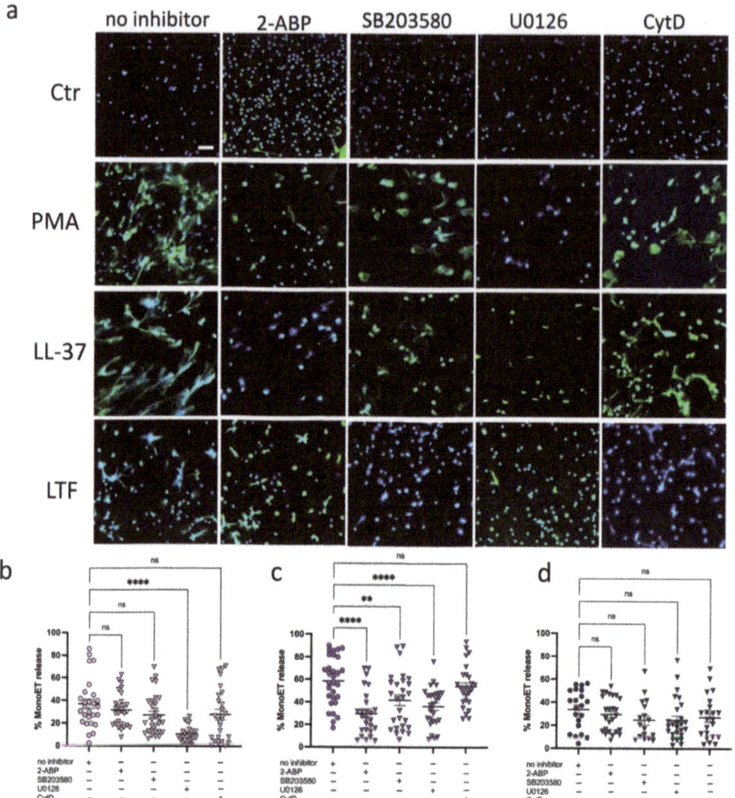

Figure 3. Different stimuli engage various pathways in ET formation. (**a**). Representative fluorescence microscopy images of CD14+ monocytes incubated with PMA, LL-37, or LTF. Blue = DAPI, Green = Histone-DNA complex. Scale bar is 50μm. (**b–d**). Quantitative analysis of CD14+ monocytes pre-incubated with inhibitors of different pathways (SOCE, MAPK, ERK1/2, actin rearrangement) and stimulated with (**b**) 25 nM PMA, (**c**) 5 μM LL-37, or (**d**) 75 nM LTF. All data represent mean +/− SEM of 6 independent experiments. ns = not significant, ** $p \leq 0.01$, **** $p \leq 0.0001$.

3.2. Neutrophil Granular Content Triggers Monocyte ET Formation

Since we observed ET formation in protein-stimulated monocytes, as a holistic approach, we were interested in the impact of a mixture of neutrophil granular content (NGC) proteins on the monocytes. For these experiments, purified blood-derived neutrophils were subjected to thermal shock to trigger degranulation in the absence of any biological or chemical stimulus. NGC from several donors ($n = 6$) was pooled and analyzed for its impact on monocytes regarding the release of TNFα, IL-6, and nucleosomes. To exclude false positive results, the TNFα, IL-6, and DNA contents of NGC were measured as a background control and the values were subtracted from the values of the monocyte samples. A significant increase in TNFα secretion was observed after incubation of the monocytes with NGC for 3 h (Figure 4a). In addition, a slight increase in IL-6 release was detected (Figure 4b). Moreover, the treatment of monocytes with NGC facilitated increased nucleosome release by two-fold (Figure 4b). While NGC co-incubation triggered the release of proinflammatory cytokines and nucleosomes, no ROS production as an indication of cell activation was observed (Figure 4c). Here, comparable to the results detected with LTF alone (Figure 1d), the fluorescent signal appeared to be quenched by the addition of NGC (Figure 4d).

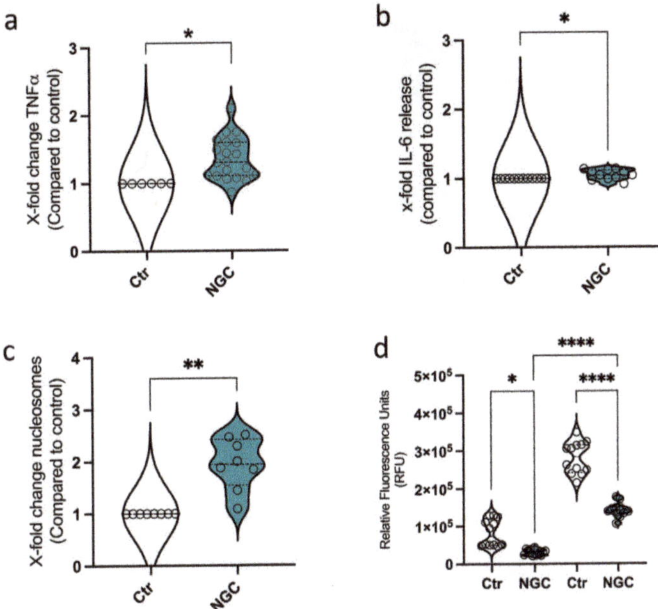

Figure 4. Interaction of primary monocytes with NGC. (**a**). Secretion of TNFα in response to NGC treatment. (**b**). IL-6 release from CD14$^+$ monocytes in response to NGC. (**c**). Release of nucleosomes as an indication for ET formation. (**d**). Analysis of ROS production in CD14$^+$ monocytes in response to NGC. All data represent mean +/− SEM of 4–5 independent experiments. * $p \leq 0.05$, ** $p \leq 0.01$, **** $p \leq 0.0001$.

To understand the impact of the NGC on the monocytes in detail and to extend our analysis beyond LL-37 and LTF, we analyzed the NGC protein fraction by quantitative proteomics mass spectrometry (MS) (Figure 5, Table S1). Moreover, the NGC protein fraction was separated by SDS-PAGE, and the five most prominent bands analyzed by MS as above (Figure 5a, Table S2). In addition to LL-37 and LTF (Figure 5b), we identified 621 proteins in the NGC fraction (Table S1). Based on the SDS-PAGE fractionation (Figure 5a), the most prominent bands are Lactoferrin, albumin, plastin-2, hemoglobin b, and protein S100-A8. The total ion count for LL-37 and LTF based on the quantitative MS analysis are displayed in Figure 5b. Other HDPs found in the NGC included azurocidin (HBP), myeloperoxidase (MPO), histidine-rich glycoprotein (HRG), and matrix metalloproteinase 9 (MMP-9), which are displayed as a selection in Figure 5c with values derived from the TIC-normalized protein abundances.

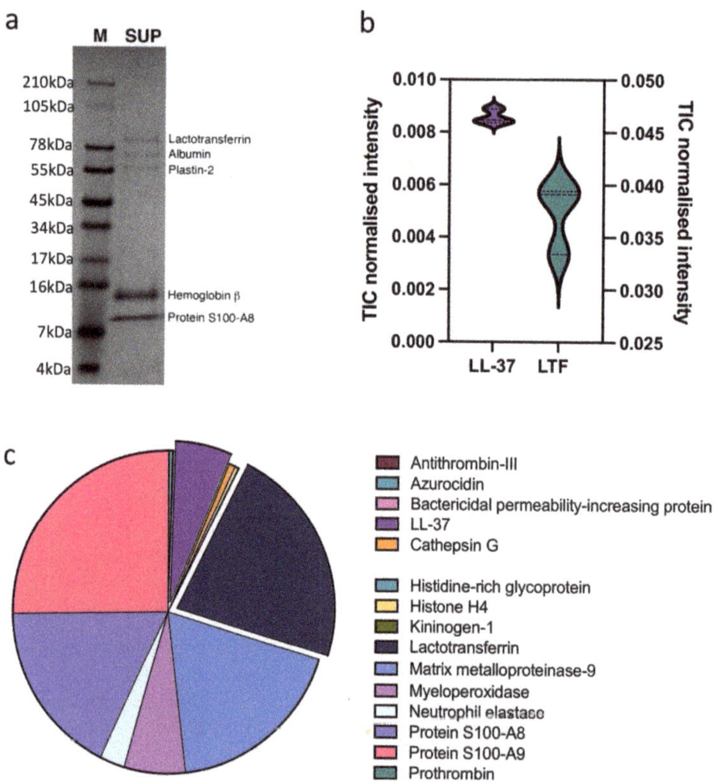

Figure 5. Quantitative MS analysis of NGC. (**a**). SDS PAGE of NGC (=SUP). The most intense gel bands correspond to 1: Lactotransferrin, 2: Albumin, 3: Plastin-2, 4: Hemoglobin β, and 5: Protein S100-A8. (**b**). Total-ion current (TIC) normalized intensity of LL-37 and LTF in NGC. (**c**). Selective representation of various HDPs found in NGC. The proteins are displayed as part of whole, and the size of each protein represents the TIC-normalized abundance. LL-37 and LTF are exploded for easier visualization.

4. Discussion

The present study identified a novel role for HDPs LL-37 and LTF, triggering the release of extracellular traps from CD14$^+$ monocytes. The release of HDPs from granules is an important process of various cell types in immune defense and intercellular communication. Janardhan et al. previously proposed a correlation between neutrophil-associated HDPs and monocyte recruitment [8], which was later confirmed by Soehnlein et al. [12]. In its role as an HDP, LL-37 contributes to innate immunity as it shows a broad range of antimicrobial activity against bacteria, such as Staphylococcus (*S.*) *aureus* and Escherichia (*E.*) *coli* [44]. In addition, it functions as a chemoattractant for monocytes, neutrophils, and other leukocytes, regulating the inflammatory response [45]. Similar to LL-37, LTF serves as a chemoattractant for monocytes [11], but it can also be used as a preventative drug against sepsis in pre-term infants [46]. Moreover, the LTF-derived peptide hLF1–11 displays LPS binding and antimicrobial characteristics while it did not affect host cells at the concentration used in the present study [29]. Autonomous production of TNFα was found to be crucial for monocyte development and survival [47], thus the release of TNFα can be used as marker for the activation of circulating monocytes [48]. In our experiments, we found that LL-37 and LTF significantly elevated the secretion of TNFα from CD14$^+$ monocytes. This contrasts with earlier findings reporting that both HDPs reduced the

LPS-triggered cytokine release [49–51]. Mortality in LPS-challenged mice was dramatically reduced when animals were treated with LTF, which was correlated with reduced TNF levels in murine serum [51]. Nonetheless, in healthy individuals, the regulatory role of LTF results in the spontaneous release of TNF from PBMCs [52].

Enhanced release of TNFα has been associated with the generation of neutrophil ETs [53]. Indeed, we found that monocytes released nucleosomes in response to LL-37 and LTF. LL-37 has previously been shown to facilitate NET formation [16] and is also found within NET structures [36]. An earlier study by Okubo and colleagues suggested that LTF acts as an intrinsic inhibitor of ET formation, since application of this exogenous protein resulted in reduced extracellular DNA fibers. However, it must be noted that in this study, LTF was always added together with PMA [42] while in our experiments, LTF was used alone. Interestingly, compared to PMA, neither HDP address ROS-dependent monocyte activation. ROS-independent pathways have been reported before in neutrophils, e.g., NET induction by *S. aureus* [54] and *Leishmania amazonensis* [55]. Depletion of cholesterol has been observed to induce NET formation [33], allowing the hypothesis that certain membrane alterations might also trigger monocyte activation or render the cells more susceptible to certain stimuli. In addition to this, Arai and colleagues found that uric acid induces ETs without activation of NADPH-oxidase with partial involvement of the NF-κB pathway [56]. Furthermore, the MAPK pathway plays a role in neutrophil ET induction when triggered by PMA, *Helicobacter pylori* or, *Neospora caninum* [57–59]. This diversity in possible pathways might also be adapted to monocyte ET formation [43].

DNA-intercalating dyes are frequently used to quantify ETs [35]. However, as demonstrated by us previously, LL-37 competes with PicoGreen™ for the binding site on DNA [36]. In this study, we hypothesized that a positive signal could be lost due to this competitive interaction. Both LL-37 and LTF are cationic molecules, prone to binding to negatively charged structures. We therefore propose the importance of utilizing additional markers in addition to DNA-intercalating dyes for ET quantification, e.g., histones, neutrophil elastase, or myeloperoxidase.

Store-operated calcium signaling (SOCE) is important in neutrophil activation, being considered as the main entry of calcium into different immune cells [60]. Various studies have shown the involvement of SOCE in the formation of monocyte extracellular traps [57,59,61]. Interaction studies of LL-37 with ion channels have revealed the importance of this HDP on calcium signaling in neutrophils and monocytes [62,63]. For neutrophils, the impact of LTF on calcium mobilization was described earlier [64]; however, in our experiments using CD14+ monocytes, no significant effect was detected.

Protein kinases play a crucial role in the formation of extracellular trap release [38,58,65]. In neutrophils, the treatment of mice with SB203580 showed reduced NET production in the lungs and bronchoalveolar lavage fluid (BALF) of poly I:C-challenged animals [66]. In the presented study, we found that LL-37-triggered monocyte ET formation was significantly decreased by the inhibition of p38 MAP kinase.

PMA is a direct activator of protein kinase C (PKC). Neeli and Radic reported that inhibition of PKC reduced ET formation in neutrophils [67]. PKC in turn activates NADPH oxidases, resulting in the production of ROS in response to PMA. In our experiments, only in PMA-treated monocytes was ROS accumulation observed, leading to the hypothesis that LL-37- and LTF-mediated ET release occurs independently of NADPH oxidases. Activation of PKC by PMA additionally triggers the activation of the Raf-MEK-ERK pathway [58]. Indeed, we showed that in the presence of U0126, PMA- and LL-37- mediated traps are significantly diminished, suggesting that these are dependent on the Raf-MEK-ERK pathway. For both intracellular calcium mobilization and kinase (ERK and p38) activation, the involvement of LL-37 has been demonstrated in human macrophages affecting Leukotriene B4 production [68]. Neither SB203580 nor U0126 (inhibitors of p38 MAPK and Raf-MEK-ERK, respectively) had an effect on LTF-triggered ET formation, even though the involvement of LTF in the ERK signaling pathway has been reported earlier [69].

Cytochalasin D is a potent inhibitor of actin filament cytoskeleton rearrangement during ETosis [70]. The application of CytD up to 15 min after the activation of neutrophils leads to a significant reduction in ET formation, as reported by Neubert and colleagues [71]. In good correlation with the findings of Granger et al., the pre-incubation of the monocytes with CytD had no effect on the PMA- (and HDP-) mediated ET formation. Taken together, our findings agree with previous reports that ET formation triggered by different stimuli occur through various signaling pathways [43].

The predominant focus on the analysis of ET formation utilizes single molecules. It should be considered, however, that purified or recombinant proteins can have different/opposite effects compared with native proteins or a mix of several proteins. Still, when we analyzed the neutrophil granular content, we observed similar effects on the monocytes compared to LL-37 and LTF regarding TNFα, nucleosome, and ROS secretion. A mixture of proteins, such as, for example, in FBS used for cell culture, affects the degradation of Mg differently compared to the single proteins contained in the FBS [72]. While contamination with LPS or other agents cannot be excluded completely, we hypothesize that it is important to consider a more holistic approach to immune cell stimulation. In this regard, it might also be interesting in the future to investigate how a mixture of immune cells, e.g., in whole blood, might affect the outcome compared to a single type of purified cells.

Here, we characterized the effects of the previously described HDPs LL-37 and Lactoferrin, the major NGC constituent identified by quantitative MS analysis, on ETosis. This study, moreover, serves as a repository of other NGC proteins, as we identified more than 600 of these (Table S1). Future studies will extend this data repository to include the protein composition of monocyte ETs triggered by various stimuli, as performed with neutrophils [73,74].

Overall, the results of this study fit the perception of neutrophils being recruiters and activators of monocytes in this complex relationship. The findings of our study suggest that neutrophils' involvement in immunity reaches even further than hitherto known. The generation of monocyte ETs triggered by neutrophil-associated HDPs could constitute a mechanism of amplifying the immune response at the site of infection. Thus, understanding the pathways and mechanisms behind monocyte ET formation still demands additional research.

Supplementary Materials: The following supporting information can be downloaded at: https://www.mdpi.com/article/10.3390/biomedicines10020469/s1; Table S1: Quantitative MS raw data of the NGC in-solution digestion; Table S2: Quantitative MS raw data of the SDS-PAGE fractionated NGC in-gel digestion; Figure S1: Determination of possible cytotoxicity. Analysis of LDH release. Triton X-100 served as positive lysis control.

Author Contributions: Conceptualization, A.N.; methodology, F.V.S., L.H., S.E., A.N.; software, L.H.; validation, F.V.S., L.H. and A.N.; formal analysis, F.V.S., L.H. and A.N.; investigation, F.V.S., L.H. and A.N.; resources, F.V.S., L.H. and A.N.; data curation, L.H. and A.N.; writing—original draft preparation, F.V.S. and A.N.; writing—review and editing, F.V.S., L.H., S.E., A.N.; visualization, F.V.S., L.H. and A.N.; supervision, A.N.; project administration, A.N.; funding acquisition, A.N. All authors have read and agreed to the published version of the manuscript.

Funding: This study was funded by grants from Knut and Alice Wallenberg Foundation (KAW 2016.0023), the Swedish Government Funds for Clinical Research (ALF; ALFSKANE-622011), the Alfred Österlund Foundation, the Royal Physiographic Society of Lund and the Wenner-Gren Foundation.

Institutional Review Board Statement: The study was conducted in accordance with the Declaration of Helsinki and approved by the Ethics Committee of Lund University (ethical permits 2021:10 and 2008/657).

Data Availability Statement: The mass spectrometry data have been deposited to the ProteomeXchange [26] consortium via the MassIVE partner repository (https://massive.ucsd.edu/ (accessed on 30 December 2021)) with the dataset identifier PXD031375.

Acknowledgments: The authors thank Ganna Petruk for help with the protein gel electrophoresis and helpful discussions and Berit Olofsson for isolation of the CD14+ monocytes. We gratefully acknowledge support from the Swedish National Infrastructure for Biological Mass Spectrometry.

Conflicts of Interest: The authors declare no conflict of interest. The funders had no role in the design of the study; in the collection, analyses, or interpretation of data; in the writing of the manuscript, or in the decision to publish the results.

References

1. Borregaard, N.; Lollike, K.; Kjeldsen, L.; Sengeløv, H.; Bastholm, L.; Nielsen, M.H.; Bainton, D.F. Human neutrophil granules and secretory vesicles. *Eur. J. Haematol.* **2009**, *51*, 187–198. [CrossRef] [PubMed]
2. Silvestre-Roig, C.; Hidalgo, A.; Soehnlein, O. Neutrophil heterogeneity: Implications for homeostasis and pathogenesis. *Blood* **2016**, *127*, 2173–2181. [CrossRef] [PubMed]
3. Teh, Y.C.; Ding, J.L.; Ng, L.G.; Chong, S.Z. Capturing the Fantastic Voyage of Monocytes Through Time and Space. *Front. Immunol.* **2019**, *10*, 834. [CrossRef] [PubMed]
4. Chong, S.Z.; Evrard, M.; Goh, C.C.; Ng, L.G. Illuminating the covert mission of mononuclear phagocytes in their regional niches. *Curr. Opin. Immunol.* **2018**, *50*, 94–101. [CrossRef]
5. Thomas, G.D.; Hamers, A.A.; Nakao, C.; Marcovecchio, P.; Taylor, A.M.; McSkimming, C.; Nguyen, A.T.; McNamara, C.A.; Hedrick, C.C. Human Blood Monocyte Subsets. *Arter. Thromb. Vasc. Biol.* **2017**, *37*, 1548–1558. [CrossRef]
6. Kapellos, T.S.; Bonaguro, L.; Gemünd, I.; Reusch, N.; Saglam, A.; Hinkley, E.R.; Schultze, J.L. Human Monocyte Subsets and Phenotypes in Major Chronic Inflammatory Diseases. *Front. Immunol.* **2019**, *10*, 2035. [CrossRef]
7. Boyette, L.B.; Macedo, C.; Hadi, K.; Elinoff, B.D.; Walters, J.; Ramaswami, B.; Chalasani, G.; Taboas, J.M.; Lakkis, F.G.; Metes, D.M. Phenotype, function, and differentiation potential of human monocyte subsets. *PLoS ONE* **2017**, *12*, e0176460. [CrossRef]
8. Kyathanahalli, S.J.; Janardhan, K.S.; Sandhu, S.K.; Singh, B. Neutrophil depletion inhibits early and late monocyte/macrophage increase in lung inflammation. *Front. Biosci.* **2006**, *11*, 1569–1576. [CrossRef]
9. Zhou, J.; Stohlman, S.A.; Hinton, D.R.; Marten, N.W. Neutrophils Promote Mononuclear Cell Infiltration During Viral-Induced Encephalitis. *J. Immunol.* **2003**, *170*, 3331–3336. [CrossRef]
10. Soehnlein, O.; Zernecke, A.; Eriksson, E.E.; Rothfuchs, A.G.; Pham, C.T.; Herwald, H.; Bidzhekov, K.; Rottenberg, M.E.; Weber, C.; Lindbom, L. Neutrophil secretion products pave the way for inflammatory monocytes. *Blood* **2008**, *112*, 1461–1471. [CrossRef]
11. De La Rosa, G.; Yang, D.; Tewary, P.; Varadhachary, A.; Oppenheim, J.J. Lactoferrin Acts as an Alarmin to Promote the Recruitment and Activation of APCs and Antigen-Specific Immune Responses. *J. Immunol.* **2008**, *180*, 6868–6876. [CrossRef]
12. Soehnlein, O.; Weber, C.; Lindbom, L. Neutrophil granule proteins tune monocytic cell function. *Trends Immunol.* **2009**, *30*, 538–546. [CrossRef] [PubMed]
13. Påhlman, L.I.; Mörgelin, M.; Eckert, J.; Johansson, L.; Russell, W.; Riesbeck, K.; Soehnlein, O.; Lindbom, L.; Norrby-Teglund, A.; Schumann, R.R.; et al. Streptococcal M Protein: A Multipotent and Powerful Inducer of Inflammation. *J. Immunol.* **2006**, *177*, 1221–1228. [CrossRef] [PubMed]
14. Wantha, S.; Alard, J.-E.; Megens, R.T.; van der Does, A.M.; Döring, Y.; Drechsler, M.; Pham, C.T.; Wang, M.-W.; Wang, J.-M.; Gallo, R.L.; et al. Neutrophil-Derived Cathelicidin Promotes Adhesion of Classical Monocytes. *Circ. Res.* **2013**, *112*, 792–801. [CrossRef] [PubMed]
15. Soehnlein, O.; Xie, X.; Ulbrich, H.; Kenne, E.; Rotzius, P.; Flodgaard, H.; Eriksson, E.E.; Lindbom, L. Neutrophil-Derived Heparin-Binding Protein (HBP/CAP37) Deposited on Endothelium Enhances Monocyte Arrest under Flow Conditions. *J. Immunol.* **2005**, *174*, 6399–6405. [CrossRef]
16. Neumann, A.; Berends, E.T.M.; Nerlich, A.; Molhoek, E.M.; Gallo, R.; Meerloo, T.; Nizet, V.; Naim, H.Y.; Von Köckritz-Blickwede, M. The antimicrobial peptide LL-37 facilitates the formation of neutrophil extracellular traps. *Biochem. J.* **2014**, *464*, 3–11. [CrossRef]
17. Granger, V.; Faille, D.; Marani, V.; Noël, B.; Gallais, Y.; Szely, N.; Flament, H.; Pallardy, M.; Chollet-Martin, S.; De Chaisemartin, L. Human blood monocytes are able to form extracellular traps. *J. Leukoc. Biol.* **2017**, *102*, 775–781. [CrossRef]
18. Brinkmann, V.; Reichard, U.; Goosmann, C.; Fauler, B.; Uhlemann, Y.; Weiss, D.S.; Weinrauch, Y.; Zychlinsky, A. Neutrophil extracellular traps kill bacteria. *Science* **2004**, *303*, 1532–1535. [CrossRef]
19. Neumann, A.; Brogden, G.; Von Köckritz-Blickwede, M. Extracellular Traps: An Ancient Weapon of Multiple Kingdoms. *Biology* **2020**, *9*, 34. [CrossRef]
20. Webster, S.J.; Daigneault, M.; Bewley, M.A.; Preston, J.A.; Marriott, H.M.; Walmsley, S.R.; Read, R.C.; Whyte, M.K.B.; Dockrell, D.H. Distinct Cell Death Programs in Monocytes Regulate Innate Responses Following Challenge with Common Causes of Invasive Bacterial Disease. *J. Immunol.* **2010**, *185*, 2968–2979. [CrossRef]
21. Pérez, D.; Muñoz, M.; Molina, J.; Muñoz-Caro, T.; Silva, L.; Taubert, A.; Hermosilla, C.; Ruiz, A. Eimeria ninakohlyakimovae induces NADPH oxidase-dependent monocyte extracellular trap formation and upregulates IL-12 and TNF-α, IL-6 and CCL2 gene transcription. *Veter. Parasitol.* **2016**, *227*, 143–150. [CrossRef] [PubMed]
22. Muñoz-Caro, T.; Silva, L.M.R.; Ritter, C.; Taubert, A.; Hermosilla, C. Besnoitia besnoiti tachyzoites induce monocyte extracellular trap formation. *Parasitol. Res.* **2014**, *113*, 4189–4197. [CrossRef]

23. Halder, L.D.; Abdelfatah, M.A.; Jo, E.A.H.; Jacobsen, I.D.; Westermann, M.; Beyersdorf, N.; Lorkowski, S.; Zipfel, P.F.; Skerka, C. Factor H Binds to Extracellular DNA Traps Released from Human Blood Monocytes in Response to Candida albicans. *Front. Immunol.* **2017**, *7*, 671. [CrossRef]

24. Neumann, A.; Björck, L.; Frick, I.-M. Finegoldia magna, an Anaerobic Gram-Positive Bacterium of the Normal Human Microbiota, Induces Inflammation by Activating Neutrophils. *Front. Microbiol.* **2020**, *11*, 65. [CrossRef] [PubMed]

25. Shevchenko, A.; Tomas, H.; Havlis, J.; Olsen, J.; Mann, M.J. In-gel digestion for mass spectrometric characterization of proteins and proteomes. *Nat. Protoc.* **2007**, *1*, 2856–2860. [CrossRef] [PubMed]

26. Vizcaíno, J.A.; Deutsch, E.W.; Wang, R.; Csordas, A.; Reisinger, F.; Ríos, D.; Dianes, J.A.; Sun, Z.; Farrah, T.; Bandeira, N.; et al. ProteomeXchange provides globally coordinated proteomics data submission and dissemination. *Nat. Biotechnol.* **2014**, *32*, 223–226. [CrossRef] [PubMed]

27. Prame Kumar, K.; Nicholls, A.J.; Wong, C.H.Y. Partners in crime: Neutrophils and monocytes/macrophages in inflammation and disease. *Cell Tissue Res.* **2018**, *371*, 551–565. [CrossRef] [PubMed]

28. Hemshekhar, M.; Choi, K.-Y.G.; Mookherjee, N. Host Defense Peptide LL-37-Mediated Chemoattractant Properties, but Not Anti-Inflammatory Cytokine IL-1RA Production, Is Selectively Controlled by Cdc42 Rho GTPase via G Protein-Coupled Receptors and JNK Mitogen-Activated Protein Kinase. *Front. Immunol.* **2018**, *9*, 1871. [CrossRef] [PubMed]

29. Stallmann, H.P.; Faber, C.; Bronckers, A.L.; de Blieck-Hogervorst, J.M.; Brouwer, C.P.; Amerongen, A.V.N.; Wuisman, P.I. Histatin and lactoferrin derived peptides: Antimicrobial properties and effects on mammalian cells. *Peptides* **2005**, *26*, 2355–2359. [CrossRef] [PubMed]

30. Lee, T.D.; Gonzalez, M.L.; Kumar, P.; Grammas, P.; Pereira, H. CAP37, a neutrophil-derived inflammatory mediator, augments leukocyte adhesion to endothelial monolayers. *Microvasc. Res.* **2003**, *66*, 38–48. [CrossRef]

31. Guimaraes-Costa, A.B.; Nascimento, M.T.C.; Froment, G.S.; Soares, R.P.P.; Morgado, F.N.; Conceicao-Silva, F.; Saraiva, E.M. Leishmania amazonensis promastigotes induce and are killed by neutrophil extracellular traps. *Proc. Natl. Acad. Sci. USA* **2009**, *106*, 6748–6753. [CrossRef]

32. Fuchs, T.A.; Abed, U.; Goosmann, C.; Hurwitz, R.; Schulze, I.; Wahn, V.; Weinrauch, Y.; Brinkmann, V.; Zychlinsky, A. Novel cell death program leads to neutrophil extracellular traps. *J. Cell Biol.* **2007**, *176*, 231–241. [CrossRef] [PubMed]

33. Neumann, A.; Brogden, G.; Jerjomiceva, N.; Brodesser, S.; Naim, H.Y.; von Köckritz-Blickwede, M. Lipid alterations in human blood-derived neutrophils lead to formation of neutrophil extracellular traps. *Eur. J. Cell Biol.* **2014**, *93*, 347–354. [CrossRef]

34. Ponath, V.; Kaina, B. Death of Monocytes through Oxidative Burst of Macrophages and Neutrophils: Killing in Trans. *PLoS ONE* **2017**, *12*, e0170347. [CrossRef]

35. Carmona-Rivera, C.; Kaplan, M.J. Induction and Quantification of NETosis. *Curr. Protoc. Immunol.* **2016**, *115*, 14–41. [CrossRef] [PubMed]

36. Neumann, A.; Völlger, L.; Berends, E.T.; Molhoek, E.M.; Stapels, D.A.; Midon, M.; Friães, A.; Pingoud, A.; Rooijakkers, S.H.; Gallo, R.; et al. Novel Role of the Antimicrobial Peptide LL-37 in the Protection of Neutrophil Extracellular Traps against Degradation by Bacterial Nucleases. *J. Innate Immun.* **2014**, *6*, 860–868. [CrossRef] [PubMed]

37. Alarcón, P.; Manosalva, C.; Conejeros, I.; Carretta, M.D.; Muñoz-Caro, T.; Silva, L.M.R.; Taubert, A.; Hermosilla, C.; Hidalgo, M.A.; Burgos, R.A. d(−) Lactic Acid-Induced Adhesion of Bovine Neutrophils onto Endothelial Cells Is Dependent on Neutrophils Extracellular Traps Formation and CD11b Expression. *Front. Immunol.* **2017**, *8*, 975. [CrossRef]

38. Muñoz-Caro, T.; Mena Huertas, S.J.; Conejeros, I.; Alarcon, P.; Hidalgo, M.A.; Burgos, R.A.; Hermosilla, C.; Taubert, A. Eimeria bovis-triggered neutrophil extracellular trap formation is CD11b-, ERK 1/2-, p38 MAP kinase- and SOCE-dependent. *Vet. Res.* **2015**, *46*, 23. [CrossRef]

39. Behrendt, J.H.; Ruiz, A.; Zahner, H.; Taubert, A.; Hermosilla, C. Neutrophil extracellular trap formation as innate immune reactions against the apicomplexan parasite Eimeria bovis. *Veter- Immunol. Immunopathol.* **2010**, *133*, 1–8. [CrossRef]

40. Han, Z.; Zhang, Y.; Wang, C.; Liu, X.; Jiang, A.; Liu, Z.; Wang, J.; Yang, Z.; Wei, Z. Ochratoxin A-Triggered Chicken Heterophil Extracellular Traps Release through Reactive Oxygen Species Production Dependent on Activation of NADPH Oxidase, ERK, and p38 MAPK Signaling Pathways. *J. Agric. Food Chem.* **2019**, *67*, 11230–11235. [CrossRef]

41. Clerk, A.; Sugden, P.H. The p38-MAPK inhibitor, SB203580, inhibits cardiac stress-activated protein kinases/c-Jun N-terminal kinases (SAPKs/JNKs). *FEBS Lett.* **1998**, *426*, 93–96. [CrossRef]

42. Okubo, K.; Kamiya, M.; Urano, Y.; Nishi, H.; Herter, J.M.; Mayadas, T.; Hirohama, D.; Suzuki, K.; Kawakami, H.; Tanaka, M.; et al. Lactoferrin Suppresses Neutrophil Extracellular Traps Release in Inflammation. *eBioMedicine* **2016**, *10*, 204–215. [CrossRef] [PubMed]

43. Kenny, E.F.; Herzig, A.; Krüger, R.; Muth, A.; Mondal, S.; Thompson, P.R.; Brinkmann, V.; Von Bernuth, H.; Zychlinsky, A. Diverse stimuli engage different neutrophil extracellular trap pathways. *eLife* **2017**, *6*, e24437. [CrossRef]

44. Turner, J.; Cho, Y.; Dinh, N.-N.; Waring, A.J.; Lehrer, R.I. Activities of LL-37, a Cathelin-Associated Antimicrobial Peptide of Human Neutrophils. *Antimicrob. Agents Chemother.* **1998**, *42*, 2206–2214. [CrossRef] [PubMed]

45. Dürr, U.H.; Sudheendra, U.; Ramamoorthy, A. LL-37, the only human member of the cathelicidin family of antimicrobial peptides. *Biochim. et Biophys. Acta (BBA) Biomembr.* **2006**, *1758*, 1408–1425. [CrossRef]

46. Gao, Y.; Hou, L.; Lu, C.; Wang, Q.; Pan, B.; Wang, Q.; Tian, J.; Ge, L. Enteral Lactoferrin Supplementation for Preventing Sepsis and Necrotizing Enterocolitis in Preterm Infants: A Meta-Analysis with Trial Sequential Analysis of Randomized Controlled Trials. *Front. Pharmacol.* **2020**, *11*, 1186. [CrossRef] [PubMed]

47. Wolf, Y.; Shemer, A.; Polonsky, M.; Gross, M.; Mildner, A.; Yona, S.; David, E.; Kim, K.-W.; Goldmann, T.; Amit, I.; et al. Autonomous TNF is critical for in vivo monocyte survival in steady state and inflammation. *J. Exp. Med.* **2017**, *214*, 905–917. [CrossRef] [PubMed]

48. Jovinge, S.; Ares, M.P.; Kallin, B.; Nilsson, J. Human Monocytes/Macrophages Release TNF-α in Response to Ox-LDL. *Arter. Thromb. Vasc. Biol.* **1996**, *16*, 1573–1579. [CrossRef] [PubMed]

49. Brown, K.; Poon, G.F.T.; Birkenhead, D.; Pena, O.M.; Falsafi, R.; Dahlgren, C.; Karlsson, A.; Bylund, J.; Hancock, R.; Johnson, P. Host Defense Peptide LL-37 Selectively Reduces Proinflammatory Macrophage Responses. *J. Immunol.* **2011**, *186*, 5497–5505. [CrossRef] [PubMed]

50. Alalwani, S.M.; Sierigk, J.; Herr, C.; Pinkenburg, O.; Gallo, R.; Vogelmeier, C.; Bals, R. The antimicrobial peptide LL-37 modulates the inflammatory and host defense response of human neutrophils. *Eur. J. Immunol.* **2010**, *40*, 1118–1126. [CrossRef]

51. Zhang, G.-H.; Mann, D.M.; Tsai, C.-M. Neutralization of Endotoxin In Vitro and In Vivo by a Human Lactoferrin-Derived Peptide. *Infect. Immun.* **1999**, *67*, 1353–1358. [CrossRef] [PubMed]

52. Lactoferrin in Health and Disease-PubMed. Available online: https://pubmed.ncbi.nlm.nih.gov/17507874/ (accessed on 2 February 2022).

53. Oyarzún, C.P.M.; Carestia, A.; Lev, P.R.; Glembotsky, A.C.; Ríos, M.A.C.; Moiraghi, B.; Molinas, F.C.; Marta, R.F.; Schattner, M.; Heller, P.G. Neutrophil extracellular trap formation and circulating nucleosomes in patients with chronic myeloproliferative neoplasms. *Sci. Rep.* **2016**, *6*, 38738. [CrossRef] [PubMed]

54. Pilsczek, F.H.; Salina, D.; Poon, K.K.H.; Fahey, C.; Yipp, B.G.; Sibley, C.D.; Robbins, S.M.; Green, F.H.Y.; Surette, M.G.; Sugai, M.; et al. A Novel Mechanism of Rapid Nuclear Neutrophil Extracellular Trap Formation in Response to *Staphylococcus aureus*. *J. Immunol.* **2010**, *185*, 7413–7425. [CrossRef] [PubMed]

55. Rochael, N.C.; Guimarães-Costa, A.B.; Nascimento, M.T.C.; DeSouza-Vieira, T.S.; De Oliveira, M.P.; Souza, L.F.G.E.; Oliveira, M.F.; Saraiva, E.M. Classical ROS-dependent and early/rapid ROS-independent release of Neutrophil Extracellular Traps triggered by Leishmania parasites. *Sci. Rep.* **2015**, *5*, 18302. [CrossRef] [PubMed]

56. Arai, Y.; Nishinaka, Y.; Arai, T.; Morita, M.; Mizugishi, K.; Adachi, S.; Takaori-Kondo, A.; Watanabe, T.; Yamashita, K. Uric acid induces NADPH oxidase-independent neutrophil extracellular trap formation. *Biochem. Biophys. Res. Commun.* **2014**, *443*, 556–561. [CrossRef] [PubMed]

57. Villagra-Blanco, R.; Silva, L.; Gärtner, U.; Wagner, H.; Failing, K.; Wehrend, A.; Taubert, A.; Hermosilla, C. Molecular analyses on Neospora caninum -triggered NETosis in the caprine system. *Dev. Comp. Immunol.* **2017**, *72*, 119–127. [CrossRef] [PubMed]

58. Hakkim, A.; Fuchs, T.A.; Martinez, N.E.; Hess, S.; Prinz, H.; Zychlinsky, A.; Waldmann, H. Activation of the Raf-MEK-ERK pathway is required for neutrophil extracellular trap formation. *Nat. Chem. Biol.* **2011**, *7*, 75–77. [CrossRef]

59. Wei, Z.; Hermosilla, C.; Taubert, A.; He, X.; Wang, X.; Gong, P.; Li, J.; Yang, Z.; Zhang, X. Canine Neutrophil Extracellular Traps Release Induced by the Apicomplexan Parasite Neospora caninum In Vitro. *Front. Immunol.* **2016**, *7*, 436. [CrossRef]

60. Clemens, R.A.; Lowell, C.A. Store-operated calcium signaling in neutrophils. *J. Leukoc. Biol.* **2015**, *98*, 497–502. [CrossRef]

61. Villagra-Blanco, R.; Silva, L.M.R.; Muñoz-Caro, T.; Yang, Z.; Li, J.; Gärtner, U.; Taubert, A.; Zhang, X.; Hermosilla, C. Bovine Polymorphonuclear Neutrophils Cast Neutrophil Extracellular Traps against the Abortive Parasite Neospora caninum. *Front. Immunol.* **2017**, *8*, 606. [CrossRef]

62. Zhang, Z.; Cherryholmes, G.; Chang, F.; Rose, D.M.; Schraufstatter, I.; Shively, J.E. Evidence that cathelicidin peptide LL-37 may act as a functional ligand for CXCR2 on human neutrophils. *Eur. J. Immunol.* **2009**, *39*, 3181–3194. [CrossRef] [PubMed]

63. Tomasinsig, L.; Pizzirani, C.; Skerlavaj, B.; Pellegatti, P.; Gulinelli, S.; Tossi, A.; Di Virgilio, F.; Zanetti, M. The Human Cathelicidin LL-37 Modulates the Activities of the P2X7 Receptor in a Structure-dependent Manner. *J. Biol. Chem.* **2008**, *283*, 30471–30481. [CrossRef] [PubMed]

64. Grigorieva, D.V.; Gorudko, I.V.; Shamova, E.V.; Terekhova, M.S.; Maliushkova, E.V.; Semak, I.V.; Cherenkevich, S.N.; Sokolov, A.V.; Timoshenko, A.V. Effects of recombinant human lactoferrin on calcium signaling and functional responses of human neutrophils. *Arch. Biochem. Biophys.* **2019**, *675*, 108122. [CrossRef] [PubMed]

65. Guo, R.; Tu, Y.; Xie, S.; Liu, X.S.; Song, Y.; Wang, S.; Chen, X.; Lu, L. A Role for Receptor-Interacting Protein Kinase-1 in Neutrophil Extracellular Trap Formation in Patients with Systemic Lupus Erythematosus: A Preliminary Study. *Cell. Physiol. Biochem.* **2018**, *45*, 2317–2328. [CrossRef] [PubMed]

66. Gan, T.; Yang, Y.; Hu, F.; Chen, X.; Zhou, J.; Li, Y.; Xu, Y.; Wang, H.; Chen, Y.; Zhang, M. TLR3 Regulated Poly I:C-Induced Neutrophil Extracellular Traps and Acute Lung Injury Partly Through p38 MAP Kinase. *Front. Microbiol.* **2018**, *9*, 3174. [CrossRef]

67. Neeli, I.; Radic, M. Opposition between PKC isoforms regulates histone deimination and neutrophil extracellular chromatin release. *Front. Immunol.* **2013**, *4*, 38. [CrossRef] [PubMed]

68. Wan, M.; Soehnlein, O.; Tang, X.; van der Does, A.M.; Smedler, E.; Uhlén, P.; Lindbom, L.; Agerberth, B.; Haeggström, J.Z. Cathelicidin LL-37 induces time-resolved release of LTB 4 and TXA 2 by human macrophages and triggers eicosanoid generation in vivo. *FASEB J.* **2014**, *28*, 3456–3467. [CrossRef]

69. Ikoma-Seki, K.; Nakamura, K.; Morishita, S.; Ono, T.; Sugiyama, K.; Nishino, H.; Hirano, H.; Murakoshi, M. Role of LRP1 and ERK and cAMP Signaling Pathways in Lactoferrin-Induced Lipolysis in Mature Rat Adipocytes. *PLoS ONE* **2015**, *10*, e0141378. [CrossRef]

70. Neeli, I.; Dwivedi, N.; Khan, S.; Radic, M. Regulation of Extracellular Chromatin Release from Neutrophils. *J. Innate Immun.* **2009**, *1*, 194–201. [CrossRef]

71. Neubert, E.; Meyer, D.; Rocca, F.; Günay, G.; Kwaczala-Tessmann, A.; Grandke, J.; Senger-Sander, S.; Geisler, C.; Egner, A.; Schön, M.P.; et al. Chromatin swelling drives neutrophil extracellular trap release. *Nat. Commun.* **2018**, *9*, 1–13. [CrossRef]
72. Hou, R.-Q.; Scharnagl, N.; Willumeit, R.; Feyerabend, F. Different effects of single protein vs. protein mixtures on magnesium degradation under cell culture conditions. *Acta Biomater.* **2019**, *98*, 256–268. [CrossRef] [PubMed]
73. Urban, C.F.; Ermert, D.; Schmid, M.; Abu-Abed, U.; Goosmann, C.; Nacken, W.; Brinkmann, V.; Jungblut, P.R.; Zychlinsky, A. Neutrophil Extracellular Traps Contain Calprotectin, a Cytosolic Protein Complex Involved in Host Defense against Candida albicans. *PLoS Pathog.* **2009**, *5*, e1000639. [CrossRef] [PubMed]
74. Thammavongsa, V.; Missiakas, D.M.; Schneewind, O. Staphylococcus aureus Degrades Neutrophil Extracellular Traps to Promote Immune Cell Death. *Science* **2013**, *342*, 863–866. [CrossRef] [PubMed]

 *biomedicines*

Article

In Vivo Evaluation of ECP Peptide Analogues for the Treatment of *Acinetobacter baumannii* Infection

Jiarui Li [1], Guillem Prats-Ejarque [1], Marc Torrent [1], David Andreu [2], Klaus Brandenburg [3], Pablo Fernández-Millán [1] and Ester Boix [1,*]

[1] Department of Biochemistry and Molecular Biology, Faculty of Biosciences, Universitat Autònoma de Barcelona (UAB), Cerdanyola del Valles, 08193 Bellaterra, Spain; jiarui.li@e-campus.uab.cat (J.L.); Guillem.Prats.Ejarque@uab.cat (G.P.-E.); Marc.Torrent@uab.cat (M.T.); Pablo.Fernandez@uab.cat (P.F.-M.)

[2] Barcelona Biomedical Research Park, Department of Experimental and Health Sciences, Universitat Pompeu Fabra, 08002 Barcelona, Spain; david.andreu@upf.edu

[3] Brandenburg Antiinfektiva GmbH, c/o Forschungszentrum Borstel, 23845 Sülfeld, Germany; kbranden@gmx.de

* Correspondence: Ester.Boix@uab.cat

Abstract: Antimicrobial peptides (AMPs) are alternative therapeutics to traditional antibiotics against bacterial resistance. Our previous work identified an antimicrobial region at the N-terminus of the eosinophil cationic protein (ECP). Following structure-based analysis, a 30mer peptide (ECPep-L) was designed that combines antimicrobial action against Gram-negative species with lipopolysaccharides (LPS) binding and endotoxin-neutralization activities. Next, analogues that contain non-natural amino acids were designed to increase serum stability. Here, two analogues were selected for in vivo assays: the all-D version (ECPep-D) and the Arg to Orn version that incorporates a D-amino acid at position 2 (ECPep-2D-Orn). The peptide analogues retained high LPS-binding and anti-endotoxin activities. The peptides efficacy was tested in a murine acute infection model of *Acinetobacter baumannii*. Results highlighted a survival rate above 70% following a 3-day supervision with a single administration of ECPep-D. Moreover, in both ECPep-D and ECPep-2D-Orn peptide-treated groups, clinical symptoms improved significantly and the tissue infection was reduced to equivalent levels to mice treated with colistin, used as a last resort in the clinics. Moreover, treatment drastically reduced serum levels of TNF-α inflammation marker within the first 8 h. The present results support ECP-derived peptides as alternative candidates for the treatment of acute infections caused by Gram-negative bacteria.

Keywords: ECP; AMPs; infection; murine model; Gram-negative bacteria; LPS

Citation: Li, J.; Prats-Ejarque, G.; Torrent, M.; Andreu, D.; Brandenburg, K.; Fernández-Millán, P.; Boix, E. In Vivo Evaluation of ECP Peptide Analogues for the Treatment of *Acinetobacter baumannii* Infection. *Biomedicines* 2022, 10, 386. https://doi.org/10.3390/biomedicines10020386

Academic Editor: Jitka Petrlova

Received: 28 December 2021
Accepted: 2 February 2022
Published: 5 February 2022

Publisher's Note: MDPI stays neutral with regard to jurisdictional claims in published maps and institutional affiliations.

1. Introduction

It is now nearly one century since the landmark discovery of penicillin. Unfortunately, the emergence of antimicrobial resistance (AMR) is demanding the development of novel antibiotics [1]. Among hundreds of active molecules, we cannot disregard the potential of biomacromolecules such as antimicrobial peptides (AMPs), which are endowed with some unique properties distinct from small molecules, widely used in the drug development field [2–4].

AMPs are peptides that can be derived from organisms of all kingdoms and display a variety of functions. Bacteria produce AMPs to fight against other bacteria while animals can take advantage of AMPs to protect themselves. AMPs, and in particular cationic AMPs, the most populated group, have common properties and mechanisms because of their physicochemical nature. In particular, AMPs will exert their roles either by disrupting the negatively charged bacterial membrane or by targeting intracellular components within bacteria cells. Some native proteins or peptides are also able to regulate innate immunity, a feature that provides them a selective superiority over traditional antibiotics [5]. In addition, AMPs can contribute to the fight against multidrug-resistant (MDR) bacteria as well as biofilm communities [6–8]. Additionally, synergism between antibiotics and AMPs has been slowly identified [9,10]. Therefore, AMPs have a great potential for drug development. Recently, thanks to novel methodologies, many natural or synthetic AMPs have been proposed as antimicrobial drug candidates [11].

However, some drawbacks remain to be solved, such as their unpredictable toxicity or bio-stability, which can hinder their unconditional applicability. In particular, many discovered AMPs are only effective in vitro due to their lower stability in vivo. To overcome this issue, unnatural amino acid or non-coded amino acids have been introduced along the whole AMP sequence or at the sites sensitive to proteases [12–15]. Interestingly, AMP with L-amino acids replaced by D-amino acids can sometimes not only be highly resistant to proteolysis but also have better antibacterial activity than the L-version [16]. On the other hand, natural AMPs may have a too-long sequence, which will increase the cost of production. Thus, it is a good option to find active fragments from original AMPs and design shorter peptides that retain the antibacterial activity [17,18].

Despite some putative drawbacks that might remain to be addressed, many AMPs are already in clinical or preclinical trials. However, to date, most AMPs under clinical trial are only for topical application and the number of AMPs for systemic administration against drug-resistant bacteria are still a minority [18–20]. One representative AMP category is polymyxins, once discarded because of nephrotoxicity, but now reconsidered as the last resort for the treatment of infection caused by multi-resistant strains [21]. It is worth noting that polymyxins as non-ribosomal peptides contain non-standard amino acids and are, therefore, protected against proteolysis. In particular, polymyxin E (colistin), the most frequently used in the clinics, has a D-Leu in its sequence at position 6 [22]. It is important to highlight that colistin is produced in vivo by non-ribosomal peptide synthetase (NRPS) in which D-amino acids are incorporated through the action of the epimerization domain [23].

Our laboratory has been long-standing working on the structure-function of secretory ribonucleases (RNases) that belong to the RNase A superfamily, a vertebrate-specific family with 13 members identified in humans with diverse biological functions, including antimicrobial activity, immune modulation and host defense [9,24,25]. Among them, RNase3, also known as the eosinophil cationic protein (ECP), stands out as the family member with the highest antimicrobial activity, where an essential domain was identified at its N-terminal, with LPS-binding, cell-agglutinating and anti-biofilm properties [26].

Considering the outstanding performance of ECP native protein and derived peptides in vitro, we designed new ECP peptide analogues based on the 30mer lead template [27], that incorporate non-natural amino acids to limit proteolysis in human serum [28]. Next, two peptides were selected for further characterization and in vivo studies. The first is the all-D version of the ECPep-L lead peptide (ECPep-D) [27], where all L-amino acids in the sequence have been replaced by D-amino acids. The second peptide incorporates Orn substitutions for Arg residues and includes a D-amino acid at position 2 (ECPep-2D-Orn), based on the previous identification of ECP peptide main proteolysis target sites [28]. In addition, recent work confirmed that Arg to Orn replacement ensured the peptide stability in serum in vitro up to more than 8 h and retained its antimicrobial properties [28].

In the present work, we explored the efficacy of the two peptides analogues (ECPep-D and ECPep-2D-Orn) in a murine infection model. Based on our previous work, in vitro characterization against representative Gram-negative bacteria species, we selected *Acinetobacter baumannii* (*A. baumannii*) [29,30], which belongs to the ESKAPE top priority WHO pathogen list, as our working reference for the present in vivo murine infection model. To note, AMPs are currently regarded as one of the most promising candidates to treat *A. baumannii* infections.

2. Materials and Methods

2.1. Bacteria Strain, Cells, Mice and Other Materials

A. baumannii strain (CECT 452, Valencia, Spain, ATCC 15308, Manassas, VA, USA) and *Pseudomonas aeruginosa* (*P. aeruginosa*) strain (CECT 4122, Valencia, Spain, ATCC 15692,

Manassas, VA, USA) were from the Colección Española de Cultivos Tipo (CECT, Valencia, Spain). The *Escherichia coli* (*E. coli*) BL21 (DE3) strain was from Novagen (Darmstadt, Germany). MRC-5 cells were from the American Type Culture Collection (ATCC, Manassas, VA, USA) and HEK293T cell line was kindly provided by Dr. Raquel Pequerul (UAB, Barcelona, Spain).

Balb/c mice (9–12 weeks old, 20–30 g, male and female, were supplied by Charles Rivers Laboratories and the study procedures have been approved by the Animal and Human Experimentation Ethics Committee at Universitat Autònoma de Barcelona (UAB, Barcelona, Spain). The animals were acclimated at least 5 days from arrival until the start of the experiment with free diet and water provided.

Lipopolysaccharides (LPS) from *E. coli* O111:B4 (O-LPS), EH100 (Ra mutant), and J5 (Rc mutant), porcine mucin, Mueller Hinton broth (MHB) and Thiazolyl Blue Tetrazolium Blue (MTT) were purchased at Sigma-Aldrich (Saint Louis, MO, USA). Colistin sulphate and Dimethyl sulfoxide (DMSO) were from Apollo Scientific (Stockport, UK). BODIPY® TR cadaverine (BC) were from Molecular Probes (Eugene, OR, USA). Mouse TNF-α ELISA Kit (MBS175787, MyBioSource, San Diego, CA, USA) and Mouse LPS ELISA Kit (MBS7700668, MyBioSource, San Diego, CA, USA) were from bioNova científica s.l. (Madrid, Spain) Human TNF-α ELISA set was from BD Biosciences (555212, BD OptEIA™, San Jose, CA, USA). Aspidasept® (pep19-2.5) is an antibacterial/anti-inflammatory peptide drug (GCKKYRRFRWKFKGKFWFWG, with C-terminal amide) [31].

2.2. Peptides Synthesis

Peptides were synthesized at the Universitat Pompeu Fabra (UPF, Barcelona, Spain) Peptide Synthesis Service as previously described [27]. Briefly, peptides were assembled in C-terminal carboxamide form on an H-Rink Amide-Chem Matrix resin using Fmoc solid-phase peptide synthesis (SPPS) protocols. After chain assembly, peptides were fully deprotected and cleaved from the resin with TFA/H_2O/Triisopropylsilane. Peptides were precipitated from the TFA solution by addition of chilled diethyl ether followed by three centrifugations at 4800 rpm, 5 min, 4 °C, taken up in water and lyophilized. Crude peptides were checked by analytical RP-HPLC and LC-MS and purified by preparative RP-HPLC as previously detailed [27]. Fractions of >95% HPLC purity and with the expected mass by LC-MS were pooled and lyophilized. Peptide purity was assessed by the area of the purified peptide peak relative to the total peak areas in the chromatogram. Peptide stock solutions were prepared in sterile deionized water and stored at −20 °C.

2.3. Circular Dichroism Assay

Far-UV CD spectra were obtained from a Jasco-715 (Jasco), as previously described [32]. The spectra were registered from 190 to 240 nm at room temperature. Data from four consecutive scans were averaged. Before reading, the sample was centrifuged at $10,000 \times g$ for 5 min. Peptide spectra were obtained at 16 µM in 5 mM Tris, pH 7.4 and 1 mM SDS, with a 0.2 cm path-length quartz cuvette.

2.4. Minimum Inhibitory Concentration (MIC) Determination

The MIC is defined as the lowest concentration of one reagent to prevent visible growth of bacteria. The MIC determination was followed by the protocol described previously [33]. Briefly, bacteria in exponential growth were used to prepare a suspension in MHB with the approximate number of 5×10^5 CFU/mL. Next, an aliquot of 90 µL of bacteria suspension was added into each well of the 96-well polypropylene plate. Immediately, 10 µL of peptide diluted by 0.01% acetic acid was added to the corresponding well to have a final concentration ranging from 20 to 0.16 µM. The plate was incubated for 24 h at 37 °C, 100 rpm. The presence or absence of bacterial growth was visually inspected and confirmed by reading OD_{600} with Tecan Microplate Reader Spark®. Each test was performed in triplicate.

2.5. LPS Affinity Assay

The LPS affinity was assessed using the fluorescent probe BC by an adaptation of the displacement assay reported [27,34]. Peptides and colistin were serially diluted in a 96-well fluorescence plate from 20 to 0.16 µM in Tris/HCl 50 mM PH 7.4. Next, LPS and BC diluted in the same buffer were added to have the final concentration of 50 µg/mL and 5 µM, respectively. Fluorescence measurements were performed on Tecan Microplate Reader Spark® at 580 nm of excitation wavelength and 620 nm of the emission wavelength at 5 nm for optimal gain. Each test was performed in triplicate.

2.6. Cytotoxicity Assay

Cytotoxicity was measured for the MRC-5 and HEK293T cell lines by MTT assay [35]. Cells were grown in 5% CO_2 at 37 °C with MRC-5 cell line maintained in MEMα and HEK293T in DMEM/F-12 media, both of which were supplemented with 10% FBS. The cells were passaged in 25 cm^2 or 75 cm^2 flasks to prepare 96-well plates with 3×10^4 cells/well, which were incubated overnight. Next, serial dilutions of ECPep-D and ECPep-2D-Orn were added to have a final concentration ranging from 100 to 0.78 µM (375 µg/mL to 3 µg/mL) and colistin was added at final concentrations ranging from 300 to 2.34 µM (380 µg/mL to 3 µg/mL) in corresponding wells. After incubation to the specified time, the medium of the plate was replaced by fresh medium containing 0.5 mg/mL MTT solution and the mixture was incubated for 2.5 h in 5% CO_2 at 37 °C. The medium was then removed, and formazan was dissolved by adding DMSO. The optical density (OD) was recorded by using a Victor^3 plate reader (PerkinElmer, Waltham, MA, USA) set at 540 nm and 620 nm as a reference. Each test was performed in triplicate.

2.7. Tolerance Study of Peptides in Mouse

Initially, the "Up and Down" protocol was followed to determine the maximum lethal dose with some modifications. Briefly, one mouse was taken per peptide and increasing doses of the peptide were injected through the intraperitoneal (*i.p.*) route every 48 h, and the concentration at which the mouse had severe clinical symptoms or died was determined (Figure 1a).

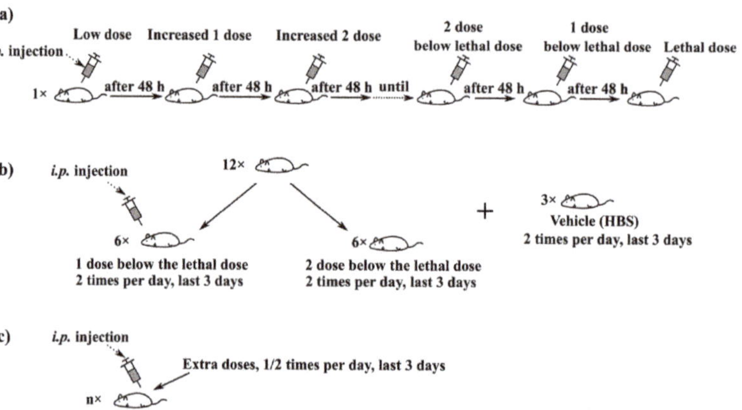

Figure 1. Toxicity study plans. (**a**) "Up and Down" protocol: 1 mouse was injected, starting at a very low peptide dose and observed at 48 h intervals. If the mouse showed no affected signs, with the same mouse, the dose was slowly increased every 48 h until it reached the lethal dose for this mouse. (**b**) Main study: mice were divided into groups randomly; 1 as well as 2 peptide doses below the lethal dose from the "Up and Down" assay was given two times per day for 3 days (the protocol was readjusted in case of emergency owing to ethical principles, as detailed in the results section and Table S1). An extra group is for the vehicle. (**c**) Last study: Peptide was tested in smaller groups at lower doses depending on the results obtained from the main study.

Once the lethal concentration had been determined, we continued with the "main study" (Figure 1b). Two groups of 6 female mice were used at two different concentrations of both peptides, one and two concentrations below the maximum considered either by lethality or by clinical signs. The toxicity was further verified by mimicking the doses and frequency foreseen in the previous study, i.e., two doses per day for 3 days as the initial plan. In all cases, the trials were stopped or readjusted owing to ethical principles when obvious suffering and sudden death of mice were observed in any of these groups.

There was also an extra group to confirm the non-toxicity of the vehicle. For extra caution, some additional doses were applied to confirm the safety dose in case of repeated injections (Figure 1c).

2.8. Mouse Systemic Infection Model

An aliquot from an *A. baumannii* stock stored in 15% glycerol at $-80\ ^\circ$C was taken to culture in 3 mL LB overnight at 250 rpm, 37 $^\circ$C. Next, OD_{600} reading of *A. baumannii* overnight culture was performed to adjust to the number of bacteria required (CFU/mL) and then diluted by half with the same volume of 10% porcine mucin, to obtain a solution of selected CFU/mL in 5% mucin for mice inoculation. Mucin was added to enhance the infectivity of *A. baumannii*, as previously established [36]. The inoculating volume was decided based on the body weight of each animal (10 mL/kg).

Following this, the evolution of the animals was supervised for 48 h by evaluating clinical signs and body weight as detailed in Section 2.10 (Table S2). When the sum of the scores reached a maximum value of 7, the mice were euthanized following ethical principles. The infection was first tested at a concentration of 10^9 CFU/kg and then set at 10^7 CFU/kg and 10^8 CFU/kg. Criteria for initial infection conditions were established as later detailed. We also performed an extra assay where mice infected by 10^8 CFU/kg of *A. baumannii* were treated with colistin, used as a positive control.

2.9. Efficacy Assay of Peptides in Infection Mice Model

In the first assay, 24 mice (12 females and 12 males) were randomly distributed in 4 groups (3 males and 3 females in each group) with different treatments: vehicle (HBS buffer, negative control), ECPep-D (10 mg/kg), ECPep-2D-Orn (10 mg/kg) and colistin (15 mg/kg), the latter as a positive control. To avoid bias, the reagents for these four groups were given to technicians in advance and renumbered randomly, designated as A, B, C, D, respectively. The treatments corresponding to these codes were not to be told to the operators until the end of the experiment. The murine acute infection model followed the same protocol as introduced before with a lethal concentration (10^8 CFU/kg) to generate sepsis. The treatment was applied 2 h later after the bacteria inoculum and once every 24 h for 3 days through *i.p.* injection. All the animals were checked for 3 days following the clinical signs described in Section 2.10 (Table S2). On study day 3, surviving animals were euthanized by using an overdose of 200 mg/kg pentobarbital given intraperitoneally.

Following the first assay, we also performed a second assay, with the introduction of some modifications. Using the same infection model and injection route, another 46 mice were divided into 4 groups randomly, in which 14 for ECPep-D (20 mg/kg), 14 for ECPep-2D-Orn (20 mg/kg), 12 for colistin (15 mg/kg) and 6 for vehicle, with gender equally divided. The treatment was also applied 2 h later after the bacteria inoculum but only once during the total study. All the animals were also checked for 3 days as in the first assay.

2.10. Evaluation of Body Weight and Clinical Symptoms

Body weight (BW, g) of mice was measured before every injection and % BW gain was calculated.

$$\% \text{ BW gain} = (\text{new BW} - \text{initial BW})/\text{initial BW} \times 100\% \tag{1}$$

In the tolerance study, for the "Up and Down" assay, the BW was recorded twice on the administration day and, for the "main study", it was measured every day at the first

3 days and then at some selected days until the 16th day, when the monitoring was finished. In both efficacy assays, BW was recorded once a day but on the first day (Day 0), a second measure was conducted at 8 h.

The clinical symptoms were assessed through a scoring system, which contains 9 items, including changes of body weight, behavior, breathing, skin, hair, eyes and gastrointestinal status (see full list at Table S2). The score for each item increases from 0 to 3 according to the severe levels, up to a total of 27. During the study, when an animal obtained a score of 3 in any of the parameters assessed with a maximum score of 3, euthanasia was carried out to apply the endpoint criteria. Euthanasia would also be practiced if the sum of several parameters that are less than 3 separately was ≥ 7.

In the tolerance study, the clinical score was recorded at about 10 min, 30 min, 1 h, 2 h, 4 h, 7 h, 24 h and 48 h after injecting the peptides for the "Up and Down" assay and in the main study, the clinical score were only monitored when the BW was measured. During the efficacy assay, the clinical score was recorded every 2–3 h on Day 0 and then twice every day.

2.11. Assay of CFU in Mice Tissues

The evaluation of CFUs in extracted tissues, spleen and lung, was performed at the end of the treatment. In both efficacy studies, one lung and half spleen were collected at the time when each animal was executed due to a high clinical score or euthanized at the end of the study. The organs were homogenized in HBS with 6 successive 1:10 dilution after being weighed on a precision scale. Each dilution was seeded in Petri dishes with LB-agar and the colonies were counted after 16 h of incubation at 37 °C. The final CFUs for each organ were calculated with respect to the weight of organs (CFU/g).

2.12. Quantification of TNF-α and LPS in Mice Serum by ELISA

In the second efficacy study, the blood of animals that either died for severe clinical symptoms or were euthanized at the end of the assay, were collected for testing TNF-α and LPS levels by ELISA. In order to check the levels of TNF-α and free LPS, only serum was analyzed, after discarding blood cells by centrifugation. The blood from mice was centrifuged at 3200 rpm for 30 min and the supernatants were collected carefully to ensure that there were no sediments left. Mouse ELISA kits for TNF-α and LPS were used.

According to the description of the manufacturer, LPS level determination is based on the principle of double antibody sandwich technology, where the microplates pre-coated with specific antibodies were used. Following addition of serum samples to the wells, the HRP-Conjugate reagent was added to form an immune complex. After incubation and washing, the unbound enzyme was removed. Then, chromogen substrates were added successively and optical density was recorded using microplate reader set at 450 nm. Sample readings were corrected using the baseline corresponding to serum of mice neither infected nor treated. A calibration standard curve was prepared and final calculated values were positively correlated with the concentration of LPS. Two replicates were performed according to the protocols for the representation of the standard curve and the sample testing.

2.13. Stimulation of Human MNC by LPS and Endotoxin Neutralization Assay

The stimulation assay of LPS R60 HL185 on the Mononuclear cells (MNC) for the TNF-α response was based on the previously used method [31]. Mononuclear cells were isolated from heparinized blood samples obtained from healthy donors as described [37]. The cells were resuspended in 1640 RPMI medium and their number was equilibrated at 5×10^6 cells/mL. For stimulation, a total of 200 µL (1×10^6) of MNC was plated in each well of a 96-well plate and the cells were then stimulated by LPS Ra (from *S. Minnesota* strain R60) alone or at selected peptide/LPS ratio, which were pre-mixed during 30 min at 37 °C. After 4 h of incubation at 37 °C with 5% CO_2, samples were centrifuged to collect the supernatants to determine the level of TNF-α by ELISA kit (BD OptEIA^{TM}). Two replicates were performed in each test.

2.14. Statistical Analysis

A Mann—Whitney test or unpaired t-test was used for statistical comparison depending on whether the sample size between groups was equal. EC_{50} or IC_{50} was calculated by nonlinear fit with normalized response. The ELISA standard curve was fitted by linear regression (Figure S1).

3. Results

3.1. Design and Structural Characterization of ECP Peptide Analogues

Two peptide analogues, named ECPep-D and ECPep-2D-Orn (Figure 2), were selected based on the ECP reference 30mer peptide (ECPep-L). ECPep-D was designed and optimized based on the identification of the main structural determinants for antimicrobial activity against Gram-negative planktonic and biofilm cultures, as previously detailed [26,27]. The 30mer sequence includes an aggregation-prone region (A6-I14) that promotes bacterial agglutination and membrane lysis and a region involved in the LPS binding (Y27-R30) [27]. In addition, the reference peptide is highly cationic and includes 6 Arg (Figure 2). Based on a target proteolysis study in serum, peptide analogues were synthesized with either all D-AA, Arg to Orn substitutions or blockage of peptide bond cleavage by D-AA introduction. Additionally, based on the previous analysis of the proteolysis digestion products [28], ECPep-2D-Orn was selected. ECPep-2D-Orn is the Orn-peptide analogue, which showed the highest half-life in serum and incorporates a D-AA at position 2 (Figure 2). The RP-HPLC and MS chromatograms of these peptides are shown in Figure S2.

Name	Sequence	a.a
ECPep-L	**RPFTRAQWFAIQHISPRTIAMRAINNYRWR**	**30**
ECPep-D	**r p f t r a q w f a i q h i s p r t i a m r a i n n y r w r**	**30**
ECPep-2D-Orn	**OpFTOAQWFAIQHISPOTIAMOAINNYOWO**	**30**

Figure 2. Sequence of the peptides. The amino acid sequence is indicated. L residues are indicated with an uppercase letter and D residues with lower case letters. Arginine or ornithine are colored in blue or red, respectively.

Structural analysis in aqueous solution and in the presence of lipid dodecyl phosphocholine (DPC) micelles was previously assessed by NMR [28]. NMR indicated that the 30mer L-reference peptide adopts in aqueous solution a defined helix from residues 5 to 13 and tends to helical structuration within the 16–27 stretch. In the presence of DPC the peptide gets more structured. A similar pattern was observed for ECPep-2D-Orn by NMR analysis [28]. In addition, a comparison of all-L and all-D peptides by CD confirmed that equivalent secondary structures are adopted by both enantiomers (Figure 3).

3.2. In Vitro Activities of ECP Peptides

First, the antibacterial activity was tested in vitro. Overall, there were no major differences among the calculated MIC for the three Gram-negative species tested, with values of 10 μM, except for ECPep-2D-Orn that displays a significant enhanced effectivity against *A. baumannii* (Table 1). All data were compared to colistin peptide, used as a positive control. In addition, the survival percentages of each species exposed to the peptides at sublethal doses were compared (Figure 4). Comparison of the bacteria survival percentage upon exposure to a range of peptide concentrations at $\leq MIC_{100}$ showed considerable reduction in the bacterial viability and highlighted significant differences between the peptides' relative activities. Overall, the results indicated that when the bacteria were exposed to the peptides at different concentrations below the MIC value, the growth was altered in a dose-dependent mode. In particular, the results highlighted the best performance for ECPep-D, which can significantly inhibit the growth of *E. coli* and *A. baumannii* between 0.625 and 1.25 μM, in comparison with the original L-version (ECPep-L) and Orn analogue (ECPep-2D-Orn) that required concentrations above 2–5 μM.

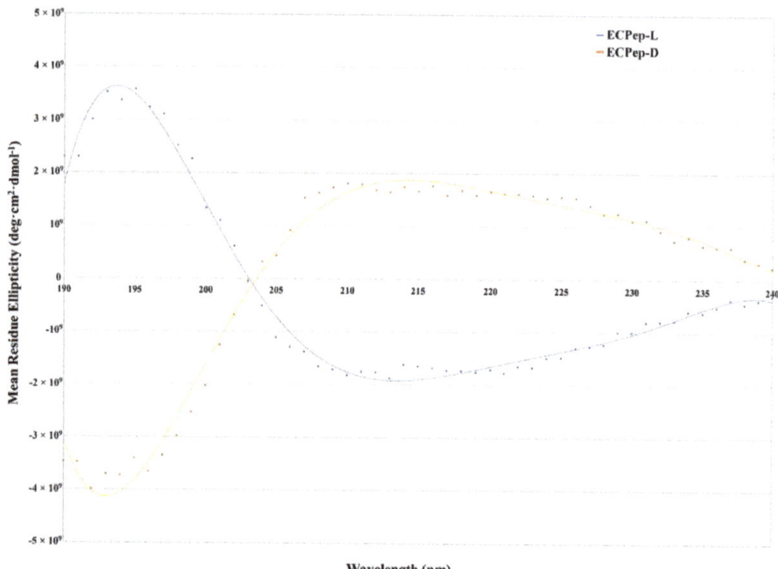

Figure 3. Circular dichroism spectra of ECPep-L and its D-enantiomer. Measures were performed in Tris 5 mM pH 7.4, 1 mM SDS and a peptide concentration of 16 μM using a Jasco J-715 spectropolarimeter.

Table 1. MIC of N-terminus derivatives of ECP and colistin.

Peptides	MIC [1]					
	E. coli		*P. aeruginosa*		*A. baumannii*	
	μM	μg/mL	μM	μg/mL	μM	μg/mL
ECPep-L	10	37.57	10	37.57	10	37.57
ECPep-D	10	37.57	10	37.57	10	37.57
ECPep-2D-Orn	>20	>70.10	>20	>70.10	5	17.53
Colistin	5	6.34	0.31	0.40	0.31	0.40

[1] MIC was the concentration with an OD_{600} value equivalent to the OD_{600} value of the bacteria-free control after 24 h incubation. Each MIC was tested in triplicate.

Following this, the relative LPS affinity of ECP peptides and colistin were compared. Comparison of relative binding affinities of the peptides with three different types of LPS extracted from *E. coli*, indicated that the ECP peptides have a lower or similar EC_{50} in comparison to colistin. Overall, ECP peptides showed better performance than colistin at the micromolar concentration range, with a higher affinity for the full O-LPS structure, followed by the LPSRa type, which lacks the O-antigen. On the other hand, both L- and D- peptide versions showed lower binding ability on LPSRc, the type with the shortest structure among the three tested LPS; where only the Orn-version was slightly better than colistin for the truncated LPS type (Table 2, Figure S3). Interestingly, no significant differences were appreciated between the calculated EC_{50} values for all-L and -D peptide versions, indicating that the all-D peptide could adopt an equivalent overall conformation essential for LPS binding.

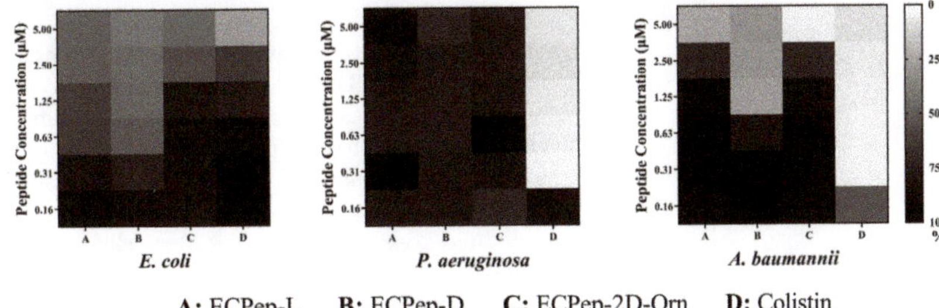

A: ECPep-L   **B:** ECPep-D   **C:** ECPep-2D-Orn   **D:** Colistin

Figure 4. Heat maps of survival percentage of bacteria after 24 h incubation exposed to sublethal doses. The average survival percentages of bacteria normalized by the bacteria-free control and no-treated control are shown. The OD_{600} value equal to that of bacteria-free control is assigned as 0% and that of non-treated control as 100%.

Table 2. LPS affinity assay of ECP-derived peptides and colistin.

	LPS Type		ECPep-L	ECPep-D	ECPep-2D-Orn	Colistin
EC_{50} [2]	O-LPS	µM	2.71 ± 0.17	2.46 ± 0.19	1.80 ± 0.08	7.87 ± 0.33
		µg/mL	10.17 ± 0.62	9.24 ± 0.71	6.30 ± 0.28	9.97 ± 0.42
	LPSRa	µM	8.87 ± 0.46	7.24 ± 0.70	9.27 ± 0.79	13.12 ± 0.40
		µg/mL	33.32 ± 1.72	27.21 ± 2.62	32.50 ± 2.77	16.63 ± 0.50
	LPSRc	µM	>20	>20	16.56 ± 1.76	19.18 ± 0.34
		µg/mL	>75.15	>75.15	58.05 ± 6.16	24.31 ± 0.42

[2] EC_{50} is the 50% peptide effective concentration, where 50% of the BC probe bound to LPS is displaced. Each group was tested in triplicate and values are presented by Mean ± SE.

Next, we tested the cytotoxicity of the peptides. For the fibroblast MRC-5 cell line, all peptides hardly showed any toxicity after 4 h of incubation. Moreover, no toxicity was observed for ECPep-2D-Orn at the highest tested concentration during all the assayed time range (up to 48 h). However, for the ECPep-D, we estimated an LD_{50} below 50 µM at 24 or 48 h exposure time (Table 3, Figure S4). In contrast, the LD_{50} of ECPep-D for kidney HEK293T cell line at 48 h was significantly higher ($LD_{50} \approx 63.60 \mu M$) than that for MRC-5 ($LD_{50} \approx 23.52 \mu M$) (Table S3, Figure S4). Overall, the values for ECP peptides are similar or slightly lower than colistin, at short and long exposure time, respectively.

Table 3. Cytotoxic activity of N-terminus derivatives of ECP and colistin.

			ECPep-L	ECPep-D	ECPep-2D-Orn	Colistin
LD_{50} [3] (MRC-5)	4 h	µM	276.23 ± 25.23	N.D.	N.D.	N.D.
		µg/mL	>1000	N.D.	N.D.	N.D.
	24 h	µM	N.D.	43.51 ± 8.84	>100	>300
		µg/mL	N.D.	163.50 ± 33.22	>350.52	>380.26
	48 h	µM	N.D.	23.52 ± 4.73	>100	>300
		µg/mL	N.D.	88.37 ± 17.77	>350.52	>380.26

[3] LD_{50} is the cytotoxic concentration of the agents to cause death to 50% of viable cells. Each assay was performed in triplicate by the MTT assay and normalized by non-treated control. Values are presented by Mean ± SE. N.D. indicates that no reduction in cell viability was detected at the highest concentration.

To sum up, we can conclude from the in vitro studies that the activities of the two analogues (ECPep-D and ECPep-2D-Orn) are similar or even better than the original version peptide (ECPep-L), indicating that these two analogues are valuable candidates for in vivo assays. In addition, previous studies demonstrated that the half-life time of ECPep-L in human serum is significantly shorter in comparison to that of ECPep-2D-Orn (12 min vs. > 480 min) [28]. Therefore, we decided to use ECPep-D and ECPep-2D-Orn to conduct the follow-up in vivo research.

3.3. Tolerance of Peptides in Mice

Next, we proceeded to evaluate the peptides potential toxicity in vivo using a mice model. Firstly, an "Up and Down" protocol was carried out to obtain an initial estimate of the maximum tolerated dose for both peptide analogues following *i.p.* injection. We started the tolerance assay for each peptide at a low dose and increased it slightly every 48 h provided that the animal did not show any obvious suffering. Both mice had a clinical score over 6 within 30 min when the dose of peptides reached 30 mg/kg, estimating the maximum tolerated dose for ECPep-D at 20 mg/kg and for ECPep-2D-Orn at 25 mg/kg with one single dose from the "Up and Down" assay (Figure S5). These results were later confirmed by the "main study" in which a single dose of 20 mg/kg for ECPep-D and 25 mg/kg for ECPep-2D-Orn did not cause any obvious clinical signs in all mice.

However, severe clinical suffering was observed when the dose was applied repeatedly in mice at concentrations from 15 mg/kg to 25 mg/kg (Figure S6). Therefore, the initial dosing plans were modified or suspended accordingly (Table S1). On the other hand, no extra death occurred and the clinical symptoms were recovered for the survival mice after the treatment suspension. The mice's healthy condition lasted till the end of the observation period (16 days).

In addition, within the extra group with low dose (7.5 or 10 mg/kg) administration, no obvious clinical symptoms were observed (Figure S6).

3.4. Murine Acute Infection Model by A. baumannii

Once the toxicity assay was finalized, we proceeded to evaluate the efficacy of the peptides in an *A. baumannii* infection model. All the mice showed very severe suffering after inoculation with 10^9 CFU/kg of *A. baumannii*, corresponding to 10^8 CFU/mL of the initial concentration of bacteria supplemented with 5% porcine mucin. Therefore, euthanasia had to be applied according to the protocol endpoint criteria after 7 h. Following this, a lower level of infection was applied at 10^7 CFU/kg in a small test group (1 male and 1 female). In this assay, both mice only experienced slight suffering during the first few hours and totally recovered on the next day. Therefore, an in-between concentration of 10^8 CFU/kg was selected for further analysis, where clinical suffering was lower than the one observed at 10^9 CFU/kg (Figure S7a). Moreover, the group of infected mice at 10^8 CFU/kg was treated with the positive control (colistin at 15 mg/kg). A slight increase in clinical signs appeared after the first 8 h and it was nearly undetectable after 24 h in the surviving mice (Figure S7b).

Thus, we decided to use 10^8 CFU/kg as the mice inoculation concentration of *A. baumannii* in the following assays and established these conditions for the infection model to test the efficacy of the peptide.

3.5. Treatment with Peptides Improves the Survival Rate and Relieves Clinical Suffering of Infected Mice

To assess the therapeutic effect of our peptides on infected mice, we performed two assays with different survival rates in peptide treatment.

In the First Efficacy Assay, we applied a peptide/mouse of 10 mg/kg once per day, a previously established safe dose, and the treatment lasted for 3 days. In the vehicle-treated group (negative control), as already observed in the set-up experiment (Figure S7), all animals presented severe clinical signs within the first 8 h post-inoculum (Figure S8) and were, therefore, euthanized after the first day. On the other hand, all the animals treated with colistin (positive control) had a 100% survival rate (Figure 5a). In this group, although a decrease in weight was observed during the first day following infection, the mice recovered weight during the assay time course, which was considered a sign of health improvement. Moreover, we also observed only a slight increase in the clinical signs score, indicative of health problems during the first 24 h, being subsequently minor or non-significant during the following observation period (up to 3 days) (Figure S8). Surviving animals treated with the ECP peptides exhibited the same behavior pattern as the colistin-treated animals (Figure S8). Although survival rates for mice treated with 10 mg/kg of the ECP peptides ranged 70–80% at the first 8 h of observation, the values drastically dropped when checked at 24 h. Overall, the survival rate after the three-day treatment was equivalent for both all D and 2D-Orn peptides, but much lower than the colistin positive control (Figure 5a). However, the weight gain of both mice was equal to the curve obtained with the colistin group. The score of the clinical signs of the surviving animals treated with ECP peptides also had similar values to the positive control group (colistin), signs of clinical recovery together with weight gain (Figure S8). In any case, no direct side-by-side comparison could be performed here between colistin and ECP peptides efficacy, as the assay concentration was not equal: 10 mg/kg for the ECP peptides versus 15 mg/kg for colistin, the latter selected to ensure a 100 % survival positive control.

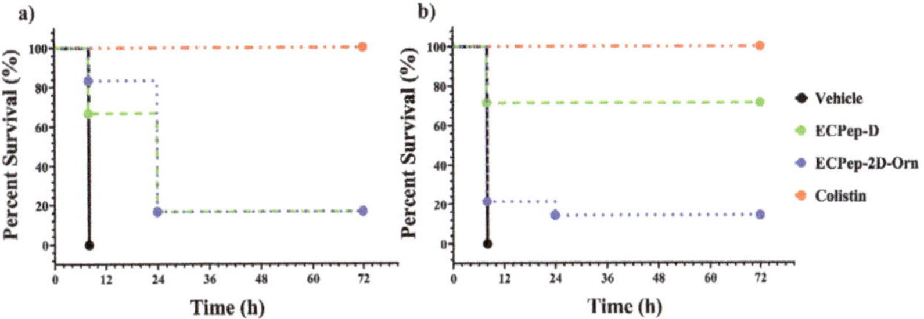

Figure 5. Survival curve of first and second efficacy assays. (**a**) Survival curve of infected mice treated with 10 mg/kg peptides. (**b**) Survival curve of infected mice treated with 20 mg/kg peptides. In both groups, each animal had been inoculated with 10^8 CFU/kg *A. baumannii* with 5% mucin before treatment and the administration of colistin set at 15 mg/kg. The peptides were administered 2 h after the bacteria inoculation. The survival mice had been monitored for 72 h and euthanized.

Subsequently, in a second efficacy assay, we decided to test a higher peptide concentration (20 mg/kg), a value where no toxicity was observed for a single dose in the previous toxicity assay (Figures S5 and S6). Additionally, the positive control (colistin) was kept at the same dose as before but with only a single injection to follow the same one-time injection protocol selected for the ECP peptides' treatment. Following an equivalent pattern as previously observed for the vehicle group, all animals without any treatment presented severe clinical signs at the first 8 h / post inoculum and were, therefore, euthanized. On the other hand, the 12 animals treated with colistin had a 100% survival rate (Figure 5b), as previously registered. When comparing the evolution of the clinical sign scores we observed first an increase during the first 8 h due to the process of infection (Figure 6). The decrease in body weight was also observed on the first night and then mice started regaining weight during the rest of the experiment time course. Interestingly, we observed a significant improvement of clinical scores and recovery of body weight in the mice groups treated with both peptides in comparison with the negative control, treated only with Vehicle (HBS) (Figure 6). Meanwhile, the score of the clinical signs of the surviving animals treated with the peptides had similar values to colistin: both signs of clinical recovery and body weight gain.

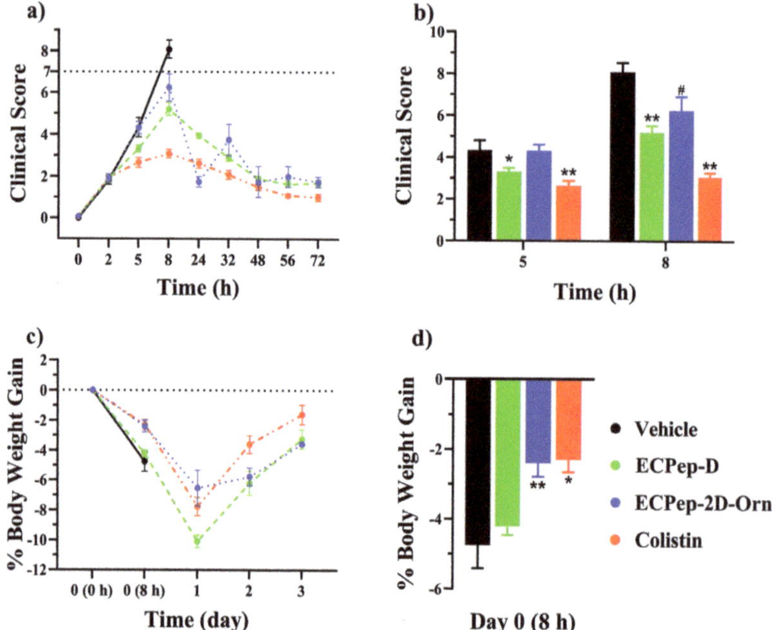

Figure 6. Changes of clinical symptoms during the 2nd efficacy assay with 20 mg/kg peptides treatment. (**a**) The average clinical scores in 3 days; (**b**) Histogram of clinical scores at 5 h and 8 h; (**c**) The time course body weight gain (%) up to 3 days; (**d**) Histogram of body weight gain (%) at 8 h. Each animal had been inoculated 10^8 CFU/kg *A. baumannii* with 5% mucin before treatment. In all cases, a single dose of the peptide was given 2 h later than the bacteria inoculation. Mann–Whitney test has been used for statistical comparison between different treatments and vehicle (** $p < 0.01$, * $p < 0.05$, # $p < 0.1$).

However, the survival curve of each ECP peptide treatments revealed a differentiated efficacy profile (Figure 5b). A significant improvement for the treatment with ECPep-D at 20 mg/kg was achieved with a survival rate of 71% at 8 h and this survival rate lasted till the end of the study (72 h). On its side, the survival rate after 3 days for ECPep-2D-Orn was only 14%, which correspond to a similar value to previous results obtained at 10 mg/kg for both peptides.

3.6. ECP Peptides Reduce Bacterial Counts in Mice Organs and TNF-α Levels after 8 h of Treatment

In both efficacy assays in *A. baumannii*-infected mice, all the animals from the vehicle-treated group (negative control) were killed after 8 h due to the bad clinical symptoms, as evaluated by the calculated clinical score (Table S2). Following this, the lung and spleen were collected and the infection level was analyzed by CFU counting. The number of calculated CFUs showed maximum values of 10^9–10^{10} CFU/g. On the contrary, in positive controls where the mice survived 3 days, colonies counted after euthanasia at the end of the study presented a minimum value in all groups, around 10^3 CFU/g (Figure S9 and Figure 7a). Complementarily, blood samples of each sacrificed animal in the second efficacy assay were also collected and levels of TNF-α and LPS in serum were evaluated. For comparative purposes, blood samples corresponding to the same timing were analyzed together: either Day 0 (8 h) or Day 3 (72 h). Due to the differences in the number of animals that could be tested in each treatment group, the Mann—Whitney test was applied.

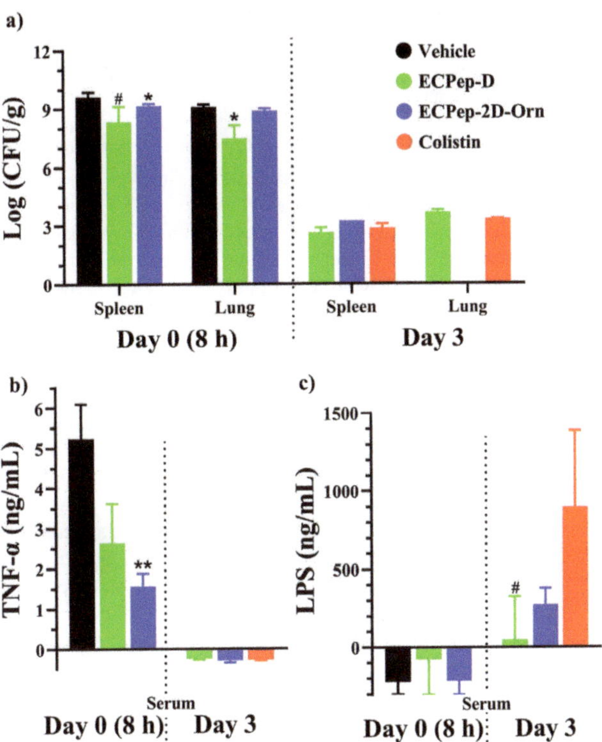

Figure 7. Evaluation of CFUs, TNF-α and LPS in the 2nd efficacy assay using 20 mg/kg of peptide treatment. (**a**) The average CFUs in mice organs (spleen and lung). (**b**) The average concentration of TNF-α in mice serum. (**c**) The average concentration of LPS in mice serum was quantified by ELISA as described in the methodology. All the animals from the colistin group survived until Day 3, so no data for organs and serum analysis can be shown at Day 0. Regarding the vehicle group, no animals survived after Day 0, so no data can be shown at Day 3. Each animal had been inoculated 10^8 CFU/Kg *A. baumannii* with 5% mucin before treatment. A single treatment was administrated 2 h after the bacteria inoculation. Figure 7b,c were obtained after the correction of OD_{450} with baseline mice, which are neither infected nor peptide treated (only injected at Day 0 with HBS). Mann—Whitney test was used for the comparison between treatments and vehicle except for the additional comparison between ECPep-D and colistin in Figure 7c at day 3 (** $p < 0.01$, * $p < 0.05$, # $p < 0.1$).

Results indicated that even if the efficacy to increase the survival rate between the two peptides differed greatly (about 70% for ECPep-D versus 20% for ECPep-2D-Orn at the first 8 h), the ability to reduce CFU counts in the studied tissues and the TNF-α inflammation marker levels in serum were similar for both peptides at the end of the assay (3 days). Therefore, we can conclude that both peptides were able to drastically reduce the CFU counts and TNF-α serum levels equivalently to colistin, the positive control. To note, only a slight significant reduction in organ CFUs was observed at Day 0 ($p < 0.05$) but CFUs counts at Day 3 were equivalent for both peptides and colistin (Figure 7a). Interestingly, both ECP peptides were able to lower serum TNF-α values at day 0 (8 h) more than twice with respect to the negative control (Figure 7b). Overall, when the mice survived at the end of the assay (Day 3) in both peptide-treatment groups, the number of colonies and the concentration of TNF-α were equivalent to the positive control.

On the other hand, regarding the LPS quantification, a different profile was observed (Figure 7c). On Day 0, the concentration of LPS in serum of sacrificed mice in all groups was undetectable. However, on Day 3, we observed an increase in the amount of LPS in both peptide-treatment groups. Moreover, the concentration of LPS was slightly increased in the colistin-treatment group respect to ECPep-D ($p < 0.1$), even though the mice seemed totally recovered with no clinical symptoms.

3.7. ECP Peptides Reduce TNF-α Levels in LPS-Stimulated Cells

Finally, we decided to evaluate the anti-endotoxin activity of the assayed peptides to complement and better interpret our in vivo results. To this end, we decided to compare the activity of ECPep-D (the peptide that showed better efficacy in vivo) with the original L-version (ECPep-L) and the colistin reference control. To evaluate the anti-endotoxin activity, we estimated the TNF-α levels in LPS-stimulated mononuclear cells. The results showed that both L- and D-peptide versions can significantly reduce the TNF-α levels when cells are exposed to LPS stimulation (Figure 8). The assay was performed at two peptide/LPS ratios and cell stimulation by three LPS concentrations. Overall, the ECP peptides' efficacy was lower than the positive control, colistin, although this difference was mostly observed when MNC cells were stimulated with a higher LPS concentration (10 ng/mL). Complementarily, we performed an additional comparative assay using an anti-endotoxin peptide (Aspidasept®), a promising peptide candidate to prevent septic shock [38]. Interestingly, at high concentrations of LPS (10 ng/mL), ECP-derived peptides present slightly better anti-inflammatory activity than Aspidasept®, which in turn shows a better performance at the lower LPS concentration (1 ng/mL) (Figure S10). Overall, all peptides (ECPep, Aspidasept® and colistin) have similar activity at the lowest LPS concentration used (Figure 8 and Figure S10a). These results back up the observed ability of ECP-derived peptides to reduce TNF-α levels in the studied mice infection model. Finally, an additional assay was performed at even higher concentrations of LPS (100 ng/mL), which induced the triple of TNF-α release (≈2300 pg/mL) in cells without peptide treatment (Figure S10b). Results highlighted that both ECPep-L and ECPep-D at 10:1 ratio with respect to LPS significantly reduced the release of TNF-α, at a similar percentage to Aspidasept®, although the latter was more effective at a 100:1 ratio.

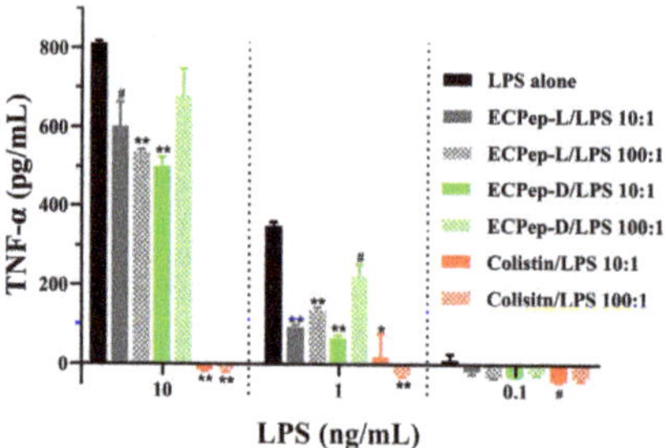

Figure 8. Inhibitory effect on LPS-induced TNF-α cytokine release. Different concentrations of LPS R60-induced secretion of TNF-α by human mononuclear cells and the inhibitory effect in the presence of the peptides ECPep-L, ECPep-D and colistin (positive control) was calculated. Significant differences were estimated in comparison to LPS alone sample (** $p < 0.01$, * $p < 0.05$, # $p < 0.1$).

4. Discussion

To achieve a proper assessment of the potentiality of novel antimicrobials as drug candidates, in vivo studies are mandatory. Following promising results in vitro, animal models are essential for monitoring both the bacterial infection process and the host response. Mice infection models can provide valuable information about how diseases or treatments may behave in humans, thanks to the similarities in the mammalian tissue structures and the functioning of their immune system. In particular, in vivo assays are essential for the evaluation of AMP agents. Firstly, many AMPs that achieve good MIC values in vitro may show bad or no effect in vivo due to their rapid degradation by proteases in the body. Secondly, the observed toxicity in vitro in cell line assays may become weak in animals thanks to the body's self-regulation. In addition, the immune response in the body can potentiate the mechanism of action of AMPs and overcome bacterial resistance systems. Moreover, by testing the AMP efficacy in animal models, we will gain an understanding of their functional role as key players of the innate host defense response [39]. Among animal studies, murine bacterial infection models are probably the best characterized for evaluation of AMP activity [40–42].

Our laboratory has long been committed to the research and development of novel antimicrobial peptides based on the structure-functional knowledge on RNase A superfamily. Extensive studies of previous work have confirmed that RNase3 (ECP) displays the highest antibacterial activity among human family members [43,44]. Based on structural analysis, proteolysis mapping and peptide synthesis we identified at the N-terminus a region that retained most of the parental protein antimicrobial activity. Following a minimization effort and the identification of the sequence determinants for bacteria cell wall binding, membrane lysis, protein aggregation and cell agglutination, a 30mer was selected as the best pharmacophore with high antibacterial activity in vitro against both planktonic and biofilm cultures of Gram-negative bacteria [26,27]. Next, ECP peptide analogues were designed to ensure protection against potential in vivo proteolysis [28]. The incorporation of non-natural amino acids has been previously reported as a successful strategy to enhance AMPs biostability while retaining their antimicrobial properties [45,46]. In the current study, aiming to develop a potential AMP drug candidate, we selected two N-terminal derivatives of ECP (5–17P24–36), intended to enhance its biological stability in vivo. Based on the original L-version of peptide ECP (5–17P24–36)], named ECPep-L, we synthesized an all D-version peptide (named ECPep-D) together with the best analogue from recent biostability assays [28], where all Arginines were substituted by Ornithine and peptide bond at position 2 was protected from proteolysis by D-amino acid substitution, named ECPep-2D-Orn (Figure 2). Our previous work indicated that shielding the peptide bond between Pro2 and Phe3 by D-replacement significantly enhanced the peptide half-life in vitro, while retaining the antimicrobial properties of the parental ECPep-L and all Orn L-version (ECPep-Orn) [28]. Here, we incorporated the characterization of the peptide all-D version, showing equivalent MIC values in comparison to the all-L version for the three tested Gram-negative species (Table 1) but potential cytotoxicity in vitro at long exposure times (Table 3). It is worth emphasizing that while the D-version has an antibacterial activity equivalent to the L-version against the three selected Gram-negative species, the 2D-Orn-version was mostly effective against *A. baumanii* but less successful against the other two bacterial species. In any case, bacterial survival curves at 24 h using sublethal doses indicated that both peptide analogues (ECPep-D and ECPep-2D-Orn) had a better performance in *A. baumannii* culture, with the highest activity for the all-D version (Figure 4). Unfortunately, as commented above, ECPep-D shows significant cytotoxicity in contrast to ECPep-2D-Orn in the tested human cell lines at 24 h and 48 h (Table 3). Therefore, we decided to compare both peptide analogues (showing either high antimicrobial activity or no toxicity at the highest tested concentration 100 µM) in an *A. baumannii* mice infection model.

Interestingly, the efficacy of the peptides in a murine acute infection model by *A. baumannii* was encouraging. More than 70% survival after 3 days of infection was achieved by a single dose of the D-peptide analogue. In addition, in vivo toxicity test indicated that only significant clinical symptoms appeared when the peptide was administered at very high concentrations, such as 25–30 mg/kg, after repeated dosing. The maximum tolerated dose and the symptomatology after repeated doses observed in our work are similar to other reported cases for AMPs administered by the intraperitoneal route. Taking colistin sulfate as a reference approved AMP, we found a reported LD_{50} by intraperitoneal administration around 20–30 mg/kg in Swiss albino mice following a single injection [47]. Additionally, at the U.S. National Library of Medicine—Toxicity information (https://chem.nlm.nih.gov/chemidplus/rn/1264-72-8, accessed on 1 December 2021), the LD_{50} of colistin sulfate for the intraperitoneal route is 21.8 mg/kg in mice. It has been reported that mice receiving 32 mg/(kg × day) became sluggish after three doses of colistin, where three of twelve mice died from neurotoxicity [48]. Thus, we can conclude that in our studies with one single dose below the maximum tolerated dose, both ECP peptide analogues are safe for mice. Nevertheless, it is worth considering further work on novel Orn-analogues to enhance their in vivo biostability while minimizing their potential toxicity.

Next, we tested ECP peptides in an acute infection model. According to the literature to achieve an acute infection of *A. baumannii* in a murine model, the recommended dose of the inoculum ranges between 10^6 CFU/mL and 10^8 CFU/mL [49–52]. Following optimization by testing an inoculum by intraperitoneal administration from 10^6 to 10^8 CFU/mL, we selected a 10^7 CFU/mL condition, where untreated mice developed severe clinical signs as well as heavy infection and had to be euthanized within a period of up to 8 to 9 h. On the other hand, a positive control was set by the administration of colistin at 15 mg/kg, where a drastic reduction in infection and recovery of clinical signs was achieved, reaching 100% mice survival. In addition, a supplement of 5% mucin was administered to boost infection and promote a proinflammatory response [36].

Subsequently, we started to test the potential efficacy of our designed peptides for acute systemic infection. Single doses from 10 mg/kg to 20 mg/kg, previously determined as safe, with no associated toxicity, were administered intraperitoneally to infected mice. A pattern of significant improvement of clinical parameters and recovery of body weight following ECP peptide administration was similar to the positive colistin-treated group. In contrast, in the negative non-treated control, none of the animals survived. The first efficacy assay at 10 mg/kg was followed by a second study at 20 mg/kg, where, in addition, a higher number of treated animals was inspected. A therapeutic effect of ECPep-D was observed with a 71.43% survival rate of mice after 3 days. In addition, a drastic reduction in CFUs was observed in the studied organs in all animals treated with both peptides, indicating that ECP peptides could be considered as potentially effective candidates for *A. baumannii* acute infection. However, although CFU levels at analyzed organs and clinical parameters recovery were similar to our positive control, a significantly lower survival ratio was reached for ECPep-2D-Orn peptide. This might be attributed to a poorer biostability in in vivo conditions of this analogue in comparison to the all-D peptide. Further work is envisaged to identify the best pharmacophore that ensures non-toxicity in vivo while retaining antimicrobial action. Complementary conjugation of ECP peptides to nanocarrier systems would be considered to reduce the effective dose and facilitate the targeted delivery, as demonstrated effective for a modified version of the parental protein [9].

Another interesting feature characteristic of ECP and its N-terminus derived peptides relies on its high binding affinity to lipopolysaccharides (LPS) present at the Gram-negative outer membrane [26]. Previous structural and functional studies characterized the protein and peptide interaction with LPS in vitro by the use of complementary approaches. The main protein residues that participate in the binding to LPS were characterized by side-directed mutagenesis, peptide-array library and NMR structural analysis [27]. The ECP targeted sequence for endotoxin binding is located at the C-terminus of the 30mer ECPep (i.e., YRWR) (see Figure 2). The present data highlights that the all-D peptide retains the same LPS binding affinity as its L counterpart (Table 2), suggesting that the overall structural determinants of the parental peptide are retained. More importantly, both ECP peptide analogues (ECPep-D and ECPep-2D-Orn) can drastically reduce TNF-α levels in mice serum and demonstrates for the first time the ability of an ECP peptide to block in vivo the release of the pro-inflammatory TNF cytokine. Nonetheless, our preliminary results on LPS determination in mice serum are less straightforward to interpret and would probably need future complementary assays. Despite the intrinsic limitations of the assay, where only a reduced number of samples could be assayed and unspecific interactions of the serum components with the ELISA kit reagents cannot be discarded, some results could be drawn when comparing the distinct analyzed groups. At Day 0 (8 h), nearly no LPS was detected within the serum of infected mice including the non-treatment group. This may be attributed to the fact that the unreleased LPS might remain on the surface of bacterial cells and would precipitate when the collected blood was centrifuged. Interesting results come out on Day 3. The amount of free LPS in the serum of mice cured with colistin is slightly higher on Day 3, which means that the LPS released from the bacteria remained in the serum and could not be neutralized after 3 days of infection. It has been reported that a large amount of LPS released from *P. aeruginosa* dead cells following treatment with colistin were still detectable, which in turn might reduce the effectiveness of this drug at the infection focus [53]. Indeed, LPS aggregates were reported as the main entities biologically active in previous studies [54]. In contrast to colistin, our designed peptides seem to be able to lower the blood circulating LPS after 3 days of treatment (Figure 7), although this issue needs further exploring. Among others, we might consider exploring the peptide anti-endotoxin activity in another animal model, such as rabbit, considered more appropriate for the study of sepsis-related symptoms [38]. In addition, a full understanding of the LPS-neutralizing process needs to take into consideration which is the biologically active conformation of blood circulating LPS [54] and how the AMP presence can alter the pro-inflammatory cell response.

Massive production of pro-inflammatory cytokines, such as TNF-α, are associated with the septic shock, a major lethal factor in Gram-negative infections in the clinics. Currently, colistin is frequently used as a last resort to treat high-risk patients that can undergo septic shock following acute infection. However, colistin is a non-ribosomal cyclic-AMP only used as a last resort due to its associated toxicity following chronic treatment [22,48]. Therefore, a new generation of antibacterial drugs that not only kill the bacteria but also neutralize LPS with no toxicity in vivo is essential. The present results corroborate the efficacy of ECP-derived peptides in vivo, showing their ability to effectively lower the number of CFUs in tested organs (lung and spleen) and the amount of TNF-α in the serum of infected mice (Figure 7). Moreover, both peptides administration can revert the severe clinical parameters during the first 8 h of infection and alleviate the mice suffering on the first day (Figure 6). In addition, the observed endotoxin neutralization activity of ECP peptides in vivo was corroborated in a human mononuclear cell assay, showing comparable values to Aspidasept® (pep19-2.5), an anti-septic AMP drug candidate (Figure 8 and Figure S10).

5. Conclusions

The current results highlight the efficacy of two N-terminal derived peptides of ECP in a murine systemic *A. baumannii* acute infection model. This is the first report of an efficacy test of an ECP peptide in an animal model. Our data indicates that the 30mer all-D peptide version (ECPep-D) successfully enhances the mice survival rate up to 71% after 3 days. In addition, both peptides can recover the body weight and supervised clinical parameters to almost normal values, following an equivalent pattern as shown for animals treated with colistin, our assay positive control. Moreover, both ECP peptides reduce the release of the pro-inflammatory cytokine TNF-α. The present in vivo results make us confident to further explore the unique advantages of ECP-derived peptides as a new generation of antimicrobial candidates.

Supplementary Materials: The following supporting information can be downloaded at: https://www.mdpi.com/article/10.3390/biomedicines10020386/s1, Table S1: Sample sizes and administration times in each group for the main toxicity study; Table S2: Assessment of general health status; Table S3: Cytotoxicity of ECP-derived peptides and colistin; Figure S1: Standard curve of TNF-α and LPS tested by ELISA; Figure S2. RP-HPLC and MS chromatograms of (a) ECPep-L, (b) ECPep-D and (c) ECPep-2D-Orn; Figure S3: BODIPY-Cadaverine displacement curves for peptides and colistin; Figure S4: Cytotoxicity curves for peptides and colistin; Figure S5: "Up and Down" assay; Figure S6: Changes of clinical score after each administration and survival percentages of animal for main toxicity assay; Figure S7: Clinical scores for the setting of murine acute infection model induced by *A. baumannii*; Figure S8: Changes clinical symptoms in 1st efficacy assays with 10 mg/kg peptides treatment; Figure S9: The average CFUs in organs (spleen and lung) of infected mice in 1st efficacy assay with 10 mg/kg peptides treatment; Figure S10: Inhibitory effect on LPS-induced TNF-α cytokine release.

Author Contributions: Conceptualization, E.B. and P.F.-M.; methodology, E.B., P.F.-M., K.B., D.A., M.T., J.L. and G.P.-E.; writing—original draft preparation.; writing—review and editing, E.B., M.T., J.L., G.P.-E. and K.B.; supervision, P.F.-M. and E.B.; project administration, E.B.; funding acquisition, E.B. All authors have read and agreed to the published version of the manuscript.

Funding: Research work was supported by Fundació La Marató de TV3 (TV3-201803-10), the Ministerio de Economía y Competitividad (PID2019-106123GB-I00) and by AGAUR, Generalitat de Catalunya (2016PROD00060; 2019 LLAV 00002), co-financed by FEDER funds. P.F.-M. was a recipient of Juan de la Cierva postdoctoral fellowship, G.P.-E. was recipient of a PIF-UAB predoctoral fellowship and J.L. was a recipient of a CSC predoctoral fellowship.

Institutional Review Board Statement: The animal study protocol was approved by the Institutional Review Board (or Ethics Committee) of Universitat Autònoma de Barcelona (protocol code 4636).

Informed Consent Statement: Not applicable.

Data Availability Statement: Not applicable.

Acknowledgments: We thank Sandra Barbosa Pérez, Gloria Costa López and Patrocinio Vergara from the Serveis Integrats de l'Animal de Laboratori (SIAL) for their technical support with the animal studies.

Conflicts of Interest: The authors declare no conflict of interest.

References

1. Ghosh, C.; Sarkar, P.; Issa, R.; Haldar, J. Alternatives to conventional antibiotics in the era of antimicrobial resistance. *Trends Microbiol.* **2019**, *27*, 323–338. [CrossRef] [PubMed]
2. Li, P.; Li, X.; Saravanan, R.; Li, C.M.; Leong, S.S.J. Antimicrobial macromolecules: Synthesis methods and future applications. *RSC Adv.* **2012**, *2*, 4031–4044. [CrossRef]
3. Sierra, J.M.; Fusté, E.; Rabanal, F.; Vinuesa, T.; Viñas, M. An overview of antimicrobial peptides and the latest advances in their development. *Expert Opin. Biol. Ther.* **2017**, *17*, 663–676. [CrossRef] [PubMed]
4. Ghosh, C.; Haldar, J. Membrane-active small molecules: Designs inspired by antimicrobial peptides. *ChemMedChem* **2015**, *10*, 1606–1624. [CrossRef] [PubMed]
5. Mahlapuu, M.; Håkansson, J.; Ringstad, L.; Björn, C. Antimicrobial peptides: An emerging category of therapeutic agents. *Front. Cell. Infect. Microbiol.* **2016**, *6*, 194. [CrossRef] [PubMed]
6. Chung, P.Y.; Khanum, R. Antimicrobial peptides as potential anti-biofilm agents against multidrug-resistant bacteria. *J. Microbiol. Immunol. Infect.* **2017**, *50*, 405–410. [CrossRef] [PubMed]
7. Feng, X.; Sambanthamoorthy, K.; Palys, T.; Paranavitana, C. The human antimicrobial peptide LL-37 and its fragments possess both antimicrobial and antibiofilm activities against multidrug-resistant *Acinetobacter baumannii*. *Peptides* **2013**, *49*, 131–137. [CrossRef]
8. Hirsch, R.; Wiesner, J.; Marker, A.; Pfeifer, Y.; Bauer, A.; Hammann, P.E.; Vilcinskas, A. Profiling antimicrobial peptides from the medical maggot *Lucilia sericata* as potential antibiotics for MDR Gram-negative bacteria. *J. Antimicrob. Chemother.* **2019**, *74*, 96–107. [CrossRef]
9. Li, J.; Fernández-Millán, P.; Boix, E. Synergism between host defence peptides and antibiotics against bacterial infections. *Curr. Top. Med. Chem.* **2020**, *20*, 1238–1263. [CrossRef]
10. Hollmann, A.; Martinez, M.; Maturana, P.; Semorile, L.C.; Maffia, P.C. Antimicrobial Peptides: Interaction with model and biological membranes and synergism with chemical antibiotics. *Front. Chem.* **2018**, *6*, 204. [CrossRef]
11. Fjell, C.D.; Hiss, J.A.; Hancock, R.E.W.; Schneider, G. Designing antimicrobial peptides: Form follows function. *Nat. Rev. Drug Discov.* **2012**, *11*, 37–51. [CrossRef] [PubMed]
12. Casciaro, B.; Cappiello, F.; Cacciafesta, M.; Mangoni, M.L. Promising approaches to optimize the biological properties of the antimicrobial peptide esculentin-1a(1-21)NH$_2$: Amino acids substitution and conjugation to nanoparticles. *Front. Chem.* **2017**, *5*, 26. [CrossRef] [PubMed]
13. Shao, C.; Zhu, Y.; Lai, Z.; Tan, P.; Shan, A. Antimicrobial peptides with protease stability: Progress and perspective. *Future Med. Chem.* **2019**, *11*, 2047–2050. [CrossRef]
14. Oddo, A.; Thomsen, T.T.; Kjelstrup, S.; Gorey, C.; Franzyk, H.; Frimodt-Møller, N.; Løbner-Olesen, A.; Hansen, P.R. An amphipathic undecapeptide with all D-amino acids shows promising activity against colistin-resistant strains of *Acinetobacter baumannii* and a dual mode of action. *Antimicrob. Agents Chemother.* **2016**, *60*, 592–599. [CrossRef] [PubMed]
15. Carmona, G.; Rodriguez, A.; Juarez, D.; Corzo, G.; Villegas, E. Improved protease stability of the antimicrobial peptide Pin2 substituted with D-amino acids. *Protein J.* **2013**, *32*, 456–466. [CrossRef] [PubMed]
16. Hamamoto, K.; Kida, Y.; Zhang, Y.; Shimizu, T.; Kuwano, K. Antimicrobial activity and stability to proteolysis of small linear cationic peptides with D-amino acid substitutions. *Microbiol. Immunol.* **2002**, *46*, 741–749. [CrossRef] [PubMed]
17. Lee, K. Development of short antimicrobial peptides derived from host defense peptides or by combinatorial libraries. *Curr. Pharm. Des.* **2002**, *8*, 795–813. [CrossRef]

18. Mookherjee, N.; Anderson, M.A.; Haagsman, H.P.; Davidson, D.J. Antimicrobial host defence peptides: Functions and clinical potential. *Nat. Rev. Drug Discov.* **2020**, *19*, 311–332. [CrossRef]

19. Koo, H.B.; Seo, J. Antimicrobial peptides under clinical investigation. *Pept. Sci.* **2019**, *111*, e24122. [CrossRef]

20. Magana, M.; Pushpanathan, M.; Santos, A.L.; Leanse, L.; Fernandez, M.; Ioannidis, A.; Giulianotti, M.A.; Apidianakis, Y.; Bradfute, S.; Ferguson, A.L.; et al. The value of antimicrobial peptides in the age of resistance. *Lancet Infect. Dis.* **2020**, *20*, e216–e230. [CrossRef]

21. Vaara, M. Polymyxins and their potential next generation as therapeutic antibiotics. *Front. Microbiol.* **2019**, *10*, 1689. [CrossRef]

22. Biswas, S.; Brunel, J.M.; Dubus, J.C.; Reynaud-Gaubert, M.; Rolain, J.M. Colistin: An update on the antibiotic of the 21st century. *Expert Rev. Anti. Infect. Ther.* **2012**, *10*, 917–934. [CrossRef] [PubMed]

23. Tambadou, F.; Caradec, T.; Gagez, A.L.; Bonnet, A.; Sopéna, V.; Bridiau, N.; Thiéry, V.; Didelot, S.; Barthélémy, C.; Chevrot, R. Characterization of the colistin (polymyxin E1 and E2) biosynthetic gene cluster. *Arch. Microbiol.* **2015**, *197*, 521–532. [CrossRef] [PubMed]

24. Gupta, S.K.; Haigh, B.J.; Griffin, F.J.; Wheeler, T.T. The mammalian secreted RNases: Mechanisms of action in host defence. *Innate Immun.* **2013**, *19*, 86–97. [CrossRef] [PubMed]

25. Schwartz, L.; Cohen, A.; Thomas, J.; Spencer, J.D. The immunomodulatory and antimicrobial properties of the vertebrate ribonuclease A superfamily. *Vaccines* **2018**, *6*, 76. [CrossRef] [PubMed]

26. Pulido, D.; Prats-Ejarque, G.; Villalba, C.; Albacar, M.; González-López, J.J.; Torrent, M.; Moussaoui, M.; Boix, E. A Novel RNase 3/ECP peptide for *Pseudomonas aeruginosa* biofilm eradication that combines antimicrobial, lipopolysaccharide binding, and cell-agglutinating Activities. *Antimicrob. Agents Chemother.* **2016**, *60*, 6313–6325. [CrossRef] [PubMed]

27. Pulido, D.; Prats-Ejarque, G.; Villalba, C.; Albacar, M.; Moussaoui, M.; Andreu, D.; Volkmer, R.; Torrent, M.; Boix, E. Positional scanning library applied to the human eosinophil cationic protein/RNase3 N-terminus reveals novel and potent anti-biofilm peptides. *Eur. J. Med. Chem.* **2018**, *152*, 590–599. [CrossRef]

28. Sandín, D.; Valle, J.; Chaves-Arquero, B.; Prats-Ejarque, G.; Larrosa, M.N.; González-López, J.J.; Jiménez, M.Á.; Boix, E.; Andreu, D.; Torrent, M. Rationally modified antimicrobial peptides from the N-Terminal domain of human RNase 3 show exceptional serum stability. *J. Med. Chem.* **2021**, *64*, 11472–11482. [CrossRef]

29. Perez, F.; Hujer, A.M.; Hujer, K.M.; Decker, B.K.; Rather, P.N.; Bonomo, R.A. Global challenge of multidrug-resistant *Acinetobacter baumannii*. *Antimicrob. Agents Chemother.* **2007**, *51*, 3471–3484. [CrossRef]

30. Falagas, M.E.; Bliziotis, I.A. Pandrug-resistant Gram-negative bacteria: The dawn of the post-antibiotic era? *Int. J. Antimicrob. Agents* **2007**, *29*, 630–636. [CrossRef]

31. Gutsmann, T.; Razquin-Olazarán, I.; Kowalski, I.; Kaconis, Y.; Howe, J.; Bartels, R.; Hornef, M.; Schürholz, T.; Rössle, M.; Sanchez-Gómez, S.; et al. New antiseptic peptides to protect against endotoxin-mediated shock. *Antimicrob. Agents Chemother.* **2010**, *54*, 3817–3824. [CrossRef] [PubMed]

32. Torrent, M.; Pulido, D.; De La Torre, B.G.; García-Mayoral, M.F.; Nogués, M.V.; Bruix, M.; Andreu, D.; Boix, E. Refining the eosinophil cationic protein antibacterial pharmacophore by rational structure minimization. *J. Med. Chem.* **2011**, *54*, 5237–5244. [CrossRef] [PubMed]

33. Wiegand, I.; Hilpert, K.; Hancock, R.E.W. Agar and broth dilution methods to determine the minimal inhibitory concentration (MIC) of antimicrobial substances. *Nat. Protoc.* **2008**, *3*, 163–175. [CrossRef] [PubMed]

34. Lakshminarayanan, R.; Tan, W.X.; Aung, T.T.; Goh, E.T.L.; Muruganantham, N.; Li, J.; Chang, J.Y.T.; Dikshit, N.; Saraswathi, P.; Lim, R.R.; et al. Branched peptide, B2088, disrupts the supramolecular organization of lipopolysaccharides and sensitizes the gram-negative bacteria. *Sci. Rep.* **2016**, *6*, 25905. [CrossRef]

35. Kósa, N.; Zolcsák, Á.; Voszka, I.; Csík, G.; Horváti, K.; Horváth, L.; Bősze, S.; Herenyi, L. Comparison of the efficacy of two novel antitubercular agents in free and liposome-encapsulated formulations. *Int. J. Mol. Sci.* **2021**, *22*, 2457. [CrossRef]

36. McConnell, M.J.; Actis, L.; Pachón, J. *Acinetobacter baumannii*: Human infections, factors contributing to pathogenesis and animal models. *FEMS Microbiol. Rev.* **2013**, *37*, 130–155. [CrossRef]

37. Jürgens, G.; Müller, M.; Koch, M.H.J.; Brandenburg, K. Interaction of hemoglobin with enterobacterial lipopolysaccharide and lipid A. Physicochemical characterization and biological activity. *Eur. J. Biochem.* **2001**, *268*, 4233–4242. [CrossRef]

38. Bárcena-Varela, S.; Martínez-De-tejada, G.; Martin, L.; Schuerholz, T.; Gil-Royo, A.G.; Fukuoka, S.; Goldmann, T.; Droemann, D.; Correa, W.; Gutsmann, T.; et al. Coupling killing to neutralization: Combined therapy with ceftriaxone/Pep19-2.5 counteracts sepsis in rabbits. *Exp. Mol. Med.* **2017**, *49*, e345. [CrossRef]

39. Lorenz, A.; Pawar, V.; Häussler, S.; Weiss, S. Insights into host–pathogen interactions from state-of-the-art animal models of respiratory *Pseudomonas aeruginosa* infections. *FEBS Lett.* **2016**, *590*, 3941–3959. [CrossRef]

40. Mardirossian, M.; Pompilio, A.; Crocetta, V.; De Nicola, S.; Guida, F.; Degasperi, M.; Gennaro, R.; Di Bonaventura, G.; Scocchi, M. In vitro and in vivo evaluation of BMAP-derived peptides for the treatment of cystic fibrosis-related pulmonary infections. *Amino Acids* **2016**, *48*, 2253–2260. [CrossRef]

41. Xiong, Y.Q.; Li, L.; Zhou, Y.; Kraus, C.N. Efficacy of ARV-1502, a proline-rich antimicrobial peptide, in a murine model of bacteremia caused by multi-Drug resistant (MDR) *Acinetobacter baumannii*. *Molecules* **2019**, *24*, 2820. [CrossRef] [PubMed]

42. Shrestha, A.; Duwadi, D.; Jukosky, J.; Fiering, S.N. Cecropin-like antimicrobial peptide protects mice from lethal *E. coli* infection. *PLoS ONE* **2019**, *14*, e0220344. [CrossRef]

43. Rosenberg, H.F.; Dyer, K.D.; Lee Tiffany, H.; Gonzalez, M. Rapid evolution of a unique family of primate ribonuclease genes. *Nat. Genet.* **1995**, *10*, 219–223. [CrossRef] [PubMed]

44. Zhang, J.; Rosenberg, H.F.; Nei, M. Positive Darwinian selection after gene duplication in primate ribonuclease genes. *Proc. Natl. Acad. Sci. USA* **1998**, *95*, 3708–3713. [CrossRef] [PubMed]

45. Hicks, R.P.; Abercrombie, J.J.; Wong, R.K.; Leung, K.P. Antimicrobial peptides containing unnatural amino acid exhibit potent bactericidal activity against ESKAPE pathogens. *Bioorgan. Med. Chem.* **2013**, *21*, 205–214. [CrossRef] [PubMed]

46. Gentilucci, L.; De Marco, R.; Cerisoli, L. Chemical modifications designed to improve peptide stability: Incorporation of non-natural amino acids, pseudo-peptide bonds, and cyclization. *Curr. Pharm. Des.* **2010**, *16*, 3185–3203. [CrossRef]

47. Nord, N.M.; Hoeprich, P.D. Polymyxin B and colistin. A Critical Comparison. *N. Engl. J. Med.* **1964**, *270*, 1030–1035. [CrossRef]

48. Eadon, M.T.; Hack, B.K.; Alexander, J.J.; Xu, C.; Dolan, M.E.; Cunningham, P.N. Cell cycle arrest in a model of colistin nephrotoxicity. *Physiol. Genom.* **2013**, *45*, 877–888. [CrossRef]

49. Huang, Y.; Wiradharma, N.; Xu, K.; Ji, Z.; Bi, S.; Li, L.; Yang, Y.-Y.; Fan, W. Cationic amphiphilic alpha-helical peptides for the treatment of carbapenem-resistant *Acinetobacter baumannii* infection. *Biomaterials* **2012**, *33*, 8841–8847. [CrossRef]

50. Bowers, D.R.; Cao, H.; Zhou, J.; Ledesma, K.R.; Sun, D.; Lomovskaya, O.; Tam, V.H. Assessment of minocycline and polymyxin B combination against *Acinetobacter baumannii*. *Antimicrob. Agents Chemother.* **2015**, *59*, 2720–2725. [CrossRef]

51. Pichardo, C.; Pachón-Ibañez, M.E.; Docobo-Perez, F.; López-Rojas, R.; Jiménez-Mejías, M.E.; Garcia-Curiel, A.; Pachon, J. Efficacy of tigecycline vs. imipenem in the treatment of experimental *Acinetobacter baumannii* murine pneumonia. *Eur. J. Clin. Microbiol. Infect. Dis.* **2010**, *29*, 527–531. [CrossRef] [PubMed]

52. Montero, A.; Ariza, J.; Corbella, X.; Doménech, A.; Cabellos, C.; Ayats, J.; Tubau, F.; Ardanuy, C.; Gudiol, F. Efficacy of colistin versus β-lactams, aminoglycosides, and rifampin as monotherapy in a mouse model of pneumonia caused by multiresistant *Acinetobacter baumannii*. *Antimicrob. Agents Chemother.* **2002**, *46*, 1946–1952. [CrossRef] [PubMed]

53. Yokota, S.-I.; Hakamada, H.; Yamamoto, S.; Sato, T.; Shiraishi, T.; Shinagawa, M.; Takahashi, S. Release of large amounts of lipopolysaccharides from *Pseudomonas aeruginosa* cells reduces their susceptibility to colistin. *Int. J. Antimicrob. Agents* **2018**, *51*, 888–896. [CrossRef] [PubMed]

54. Mueller, M.; Lindner, B.; Kusumoto, S.; Fukase, K.; Schromm, A.B.; Seydel, U. Aggregates are the biologically active units of endotoxin. *J. Biol. Chem.* **2004**, *279*, 26307–26313. [CrossRef] [PubMed]

Review

Chemical Barrier Proteins in Human Body Fluids

Gergő Kalló [1,2,3,*], Ajneesh Kumar [1,2,3], József Tőzsér [1,2,3,4] and Éva Csősz [1,2,3]

1   Proteomics Core Facility, Department of Biochemistry and Molecular Biology, Faculty of Medicine, University of Debrecen, Egyetem tér 1, 4032 Debrecen, Hungary; kumar.ajneesh@med.unideb.hu (A.K.); tozser@med.unideb.hu (J.T.); cseva@med.unideb.hu (É.C.)
2   Biomarker Research Group, Department of Biochemistry and Molecular Biology, Faculty of Medicine, University of Debrecen, Egyetem tér 1, 4032 Debrecen, Hungary
3   Doctoral School of Molecular Cell and Immune Biology, University of Debrecen, Egyetem tér 1, 4032 Debrecen, Hungary
4   Laboratory of Retroviral Biochemistry, Department of Biochemistry and Molecular Biology, Faculty of Medicine, University of Debrecen, Egyetem tér 1, 4032 Debrecen, Hungary
*   Correspondence: kallo.gergo@med.unideb.hu; Tel.: +36-52-416432

Abstract: Chemical barriers are composed of those sites of the human body where potential pathogens can contact the host cells. A chemical barrier is made up by different proteins that are part of the antimicrobial and immunomodulatory protein/peptide (AMP) family. Proteins of the AMP family exert antibacterial, antiviral, and/or antifungal activity and can modulate the immune system. Besides these proteins, a wide range of proteases and protease inhibitors can also be found in the chemical barriers maintaining a proteolytic balance in the host and/or the pathogens. In this review, we aimed to identify the chemical barrier components in nine human body fluids. The interaction networks of the chemical barrier proteins in each examined body fluid were generated as well.

Keywords: body fluid; AMP; interaction network; chemical barrier

Citation: Kalló, G.; Kumar, A.; Tőzsér, J.; Csősz, É. Chemical Barrier Proteins in Human Body Fluids. *Biomedicines* 2022, 10, 1472. https://doi.org/10.3390/biomedicines10071472

Academic Editor: Jitka Petrlova

Received: 31 May 2022
Accepted: 20 June 2022
Published: 22 June 2022

Publisher's Note: MDPI stays neutral with regard to jurisdictional claims in published maps and institutional affiliations.

1. Introduction

At those sites where the human body can make contact with potential pathogens, well-defined chemical barriers exist as part of the immune system. These chemical barriers provide passive protection against infections by diluting the colony number of pathogens and they can also actively inhibit bacterial growth due to the secretion of antimicrobial and immunomodulatory proteins/peptides (AMPs) [1]. The human body contains several contact sites: the eye, the oral cavity, the nose, the skin, the intestinal surface, and the urogenital tract. Each of these sites is protected by a chemical barrier maintained by different body fluids such as tears, sweat, saliva, nasal secretion, urine, intestinal mucus, and cervicovaginal fluid. These chemical barriers are made up of the secretion of various glands and epithelial cells and the characteristic composition of the chemical barrier makes the secreted AMP cocktail specific for a given body fluid [2]. The secreted AMP cocktail can adapt to various conditions [3,4]; therefore, the composition of the chemical barrier is continuously changing. Regarding the protein composition of the body fluids providing chemical barriers, it has been observed that the highly abundant proteins characteristic of each body fluid are part of the immune system and have protective roles. In the case of tears and sweat, it has been demonstrated that more than 90% of the secreted proteins have a role in the host defense mechanisms [5,6]. In this review, we aimed to examine the proteins identified in nine body fluids to identify the chemical barrier components. We also generated the interaction networks of these chemical barrier proteins in the examined body fluids.

2. Members of the Chemical Barriers

While some proteins/peptides such as defensins and LL-37 cathelicidin were first isolated due to their antimicrobial properties (so-called prototypic AMPs), other highly abundant proteins were initially recognized for their other functions, such as serum albumin [7], hemoglobin [8] or thrombin [9], and later, they or their peptides were found to have antimicrobial activity. While AMPs are parts of the innate immune system, members of the adaptive immune system, such as secreted immunoglobulin A (IgA), can be found in the chemical barriers as well. IgA is capable of blocking pathogens from attaching to intestinal epithelial cells [10], while the different AMPs are responsible for pathogen killing, by limiting their growth and modulating the immune reactions [2].

2.1. Prototypic AMPs in the Chemical Barriers

In the human body, the chemical barriers contain several prototypic AMPs, such as defensins, dermcidin and LL-37 cathelicidin [2]. Prototypic AMPs adsorb onto the bacterial cell membrane by electrostatic attraction or hydrophobic interactions [11] and insert into the membrane leading to the formation of channels and transmembrane pores or creating extensive membrane ruptures [2].

Defensins are small cationic peptides produced by epithelial cells forming a very stable 3D structure called defensin fold because of their cysteine-rich sequence [12]. The two subfamilies of human defensins (α- and β-defensins) show broad antimicrobial, antifungal, and antiviral activities [2,13]. Human α-defensins 1–4 are expressed by neutrophils and have been shown to protect against mycobacterium and various viruses including herpes simplex virus 1 and 2 (HSV-1, HSV2), cytomegalovirus (CMV), and also influenza virus; however, their activity may depend on the lipid composition of the viral envelope [12,14]. The α-defensins 1–3 also possess chemotactic activity for monocytes [12], α-defensins 5–6 are human enteric defensins constitutively expressed by Paneth cells and α-defensin 5 has been detected preferentially in the cervical mucosa and the oviduct [15]. Human β-defensins (hBDs) are coded on 11 genes but not all transcripts have been identified so far [16]. They can be considered as potential AMPs in epithelial cells providing a protective barrier against Gram-negative bacteria and *Candida* species [2]. Another important function of hBDs is their chemotactic activity toward various immune cells [17]. While some defensins, such as hBD1, show constitutive expression pattern, other members of this family have been found to be induced upon pathogenic or inflammatory stimuli [18]. In addition to their antimicrobial and immunomodulatory effects, some defensins have been identified as cancer-associated molecules with anti-tumor effects [19].

LL-37 cathelicidin is an α-helix type AMP of the body fluids mainly produced by epithelial cells and immune cells [2,20]. Similar to defensins, it exerts various antimicrobial activities against different types of pathogens, such as *Listeria monocytogenes*, *Staphylococcus aureus*, *Pseudomonas aeruginosa* or *Candida albicans* [21]. In addition, LL-37 cathelicidin has an important role in re-epithelialization during wound healing [22]. Besides the antimicrobial activity and wound healing, LL-37 cathelicidin has a neutralizing effect on lipopolysaccharide (LPS), it has chemotactic activity for different immune cells, modulates the transcriptional response of macrophages, stimulates vascularization, and exerts antitumor activity [21]. It should be noted that besides the LL-37 form, there are several cathelicidin-derived peptides with higher abundance in the different body fluids. These peptides are generated by proteolytic cleavage of the propeptide by proteases belonging to the kallikrein family [20]. These cathelicidin peptides exert various antimicrobial activity but they have no chemotactic activity [23].

2.2. Highly Abundant Body Fluid Proteins Are Constituents of the Chemical Barriers

Besides the prototypic AMPs, there are several proteins with much higher concentration as compared to those of prototypic AMPs. These proteins, e.g., lactotransferrin, lipocalins, lysozyme-C, extracellular glycoprotein lacritin, and prolactin-inducible protein are the highly abundant body fluid proteins with various defense functions.

It has been shown that lactotransferrin found in all body fluids is an active AMP against bacteria and parasites, and has been implicated in protection against cancer [2]. Because of its iron-sequestering activity, lactotransferrin has an important role in the prevention of bacterial colonization. Due to proteolytic cleavage, different lactotransferrin-derived peptides are produced such as Lf(1-11), lactoferricin, and lactoferrampin having stronger antimicrobial activity than the intact protein [24]. Besides its antimicrobial activity, lactotransferrin can promote autophagy [25] and it has also been shown that via its DNA-binding activity, the protein can alter the transcriptional machinery of different cells [26].

Lysozyme-C is a ubiquitous hydrolytic enzyme that exerts muraminidase activity, required for the peptidoglycan degradation of the bacterial cell wall [2]. Shimada et al., have shown that along with the degradation of the bacterial cell wall, the muraminidase activity is also important in the cleavage of pro-interleukin-1β (pro-IL-1β) to active IL-1β [27]. Besides its antibacterial activity, lysozyme-C has various functions including antifungal activity [28] and protection against HIV infection [29]. Lysozyme-C isolated from hen egg white also shows an immunomodulatory effect by suppressing the inflammatory response induced by LPS via inhibition of the phosphorylation of c-jun N-terminal kinase (JNK) in a murine model [30].

Lipocalins are a family of lipid and other small molecule binding proteins with protease inhibitor activity and with iron sequestrating capability that can limit the growth of pathogenic bacteria [31,32]. Lipocalins typically transport and/or store different small molecules, including vitamins, steroid hormones, and various secondary metabolites [33]. Studies have shown that lipocalins have roles in cancer development [33–35] and a study published by Mesquita et al. showed that the expression of lipocalin-2 is stimulated by amyloid-beta (Aβ) 1-42 in Alzheimer's disease (AD) [36].

Lacritin is a secreted extracellular glycoprotein found in tears and saliva. The protein has various functions, including the modulation of lacrimal gland secretion [37], epithelial cell proliferation [38], and corneal wound healing [39]. Additionally, the C-terminal fragment of extracellular glycoprotein lacritin has proved bactericidal activity [40]. Regarding its effect on tear secretion rate, lacritin is a potent therapeutical target in dry eye disease [41].

Prolactin-inducible protein is an aspartyl protease [42] identified in various body fluids as part of the host defense system. Besides the protease activity, prolactin-inducible protein can modulate immune reaction by binding to immunoglobulin G and Zn-α-2-glycoprotein [43,44] and its elevated expression has been associated with breast cancer progression [42,45]. Interestingly, Edechi et al. proved that prolactin-inducible protein modulates antitumor immune response as well, suggesting a bifunctional role of this protein [46].

Dermcidin is the main skin AMP, which is also present in tears and exerts broad-spectrum antimicrobial activity [2,47]. Dermcidin is constitutively secreted by eccrine sweat glands and epithelial cells, and its secretion cannot be further induced either by skin injury or inflammation [48]. By post-secretory proteolytic processing, dermcidin is cleaved to several truncated dermcidin-derived peptides which differ in length and net charge. Reduced levels of dermcidin-derived peptides has been detected in the sweat of patients with atopic dermatitis in association with an impaired cutaneous antimicrobial defense [49], while the increased expression of dermcidin has been demonstrated in lung, prostate and pancreatic cancer cells [50,51].

2.3. AMPs with Lower Abundance in the Chemical Barriers

Besides the highly abundant AMPs, the different body fluids contain various proteins with antimicrobial activity in a lower concentration, such as members of the S100 family, RNase7, bactericidal/permeability increasing protein, azurocidin, cathepsin G, histatins, and hemocidins.

The members of the S100 family are Ca^{2+}-binding proteins with EF-hand domains [52]. The family consists of 24 members with various intracellular and extracellular functions [53]. In the intracellular space, they can regulate cell proliferation, differentiation and apoptosis, cytoskeletal organization and Ca^{2+} homeostasis [53,54]. The secreted forms of S100 proteins have a paracrine effect on the nearby cells by receptor activation and regulation of different cell types such as immune cells, endothelial cells, and muscle cells [53,54]. S100A7 or psoriasin can be found on the surface of the skin secreted by the sweat and is a potent AMP against different pathogenic bacteria [2].

RNase7 mainly produced by keratinocytes and also released to the skin surface exerts a broad range of antimicrobial activity [55]. RNase7 is an inducible AMP, its expression can be enhanced by different stimuli such as inflammatory cytokines, growth factors, or pathogenic bacteria [56]. Besides the skin surface, RNase7 was also identified in the urinary tract acting as a potent AMP in the protection against different pathogens [57].

There are several AMPs, such as bactericidal/permeability increasing protein (BPI), azurocidin, myeloperoxidase or cathepsin G that are stored in the granules of neutrophil granulocytes and released upon the activation of the cell triggered by different pathogenic stimuli. BPI can be released to various sites of the human body and can initiate the permeabilization of the bacterial cell wall; therefore, it promotes the lysis of the bacteria [2,58]. Azurocidin acts against Gram-positive and negative bacterial strains and fungi as well [2] and it has been shown to have binding affinity to heparin [59]. Cathepsin G is a serine protease involved in the innate immunity, regulation of inflammatory pathways, degradation of extracellular matrix components and also in antigen presentation [60].

Histatins represent a group of 12 different histidine-rich peptides with antimicrobial activity in saliva [2]. Along with the defensins and LL-37 cathelicidin, histatins exert a broad range of antibacterial and antifungal activity but on the other hand, histatins can also act as protease inhibitors and they are involved in wound healing [61]. Moreover, these proteins can bind metal ions such as Cu^{2+} and Zn^{2+} that can modulate the activity of these proteins [61].

Hemocidins are a recently discovered group of AMPs that are derived from heme-binding proteins, such as hemoglobin and myoglobin [62]. Hemoglobin is the major protein of the red blood cells and has a crucial role in the oxygen transport via the cardiovascular system. Hemocidins are released from the hemoglobin α and β chains by limited proteolysis and exert antimicrobial activity against various pathogens: α (1–32), α (33–76), α (1–76), α (77–144), β (1–55), β (56–146), and β (116–146) hemocidins showing an antimicrobial effect against *Staphylococcus aureus*, *Escherichia coli*, *Streptococcus faecalis*, and *Candida albicans*, while β (56–72) hemocidin shows antimicrobial effect against *Escherichia coli* and *Streptococcus faecalis* [8]. It was recently demonstrated that hemoglobin-derived peptides can be generated by oxidative stress in atherosclerotic regions and intraventricular hemorrhage [63]. Myoglobin is an oxygen-binding protein located primarily in muscles serving as a local oxygen reservoir that can temporarily provide oxygen when blood oxygen delivery is insufficient during periods of intense muscular activity. Similar to hemoglobin, hemocidins can also be released from myoglobin: (1–55), (56–131), and (132–153) hemocidins exert antimicrobial activity against the above-mentioned pathogens [64].

2.4. Proteases and Protease Inhibitors

Proteases and their inhibitors are constitutive parts of the chemical barriers. A variety of cells involved in the defense against pathogenic microorganisms are expressing and secreting a wide range of proteolytic enzymes in order to degrade the bacterial/viral/fungal proteins involved in the life cycle of the microorganism. The proteolytic activity of different aminopeptidases, carboxypeptidases, and endopeptidases such as serine proteases, aspartyl proteases, and metalloproteases [65] as part of the chemical barriers, can modulate the homeostatic functions and control the microorganisms entering the human body [66–68].

Since the proteolytic activity is a double-edge sword capable of the degradation of the host proteins as well, the presence of protease inhibitors is crucial for the host. The human body expresses many different protease inhibitors with different specificity. Several protease inhibitors such as alpha-2-macroglobulin [69] and Kunitz-type protease inhibitors [70] have a broad inhibitory effect, while specific inhibitors such as alpha-1-antitrypsin [71] and antithrombin-III [72] act only on well-defined proteases. While the protease inhibitors of the host provide the defense against their own proteases, they can also inhibit the proteases secreted by pathogenic microorganisms [73].

2.5. Role of AMPs in Nosocomial Infections

Nosocomial infection, also known as a hospital-acquired infection, is contracted from the environment or staff of a healthcare facility affecting nearly 9% of patients according to the WHO [74]. The most common nosocomial infections are caused by common bacteria such as *Pseudomonas aeruginosa*, *Staphylococcus aureus*, *Acinetobacter baumannii*, and *Enterococci* species [75–77] usually leading to milder diseases. Bacterial infections are treated with antibiotics that kill or suppress the bacterial species sensitive to the antibiotic, while the resistant strains survive and may spread in the hospital [78,79]. Based on the literature data, the major mechanisms of antibiotic resistance include the hydrolysis of antibiotics, avoiding antibiotics targeting, prevention of antibiotics permeation, and the active efflux of antibiotics from bacteria [74]. The development of new antibiotics is far behind the increasing emergence of drug-resistant bacteria; therefore, alternative treatment for the infection of multidrug-resistant bacteria is an urgent task of biomedical research.

AMPs are promising candidates due to their fast killing kinetics, pharmacodynamic properties, and mechanisms of killing that overcome the common resistance mechanisms of pathogens [75]. The biofilm-disrupting properties of AMPs may also confer efficacy against multidrug-resistant bacterial infections associated with wounds and/or medical implants [75]. While preclinical studies were promising for numerous AMPs, most of the investigated AMPs failed in clinical studies [80–82]. Combination therapies of AMPs with antibiotics have also been investigated. The synergistic effect of proline-rich peptides with polymixin E was revealed in in vitro and in *Klebsiella pneumonia* infected mice models [83]. Besides the combination therapies, many drug delivery systems are investigated to deliver AMPs, such as nanocarriers and star polymers [84].

Wounds and other lesions are highly exposed to nosocomial infections due to the possible contamination with bacteria from the surface of the body or the environment. The wound healing process involves the activation of a variety of protease enzymes that can be altered by invading pathogens [85]. The analysis of wound fluids from the normal healing process and infected wounds revealed differences in the secreted proteins and processed peptides between the studied groups [85,86] that can lead to deteriorated healing. Since the proteolytic events are also crucial for the processing of AMPs [9,87], nosocomial wound infections may alter the host defense system by altering the production of AMPs in the wound fluid.

3. Chemical Barrier Proteins in Human Body Fluids

3.1. The Composition of the Chemical Barrier in Serum

Serum is the most widely used body fluid in biomedical sciences and contains various proteins, lipids, and small molecules. Serum is often used for the diagnosis of different pathological conditions and also for monitoring applied therapies. The serum proteins mainly originate from the liver and other tissues but the secretory activity of blood cells can also increase the number of serum proteins [88]. The protein concentration of healthy human serum varies between 60–80 mg/mL [89]. To date, more than 12,000 proteins have been found in human serum [90] with several highly abundant proteins such as albumin, immunoglobulins, transferrin, haptoglobin, and apoproteins that constitute approximately 90% of the total protein content [91].

Albumin is one of the most important transport proteins in the human body that can bind a variety of ligands such as ions, fatty acids, vitamins, hormones, and has a role in the maintenance of blood viscosity, in the regulation of cholesterol transport and coagulation events [92]. As a negative acute-phase protein, albumin has a role in the inflammatory response of the body [93] and Gum et al., have shown that albumin has antioxidant activity due to its N-terminal DAHK tetrapeptide [94]. Moreover, albumin exerts antimicrobial effect against *Candida albicans, Cryptococcus neoformans, Escherichia coli,* and *Staphylococcus aureus* [7].

Transferrin is the major iron transport protein having a role in cell growth and differentiation [95]. As an iron-binding molecule, transferrin can act against bacterial infections by sequestering the free iron from the serum, therefore, inhibits bacterial growth [96]. Furthermore, Ardehali. et al. proved that apotransferrin can reduce the surface adhesion of both Gram-positive and Gram-negative bacteria [97].

Haptoglobin can bind to hemoglobin thus preventing the iron loss from the free hemoglobin. The hemoglobin–haptoglobin complex can be removed via the reticuloendothelial system and by receptor-mediated endocytosis [98,99]. Haptoglobin has been shown to have immunoregulatory function via the regulation of T-cells and the expression of IL-6 and IL-10 cytokines [100]. The binding of hemoglobin can be considered as an iron-sequestering effect; thus, haptoglobin has a role in the defense against reactive oxygen species (ROS), that can be formed due to the free iron [100].

The major function of apoproteins is the construction of lipoproteins, such as chylomicron, very low density lipoprotein (VLDL), low-density lipoprotein (LDL), or high-density lipoprotein (HDL) that carry triglycerides, cholesterol, cholesterol esters, and other type of lipids in the circulation system [101]. Apoproteins also have a role in the host defense mechanisms; antimicrobial activity of Apo A1 against *Staphylococcus epidermidis* has been described [102] and antimicrobial peptides derived from Apo B acting against *Salmonella* strains have also identified [103].

From the approximately 12,000 identified serum proteins, 422 can act in the first line of host defense (Table S1). The identified chemical barrier proteins were subjected to network analysis, and the interaction network of these proteins was generated by CluePedia v.1.5.7 as we described earlier by our group [104]. The interaction network of the chemical barrier proteins in serum is visualized in Figure 1.

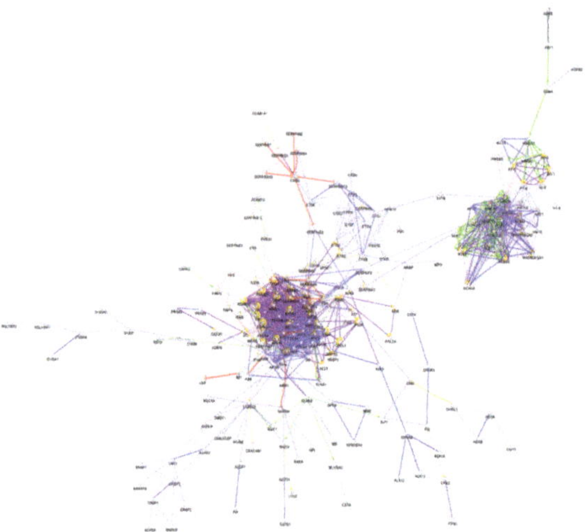

Figure 1. Partial network view of the interaction network of the chemical barrier proteins in serum. Each circle represents a protein and the lines indicate interactions. The lines with an arrow represent activation, blocking lines represent inhibition, and simple lines represent protein–protein interaction. Line color indicates the type of interaction: green color refers to activation, red color to inhibition, blue color to binding, yellow color to co-expression, and purple color to catalysis. The proteins are labeled with their gene name. The high resolution and complete network image of this network is presented in Figure S1.

The network analysis revealed that most of the chemical barrier proteins are part of two core clusters that interact with each other. The identified histone proteins form one of the clusters, while more than 80% of the other identified chemical barrier proteins form the other highly interconnected cluster. In this cluster, we can find many different hub proteins such as alpha-1-antitrypsin, metalloproteinase inhibitor 1, apolipoprotein AI, apolipoprotein E, serotransferrin, and cystatin-C with a wide range of interactions. On the other hand, we identified seven additional clusters with a small number of proteins without interactions with the core hubs (Figure S1). The network architecture is serum showed the highest complexity among the examined body fluids.

3.2. Tears, the Chemical Barrier of the Eye

Tear is a protein-, metabolite-, and salt-rich fluid produced by the lacrimal glands, Meibomian glands, and conjunctival goblet cells. The normal tear production rate is approximately 2 μL/min [105] and its typical protein concentration is 5–7 μg/μL [106]. The functions of the tear film are the lubrication of the eye, delivery of nutrients, and maintaining the refractivity of the cornea [107]. Besides these roles, tears create an effective chemical barrier on the surface of the eye via secreted AMPs, which provide protection against pathogens [108].

Currently, more than 1800 proteins have been identified in tears by state-of-the-art proteomics techniques [109]. The major tear proteins, such as lysozyme-C, prolactin-inducible protein, lactotransferrin, and lacritin, have antimicrobial activity; therefore, they are involved in the defense against pathogens [5]. While many of the tear proteins are produced by the lacrimal glands, some of them originate from epithelial cells, such as dermcidin and defensins, and there are also proteins, such as albumin, originating from the blood [110].

The examination of the tear proteome revealed 200 proteins secreted to the surface of the eye and building the chemical barrier against pathogens (Table S2). Overall, 97% of the AMPs found in tears are also components of the chemical barrier of the serum. Regarding the protein–protein interaction network of the chemical barrier proteins in tears (Figure 2), we can say that there are similarities at the level of small, unrelated clusters, but in case of the tears, the majority of AMPs are organized only into one core cluster. Most of the histone-derived AMPs found in serum are missing from tears.

Compared to serum, the number of hub proteins such as protease inhibitors, transferrin, kininogen-1, amyloid-beta precursor protein, and alpha-2-HS-glycoprotein were similar. We also identified three additional small clusters with no connections to the core cluster (Figure S2). Our results demonstrate that in spite of the similarities, the networks characteristic to serum and tear AMPs are different from each other.

3.3. Salivary Proteins in the Defense of the Oral Cavity

Saliva is a complex mixture of organic and inorganic compounds secreted from major and minor salivary glands and the gingival crevice [111]. Saliva is a very dilute body fluid composed of approximately 99% water with 0.7–2.4 μg/μL protein concentration [112,113]. The protein concentration shows high variability between the individuals depending on the age, sex, sample collection time, and the health status of the oral cavity. Saliva contains more than 2700 proteins [109] and the most abundant ones are α-amylase, mucins, cystatins, proline-rich peptides, and serum albumin [114]. Similar to tears, the abundant salivary proteins are part of the immune system due to their antimicrobial activity, antioxidant function, and protective role against microbial proteases [115].

Amylases are mostly known by their hydrolytic activity on polysaccharides, and due to their hydrolytic activity, amylases can inhibit biofilm formation by cleaving the polysaccharide backbone of extracellular polymeric substances [116]. At the same time, there is evidence showing that amylase can bind to the amylase-binding protein of *Streptococcus* species and induce biofilm formation [116]. Therefore, the function of amylase in the biofilm formation is still unclear.

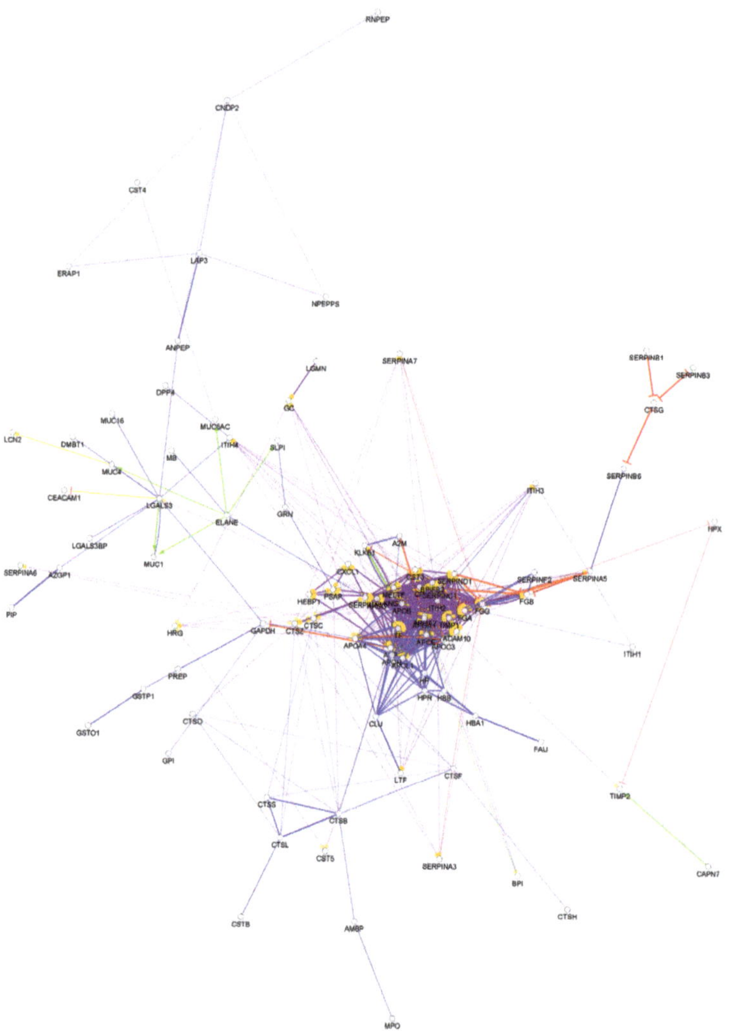

Figure 2. Partial network view of the interaction network of the chemical barrier proteins in tears. Each circle represents a protein and the lines indicate interactions. The lines with an arrow represent activation, blocking lines represent inhibition, and simple lines represent protein–protein interaction. Line color indicates the type of interaction: green color refers to activation, red color to inhibition, blue color to binding, yellow color to co-expression, and purple color to catalysis. The proteins are labeled with their gene name. The high resolution and complete network image of this network is presented in Figure S2.

Mucins are high-molecular-weight glycoproteins composed of a core protein coupled to carbohydrate chains and sulfate groups. They have characteristic amino acid repeats and a Cys-rich domain that has a key role in multimerization and mucus function by forming disulfide bridges [117]. Acting in the host defense, salivary MUC5B and MUC7 can interact with numerous bacteria such as the *Streptococcus* species or *Pseudomonas aeruginosa* and pathogenic fungi such as *Candida albicans* [118].

Cystatins represent a family of structurally conserved, low-molecular-weight cysteine protease inhibitors found in nearly everywhere in the human body and in body fluids.

Secreted cystatins regulate extracellular proteases such as a papain-like protease and matrix metalloproteases, while intracellularly they can regulate the activity of cathepsin C [119]. Along with the regulation of human proteases, cystatins can also inhibit the proteases secreted by pathogenic microorganisms, being in this way important members of the chemical barrier.

Proline-rich proteins (PRPs) and their peptide fragments represent a major fraction of salivary proteome produced by the parotid and submandibular glands [120]. PRPs can be divided into two subgroups: acidic PRPs that can adhere to the surface of the teeth and basic PRPs that cannot adhere to teeth but can bind polyphenols and bacteria [121]. By binding specific bacterial strains such as *Streptococci*, PRPs play an important role in preventing caries formation [121].

Among the more than 2700 so far identified salivary proteins, 319 proteins take part in the composition of the chemical barrier of the oral cavity (Table S3). The interaction network of the salivary chemical barrier proteins was generated (Figure 3).

Based on the network analysis, we can observe that the chemical barrier proteins in saliva are organized in two core clusters interacting with each other. Similar to the interaction network of serum AMPs, the histones are organized in one core cluster and the majority of the other chemical barrier proteins form the other core cluster with multiple hub proteins such as alpha-1-antitrypsin, metalloproteinase inhibitor 1, fibrinogen alpha and gamma chains, apolipoprotein AI, apolipoprotein E, and serotransferrin. Those proteins which are not part of these two core clusters are organized in five additional small clusters (Figure S3). The data indicate that the chemical barrier of saliva shows high similarity to the chemical barrier of serum and some similarity to the AMPs characteristic for tears.

3.4. Sweat–The Chemical Barrier of the Skin

The human skin creates an effective barrier against pathogens as a first line of host defense of the body. Besides the provided physical barrier by the cornified envelope, the skin also creates a chemical barrier via AMPs secreted by epithelial cells, sebocytes, and keratinocytes [122]. Sweat is composed of more than 99% water, making it a very dilute body fluid [123]. To date, more than 1200 sweat proteins have been identified by proteomic analyses [109]. Along with the saliva and tear fluid, the abundant sweat proteins are part of the innate immune system. Our workgroup demonstrated that 91% of the secreted sweat proteome is made up of six highly abundant proteins which are dermcidin, clusterin, ApoD, prolactin-inducible protein, and serum albumin [6]. Sweat protein content provides an effective defense against pathogens, and is involved in tissue regeneration after injury [2]. Some AMPs have been shown to be expressed constitutively (e.g., dermcidin) while others have been found to be inducible upon pathogenic stimuli (e.g., LL-37 cathelicidin, hBD2, and hBD3) [2,122,124]. Besides these prototypic AMPs, the presence of lysozyme-C and lactotransferrin in the sweat has been reported as well [2,125].

The secreted form of clusterin belongs to the family of extracellular chaperones and can be found in almost all body fluids. The major functions of these proteins are the maintenance of fluid–epithelial interface homeostasis, and the prevention of the onset of inflammation [126,127]. The study of Jeong et al. suggests that clusterin can interact with matrix metalloproteinases inhibiting their enzymatic activity and it is also suggested that the protein is able to reduce keratinocyte damage and inflammation of the skin [126].

By checking the function of the identified sweat proteins, 128 sweat AMPs were identified (Table S4). The interaction network of the sweat chemical barrier proteins shows a distinct feature compared to the networks observed in the case of other body fluids (Figure 4).

Nearly half of the identified chemical barrier proteins did not interact with each other (Figure 4). Unlike in the other networks, few hub proteins such as antithrombin-III, disintegrin, and metalloproteinase domain-containing protein 10, fibrinogen alpha and gamma chains, serotransferrin, and lactotransferrin are holding most of the interactions. Besides the core cluster, the interacting AMPs formed three additional small clusters

(Figure S4). The data indicate that the sweat chemical barrier is formed in a completely different way than the other examined body fluids highlighting the uniqueness of sweat.

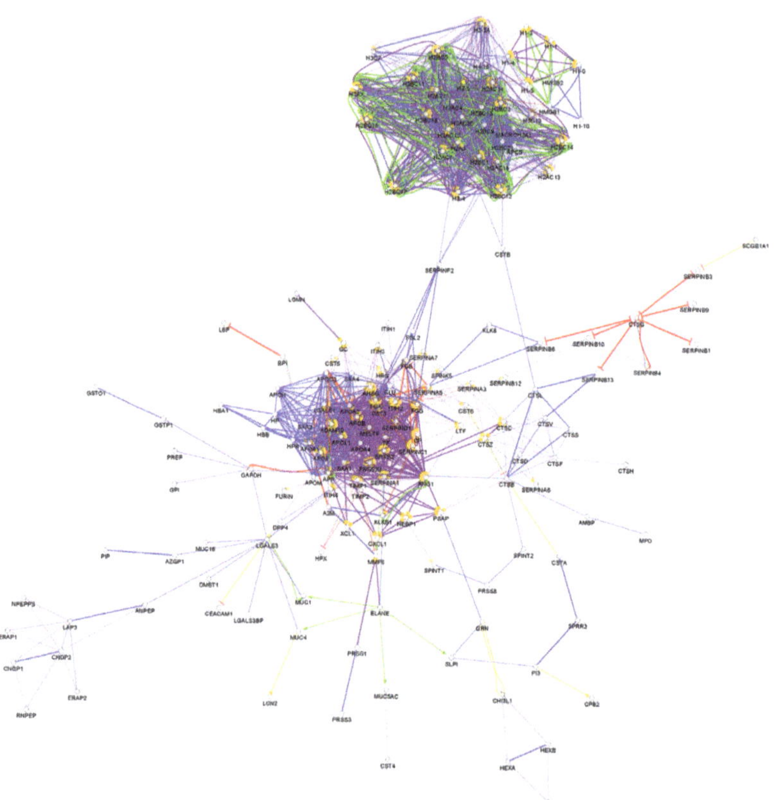

Figure 3. Partial network view of the interaction network of the chemical barrier proteins in saliva. Each circle represents a protein and the lines indicate interactions. The lines with an arrow represent activation, blocking lines represent inhibition, and simple lines represent protein–protein interaction. Line color indicates the type of interaction: green color refers to activation, red color to inhibition, blue color to binding, yellow color to co-expression, and purple color to catalysis. The proteins are labeled with their gene name. The high resolution and complete network image of this network is presented in Figure S3.

3.5. The Chemical Barrier of the Nasal Secretion

The nasal secretion provides protection in the upper respiratory tract by creating an effective chemical barrier in the nasal cavity. The production rate and the protein concentration of nasal secretion show high variability. The protein concentration varies between 0.8 mg/mL and 32.7 mg/mL [128,129]. Nasal secretion contains AMPs such as lactotransferrin, lysozyme-C [130], and immunoglobulin molecules [131]. As a result of various studies, more than 2000 proteins have been identified in the nasal mucus and many of them are related to the host defense system [132–135].

The analysis of the available data showed that 164 AMPs are present in the nasal secretion (Table S5). The protein–protein interactions among the AMPs are mainly organized into two core clusters (Figure 5).

Our analysis of the interactions among the nasal AMPs highlighted a similar pattern to the interactions found in saliva. The majority of the chemical barrier proteins in the nasal secretion were organized into two core clusters, one core cluster composed of the histone proteins and the other containing the majority of the other chemical barrier proteins. The hub proteins (alpha-1-antitrypsin, fibrinogen alpha and gamma chain, apolipoprotein AI, apolipoprotein E, and serotransferrin) were similar in the above-mentioned body fluids. We also observed that there is no interaction between the core clusters in the nasal secretion. We also identified four small clusters (Figure S5). The nasal and oral cavity are connected to each other and in some cases similar factors may influence the composition of the secreted AMP cocktail, the interaction pattern of the AMPs observed in these two body fluids show similarities but at the same time clearly indicate the differences as well.

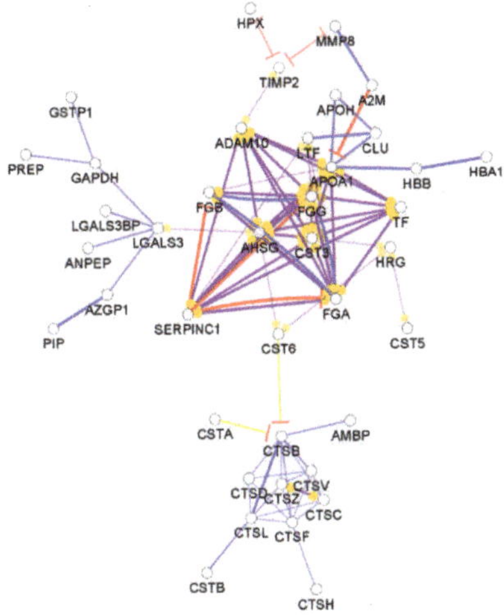

Figure 4. Partial network view of the interaction network of the chemical barrier proteins in sweat. Each circle represents a protein and the lines indicate interactions. The lines with an arrow represent activation, blocking lines represent inhibition, and simple lines represent protein–protein interaction. Line color indicates the type of interaction: green color refers to activation, red color to inhibition, blue color to binding, yellow color to co-expression, and purple color to catalysis. The proteins are labeled with their gene name. The high resolution and complete network image of this network is presented in Figure S4.

3.6. AMPs Secreted into the Urine

Urine is formed by the kidneys as a result of ultrafiltration of the plasma to eliminate waste products such as urea, metabolites, or xenobiotics from the body. The protein concentration in urine under physiological conditions is very low, nearly 1000 times less compared to other body fluids such as plasma [136]. Urine contains more than 7000 proteins; unsurprisingly, many of these proteins are part of the defense system in the body [109]. The most abundant urine proteins have been found to be serum albumin, uromodulin, α-1-microglobulin, kininogen and various immunoglobulin chains [137].

Uromodulin (or Tamm-Horsfall protein) is produced by the renal epithelial cells and has various roles in the body, such as the regulation of salt transport, protection against kidney stones, and an antimicrobial effect against pathogens, as well as immunomodulatory functions [138]. It has been shown that uromodulin can form filaments that act as a multivalent decoy for pathogenic bacteria [139]. The resulting uromodulin pathogen aggregates prevent bacterial adhesion to epithelial glycoproteins and promote pathogen clearance as they are excreted with urine [140]. Uromodulin also shows immunomodulatory effects by binding to EGF-like receptors, cathepsin G and lactoferrin, to enhance the phagocytic activity of polymorphonuclear leukocytes, production of proinflammatory cytokines by monocytes/macrophages, and lymphocyte proliferation via the activation of mitogen-activated protein kinase signaling pathways [141].

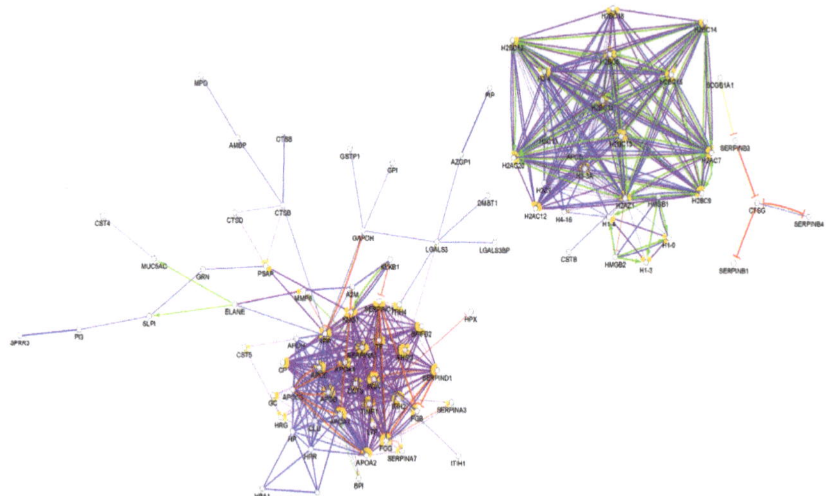

Figure 5. Partial network view of the interaction network of the chemical barrier proteins in the nasal secretion. Each circle represents a protein and the lines indicate interactions. The lines with an arrow represent activation, blocking lines represent inhibition, and simple lines represent protein–protein interaction. Line color indicates the type of interaction: green color refers to activation, red color to inhibition, blue color to binding, yellow color to co-expression, and purple color to catalysis. The proteins are labeled with their gene name. The high resolution and complete network image of this network is presented in Figure S5.

A-1-microglobulin is a member of the lipocalin family responsible for the protection against free radicals and is involved in natural tissue repair [142]. The protein is expressed by the liver and after secretion into the blood, the protein can form complexes with other macromolecules or can exist in a free form. The free form can pass through the glomeruli filter to the urine [142]. As an antioxidant molecule, α-1-microglobulin has reductase activity [143], exerts free radical scavenging activity [144], and can bind to free hem [145].

Kininogens represent a group of multifunctional glycoproteins synthesized by the liver and predominantly circulate in the blood but they can also be found in other body fluids such as urine [146]; however, the function of the urinary kininogens are still unclear. In the blood, kininogen can interact with plasma kallikrein to produce bradykinin in the contact activation of the coagulation system [147]. It was demonstrated in the early 1980s that kininogen has a role in the regulation of blood pressure and in the modulation of water and salt transport in the kidney [148]. Besides the above-mentioned functions, kininogen is able to bind to the surface of a variety of pathogen microorganisms such as *Streptococcus pyogenes* [149] or *Candida albicans* [150] leading to activation of the contact system in the blood. In addition, Sonesson et al. demonstrated that due to the proteolytic process of kininogen, peptide fragments with antifungal activity are generated [151].

From the so-far identified urine proteins, 396 have antimicrobial and/or immunomodulatory activity (Table S6). The interaction network of the identified AMPs in urine was generated and visualized, as can be seen in Figure 6.

The network analysis of the urine chemical barrier proteins indicates a similar pattern to the one identified in serum: one core cluster with the histone proteins and another one with the majority of other AMPs. Similar to saliva and serum, in the case of urine, there are connections between the two major clusters. In the case of saliva and urine, the SERPINF2 provides a hub linking the two clusters. Six small clusters were observed in case of urine as well (Figure S6). Since the urine is filtered from the serum and low-molecular-weight proteins can be transferred through the glomeruli, the similarity between the interaction network of serum and urine is not surprising. However, differences between the organization of the connections and hub proteins are also observed, indicating that the urine has a distinct chemical barrier.

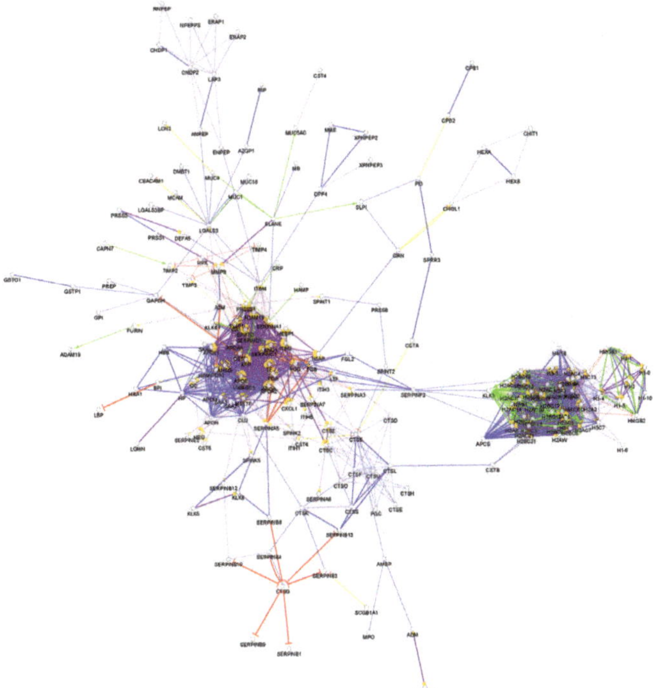

Figure 6. Partial network view of the interaction network of the chemical barrier proteins in urine. Each circle represents a protein and the lines indicate interactions. The lines with an arrow represent activation, blocking lines represent inhibition, and simple lines represent protein–protein interaction. Line color indicates the type of interaction: green color refers to activation, red color to inhibition, blue color to binding, yellow color to co-expression, and purple color to catalysis. The proteins are labeled with their gene name. The high resolution and complete network image of this network is presented in Figure S6.

3.7. Antimicrobial and Immunomodulatory Properties of Cervicovaginal Fluid

Cervicovaginal fluid is constitutively secreted from the vagina, cervix, and the upper genital tract [152] containing water, nutrients, electrolytes, proteins, and different cells [153]. The mucosa creates an effective physical barrier against invading microorganisms due to the hindrance of the adherence of bacteria to the surface of epithelial cells [154]. The protein composition of the vaginal fluid can be influenced by various factors, such as the varying ratio of estrogens and progesterone that can cause changes in the amount of secreted proteins [154]; thus, the determination of the normal protein concentration is practically impossible. To date, more than 900 proteins have been identified in the cervicovaginal fluid [109]. The majority of the proteins are proteases including complement factors, coagulation components and members of the kallikrein family, and protease inhibitors such as SERPIN and Kazal type (SPINK) serine protease inhibitor proteins [155,156]. Besides the proteases and protease inhibitors, various AMPs are also found in the vaginal fluid such as lysozyme-C, lactotransferrin, LL-37 cathelicidin, S100A7, BPI, and α-defensins [156].

By examining the identified vaginal fluid proteins, we identified 224 AMPs among them (Table S7). The interaction network of the identified AMPs in cervicovaginal fluid was generated (Figure 7).

Since only a few histone proteins were identified in the cervicovaginal fluid, the pattern of this interaction network was different from those of the so far-mentioned body fluids (Figure 8). Most of the AMPs are clustered to one core cluster, while the histones form a smaller cluster. Several major hub proteins were identified such as alpha-1 antitrypsin, kininogen-1, serotransferrin, apolipoprotein AI apolipoprotein AIV, and fibrinogen alpha and gamma chain. In addition, five small clusters were identified (Figure S7). This network was only slightly similar to the network of AMPs observed in serum, saliva, nasal secretion, and urine.

3.8. Chemical Barrier Proteins in the Seminal Fluid

The seminal fluid contains spermatozoa and seminal plasma, and its primary function is the fertilization of the oocyte. It contains many different bioactive signaling factors, including cytokines, prostaglandins, sex hormones, glycans, nucleic acids, and other small molecules that elicit molecular and cellular changes in the female reproductive tract [157]. These bioactive molecules exist either in soluble form, encapsulated within seminal plasma extracellular vesicles, or associated with spermatozoa [157]. Besides the interaction with the female reproductive tract, seminal fluid has also been found to interact with the female immune system [158]. Currently, more than 4000 proteins have been identified in human seminal plasma [109] and the most abundant proteins are fibronectin, semenogelins, lactotransferrin, laminin, and serum albumin [159].

Fibronectin is an extracellular glycoprotein ubiquitously found in the extracellular matrix and in biological fluids with a major function to connect cells to the extracellular matrix [160]. Bacteria can also bind to fibronectin through their receptors; therefore, fibronectin may play an important role in the attachment of bacteria and the infection of the host cells [161]. Laminin is also a part of the extracellular matrix and is a major component of the basal membrane [162], participating in the adhesion of sperms to the oocyte [163].

Semenogelins (semenogelin I and semenogelin II) together with fibronectin, are maintaining the gel-like coagulum of newly ejaculated semen [164]. Semenogelins have been found to inhibit the motility of the sperms but due to their proteolytic processing by prostate-specific antigen, the movement of sperms can be initiated [165]. In addition, semenogelins and their processed forms participate in Zn^{2+} shuttling, hyaluronidase activation, and exert antimicrobial activity [165]. Peptides derived from semenogelin I can act against various pathogenic bacteria, such as *Escherichia coli* and *Pseudomonas aeruginosa* [166,167].

Among the more than 4000 seminal fluid proteins, 284 are AMPs (Table S8).

The interaction network of the proteins identified as chemical barrier components in the seminal fluid shows a similar pattern to the protein–protein interaction network observed in the urine (Figure 8).

We observed the two major clusters composed of the histone proteins and the majority of the remaining AMPs, but this later one did not contain that many interactions as in the case of urine. We also observed that the core clusters were interacting with each other. The hub proteins identified were very similar in the examined body fluids. Besides the proteins part of the core clusters, we identified proteins forming six small additional clusters (Figure S8).

3.9. AMPs in the Cerebrospinal Fluid

Cerebrospinal fluid (CSF) is secreted into the brain ventricles and the cranial and spinal subarachnoid spaces predominantly by the choroid plexus [168]. The secretion rate of CSF varies between 400 and 600 mL/day with a protein concentration of 15–45 mg/dL [169]. More than 4000 proteins have been identified in the human CSF so far [109]. Approximately 80% of the total proteins in CSF originate from the blood by size-dependent filtration across the blood–brain barrier, the other proteins originate from the drainage of interstitial fluid from the central nervous system [170]. The most abundant blood-derived proteins in CSF are serum albumin, immunoglobulins, transthyretin, transferrin, and α1-antitrypsin [171,172]. Several blood-derived proteins, such as prostaglandin D2 synthase, S-100B, tau protein, and cystatin C have higher concentration in the CSF than in serum [173,174].

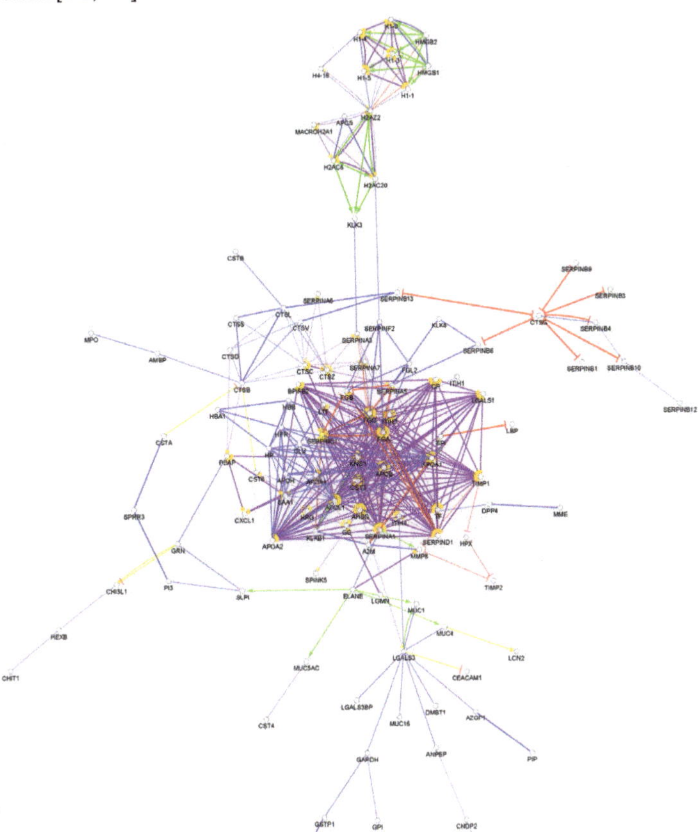

Figure 7. Partial network view of the interaction network of the chemical barrier proteins in cervicovaginal fluid. Each circle represents a protein and the lines indicate interactions. The lines with an arrow represent activation, blocking lines represent inhibition, and simple lines represent protein–protein interaction. Line color indicates the type of interaction: green color refers to activation, red color to inhibition, blue color to binding, yellow color to co-expression, and purple color to catalysis. The proteins are labeled with their gene name. The high resolution and complete network image of this network is presented in Figure S7.

Transthyretin is a transport protein in the plasma and in CSF transporting thyroxine and retinol to various parts of the body [175]. Studies from animal models and human samples also suggest the importance of transthyretin in the preservation and regulation of memory function and behavior, protection against neurodegeneration, neuroprotection in response to ischemic injury, and nerve regeneration [175]. Sharma et al. also suggested the role of this protein in the oxidative stress response [176]. It has been described that transthyretin inhibits the biofilm formation of *Escherichia coli* [177].

α1-antitrypsin is one of the most abundant serine protease inhibitors in the human body and is a well-known acute phase protein [178]. The protein is mainly synthesized in the liver, but it is also synthesized by monocytes, macrophages, pulmonary alveolar cells, and by intestinal and corneal epithelium [179]. As a potent protease inhibitor, α1-

antitrypsin protects the host cells from the activity of proteases during inflammation, and by inhibiting proteases released from pathogenic bacteria, the protein has an important role in the inhibition of bacterial colonization [178,179].

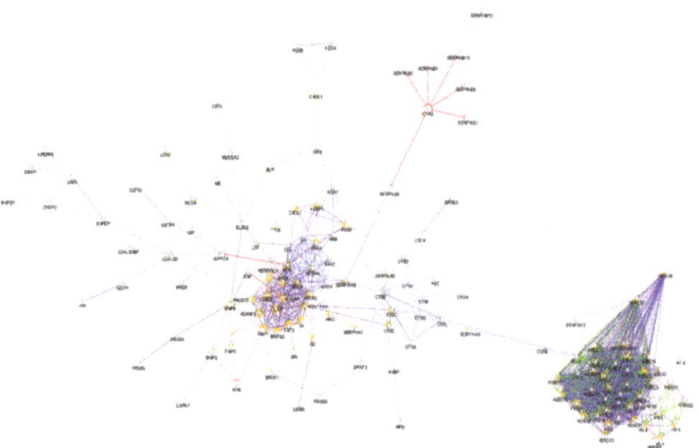

Figure 8. Partial network view of the interaction network of the chemical barrier proteins in the seminal fluid. Each circle represents a protein and the lines indicate interactions. The lines with an arrow represent activation, blocking lines represent inhibition, and simple lines represent protein–protein interaction. Line color indicates the type of interaction: green color refers to activation, red color to inhibition, blue color to binding, yellow color to co-expression, and purple color to catalysis. The proteins are labeled with their gene name. The high resolution and complete network image of this network is presented in Figure S8.

Prostaglandin D2 synthase catalyzes the conversion of prostaglandin H_2 (PGH_2) generated by cyclooxygenases to PGD_2. PGD_2 stimulates G-protein coupled receptors leading to the regulation of the circadian rhythm, food intake, pain perception, myelination, and adipocyte differentiation [180]. PGD_2 has been recognized as a proinflammatory molecule that can initiate IgE-mediated type I acute allergic response [181]. On the contrary, the anti-inflammatory effects of PGD_2 have also been described [181]. PGD_2 can also induce the expression of hBD3; thus, it can initiate the antimicrobial response against pathogenic microorganisms [182].

Tau protein is a microtubule-associated protein in the nervous system regulating the axonal transport and signaling pathways within and between neurons [183]. In addition, tau protein has an important role in the maintenance of the blood-brain barrier, since tauopathies can disrupt this barrier [184]. Interestingly, tau protein can serve as a scaffold for the binding of peptides that exert antimicrobial properties [185].

The examination of the identified CSF proteins revealed that 338 take part in the chemical barrier (Table S9) and their protein–protein interaction shows a similar architecture to the ones observed in serum, saliva, nasal secretion, urine, and seminal fluid (Figure 9).

The majority of the CSF chemical barrier proteins are organized into two core clusters, one cluster of the histones and related proteins and another cluster for the majority of the remaining chemical barrier proteins with several hubs, such as apolipoprotein AII, apolipoprotein L1, alpha-1 antitrypsin, Disintegrin and metalloproteinase domain-containing protein 10, serotransferrin and fibrinogen alpha and gamma chains. We also observed that the core clusters are interacting with each other. Compared to the serum, the pattern of the network is slightly different, and different connections and sub-clusters can be observed. Besides the core clusters, five additional small clusters were identified (Figure S9).

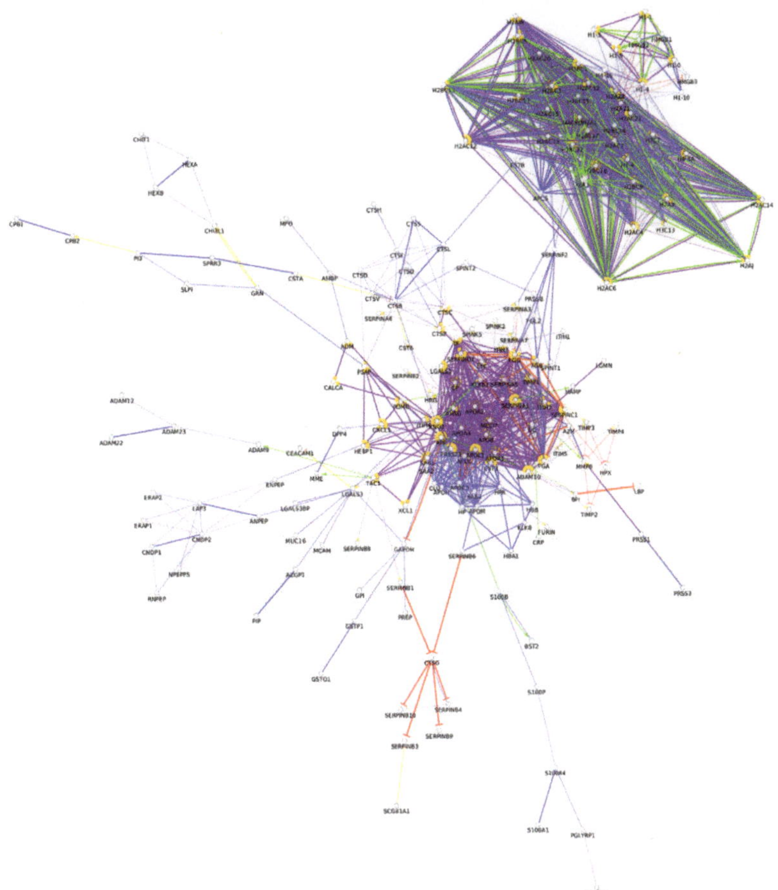

Figure 9. Partial network view of the interaction network of the chemical barrier proteins in CSF. Each circle represents a protein and the lines indicate interactions. The lines with an arrow represent activation, blocking lines represent inhibition, and simple lines represent protein–protein interaction. Line color indicates the type of interaction: green color refers to activation, red color to inhibition, blue color to binding, yellow color to co-expression, and purple color to catalysis. The proteins are labeled with their gene name. The high resolution and complete network image of this network is presented in Figure S9.

4. Comparison of the Protein–Protein Interaction Network in the Examined Body Fluids

The examination of the interaction networks revealed similar network architecture in the case of the different body fluids. Most of the proteins were organized into two core clusters: the cluster of histone proteins and the cluster of the other AMPs in networks except the one characteristic to sweat. These two clusters were connected to each other in all of the examined body fluids except the nasal secretion. The network generated from the sweat chemical barrier proteins was different from all the other networks highlighting the distinct feature of sweat.

In order to examine the hub proteins, we identified the top 20 hub proteins [104] in all of the examined networks (Figure 10).

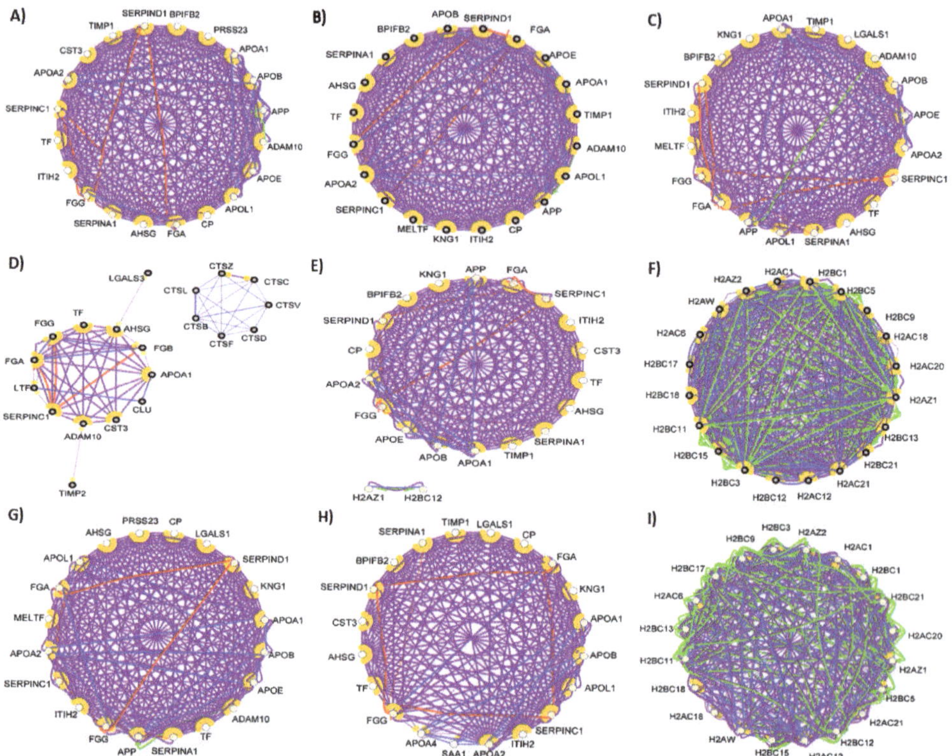

Figure 10. Interaction between the top 20 hub proteins in (**A**) serum, (**B**) tears, (**C**) saliva, (**D**) sweat, (**E**) nasal secretion, (**F**) urine, (**G**) CSF, (**H**) cervicovaginal fluid, and (**I**) seminal fluid. The blue lines represent binging, green lines represent activation, red lines represent inhibition, and purple lines represent catalysis. The proteins are labeled with their gene name.

Our analysis revealed that there is no major difference between the 20 hub proteins with the highest number of connections in serum, tears, saliva, nasal secretion, CSF, and cervicovaginal fluid. The chemical barrier of the sweat was different from the others regarding the top hub proteins, while in case of urine and seminal fluid, the histone proteins were found to be the proteins with most of the interactions. Considering the anatomical and physiological characteristics, the similarity between these two body fluids regarding chemical barrier is expected. Very likely, both body fluids have to cope with similar challenges regarding the antimicrobial protection, similarity reflected at the level of AMPs and their interaction networks.

Our results indicate that the composition and the interaction network of the examined human body fluids show high similarity. There is a considerable overlap between the protein composition of the chemical barrier, indicating the importance of the secreted AMPs and the proteases and protease inhibitors at the contact sites of the human body. We identified that approximately 9% of the AMPs can be found in all of the examined body fluids composing the core protein content of the chemical barriers and the other proteins make the body fluids specific for their localization and function.

5. Conclusions

In this review, we collected information about proteins and peptides taking part in the formation of chemical barriers of nine human body fluids and we present a comprehensive list of these components. We also generated the protein–protein interaction networks of the chemical barrier proteins and we highlighted that the interaction pattern is similar to most of the body fluids with slight differences. The network analysis proved that in case of saliva, nasal secretion, urine, and CSF, the interaction pattern is similar to the one observed in the case of serum. Regarding the networks of tears and cervicovaginal fluid, differences were observed compared to the other networks, while the network of the sweat proteins was completely different from the others. The networks of the urine and seminal fluid proteins, respectively, show high similarity in the organization of the connections and in the top hub proteins, indicating a strong relationship between these body fluids.

Supplementary Materials: The following supporting information can be downloaded at: https://www.mdpi.com/article/10.3390/biomedicines10071472/s1, Figure S1: Interaction network of the chemical barrier proteins in serum; Figure S2: Interaction network of the chemical barrier proteins in tears; Figure S3: Interaction network of the chemical barrier proteins in saliva; Figure S4: Interaction network of the chemical barrier proteins in nasal secretion; Figure S5: Interaction network of the chemical barrier proteins in sweat; Figure S6: Interaction network of the chemical barrier proteins in urine; Figure S7: Interaction network of the chemical barrier proteins in cervicovaginal fluid; Figure S8: Interaction network of the chemical barrier proteins in seminal fluid; Figure S9: Interaction network of the chemical barrier proteins in the CSF; Table S1: Proteins involved in the first line of host defense in serum; Table S2: Proteins involved in the first line of host defense in tears; Table S3: Proteins involved in the first line of host defense in saliva; Table S4: Proteins involved in the first line of host defense in the nasal secretion; Table S5: Proteins involved in the first line of host defense in sweat; Table S6: Proteins involved in the first line of host defense in urine; Table S7: Proteins involved in the first line of host defense in CSF; Table S8: Proteins involved in the first line of host defense in cervicovaginal fluid; Table S9: Proteins involved in the first line of host defense in the seminal fluid.

Author Contributions: Conceptualization, G.K. and É.C.; methodology, G.K. and A.K.; software, A.K.; formal analysis, A.K.; investigation, G.K.; resources, É.C. and J.T.; data curation, G.K.; writing—original draft preparation, G.K.; writing—review and editing, É.C. and J.T.; visualization, A.K.; supervision, É.C.; project administration, G.K. and É.C.; funding acquisition, É.C. and J.T. All authors have read and agreed to the published version of the manuscript.

Funding: This research was funded by GINOP-2.3.3-15-2016-00020, FK134605.

Institutional Review Board Statement: Not applicable.

Informed Consent Statement: Not applicable.

Data Availability Statement: Data sharing not applicable.

Acknowledgments: We are grateful for János András Mótyán for the critical revision of our manuscript.

Conflicts of Interest: The authors declare no conflict of interest.

References

1. Sperandio, B.; Fischer, N.; Sansonetti, P.J. Mucosal physical and chemical innate barriers: Lessons from microbial evasion strategies. *Semin. Immunol.* **2015**, *27*, 111–118. [CrossRef] [PubMed]
2. Wiesner, J.; Vilcinskas, A. Antimicrobial peptides: The ancient arm of the human immune system. *Virulence* **2010**, *1*, 440–464. [CrossRef] [PubMed]
3. Kalló, G.; Emri, M.; Varga, Z.; Ujhelyi, B.; Tőzsér, J.; Csutak, A.; Csősz, É. Changes in the Chemical Barrier Composition of Tears in Alzheimer's Disease Reveal Potential Tear Diagnostic Biomarkers. *PLoS ONE* **2016**, *11*, e0158000. [CrossRef]
4. Brandwein, M.; Bentwich, Z.; Steinberg, D. Endogenous Antimicrobial Peptide Expression in Response to Bacterial Epidermal Colonization. *Front. Immunol.* **2017**, *8*, 27. [CrossRef]
5. Zhou, L.; Zhao, S.Z.; Koh, S.K.; Chen, L.; Vaz, C.; Tanavde, V.; Li, X.R.; Beuerman, R.W. In-depth analysis of the human tear proteome. *J. Proteom.* **2012**, *75*, 3877–3885. [CrossRef]
6. Csősz, É.; Emri, G.; Kalló, G.; Tsaprailis, G.; Tőzsér, J. Highly abundant defense proteins in human sweat as revealed by targeted proteomics and label-free quantification mass spectrometry. *J. Eur. Acad. Dermatol. Venereol.* **2015**, *29*, 2024–2031. [CrossRef] [PubMed]
7. Arzumanyan, V.G.; Ozhovan, I.M.; Svitich, O.A. Antimicrobial Effect of Albumin on Bacteria and Yeast Cells. *Bull. Exp. Biol. Med.* **2019**, *167*, 763–766. [CrossRef]

8.   Parish, C.A.; Jiang, H.; Tokiwa, Y.; Berova, N.; Nakanishi, K.; McCabe, D.; Zuckerman, W.; Xia, M.M.; Gabay, J.E. Broad-spectrum antimicrobial activity of hemoglobin. *Bioorg. Med. Chem.* **2001**, *9*, 377–382. [CrossRef]

9.   Petrlova, J.; Petruk, G.; Huber, R.G.; McBurnie, E.W.; van der Plas, M.J.A.; Bond, P.J.; Puthia, M.; Schmidtchen, A. Thrombin-derived C-terminal fragments aggregate and scavenge bacteria and their proinflammatory products. *J. Biol. Chem.* **2020**, *295*, 3417–3430. [CrossRef] [PubMed]

10.   Mantis, N.J.; Rol, N.; Corthésy, B. Secretory IgA's complex roles in immunity and mucosal homeostasis in the gut. *Mucosal Immunol.* **2011**, *4*, 603–611. [CrossRef] [PubMed]

11.   Von Deuster, C.I.E.; Knecht, V. Competing interactions for antimicrobial selectivity based on charge complementarity. *Biochim. Biophys. Acta-Biomembr.* **2011**, *1808*, 2867–2876. [CrossRef] [PubMed]

12.   Jarczak, J.; Kościuczuk, E.M.; Lisowski, P.; Strzałkowska, N.; Jóźwik, A.; Horbańczuk, J.; Krzyzewski, J.; Zwierzchowski, L.; Bagnicka, E. Defensins: Natural component of human innate immunity. *Hum. Immunol.* **2013**, *74*, 1069–1079. [CrossRef]

13.   Ganz, T. Defensins: Antimicrobial peptides of innate immunity. *Nat. Rev. Immunol.* **2003**, *3*, 710–720. [CrossRef] [PubMed]

14.   Andreas, B.; Heiner, S. Human alpha-defensin 1 (HNP-1) inhibits adenoviral infection in vitro. *Regul. Pept.* **2001**, *101*, 157–161.

15.   Lehrer, R.I.; Ganz, T. Defensins of vertebrate animals. *Curr. Opin. Immunol.* **2002**, *14*, 96–102. [CrossRef]

16.   Chen, H.; Xu, Z.; Peng, L.; Fang, X.; Yin, X.; Xu, N.; Cen, P. Recent advances in the research and development of human defensins. *Peptides* **2006**, *27*, 931–940. [CrossRef]

17.   Yang, D.; Chertov, O.; Bykovskaia, S.N.; Chen, Q.; Buffo, M.J.; Shogan, J.; Anderson, M.; Schröder, J.M.; Wang, J.M.; Howard, O.M.; et al. Beta-defensins: Linking innate and adaptive immunity through dendritic and T cell CCR6. *Science* **1999**, *286*, 525–528. [CrossRef]

18.   Radek, K.; Gallo, R. Antimicrobial peptides: Natural effectors of the innate immune system. *Semin. Immunopathol.* **2007**, *29*, 27–43. [CrossRef]

19.   Li, D.; Wang, W.; Shi, H.; Fu, Y.; Chen, X.; Chen, X.; Liu, Y.; Kan, B.; Wang, Y. Gene therapy with beta-defensin 2 induces antitumor immunity and enhances local antitumor effects. *Hum. Gene Ther.* **2014**, *25*, 63–72. [CrossRef]

20.   Auvynet, C.; Rosenstein, Y. Multifunctional host defense peptides: Antimicrobial peptides, the small yet big players in innate and adaptive immunity. *FEBS J.* **2009**, *276*, 6497–6508. [CrossRef]

21.   De Smet, K.; Contreras, R. Human antimicrobial peptides: Defensins, cathelicidins and histatins. *Biotechnol. Lett.* **2005**, *27*, 1337–1347. [CrossRef]

22.   Heilborn, J.D.; Nilsson, M.F.; Kratz, G.; Weber, G.; Sørensen, O.; Borregaard, N.; Ståhle-Bäckdahl, M. The cathelicidin antimicrobial peptide LL-37 is involved in re-epithelialization of human skin wounds and is lacking in chronic ulcer epithelium. *J. Investig. Dermatol.* **2003**, *120*, 379–389. [CrossRef]

23.   Yamasaki, K.; Schauber, J.; Coda, A.; Lin, H.; Dorschner, R.A.; Schechter, N.M.; Bonnart, C.; Descargues, P.; Hovnanian, A.; Gallo, R.L. Kallikrein-mediated proteolysis regulates the antimicrobial effects of cathelicidins in skin. *FASEB J.* **2006**, *20*, 2068–2080. [CrossRef] [PubMed]

24.   Bruni, N.; Capucchio, M.T.; Biasibetti, E.; Pessione, E.; Cirrincione, S.; Giraudo, L.; Corona, A.; Dosio, F. Antimicrobial Activity of Lactoferrin-Related Peptides and Applications in Human and Veterinary Medicine. *Molecules* **2016**, *21*, 752. [CrossRef]

25.   Aizawa, S.; Hoki, M.; Yamamuro, Y. Lactoferrin promotes autophagy via AMP-activated protein kinase activation through low-density lipoprotein receptor-related protein 1. *Biochem. Biophys. Res. Commun.* **2017**, *493*, 509–513. [CrossRef] [PubMed]

26.   He, J.; Furmanski, P. Sequence specificity and transcriptional activation in the binding of lactoferrin to DNA. *Nature* **1995**, *373*, 721–724. [CrossRef]

27.   Shimada, T.; Park, B.G.; Wolf, A.J.; Brikos, C.; Goodridge, H.S.; Becker, C.A.; Reyes, C.N.; Miao, E.A.; Aderem, A.; Götz, F.; et al. Staphylococcus aureus evades lysozyme-based peptidoglycan digestion that links phagocytosis, inflammasome activation, and IL-1beta secretion. *Cell Host Microbe* **2010**, *7*, 38–49. [CrossRef] [PubMed]

28.   Marquis, G.; Garzon, S.; Strykowski, H.; Auger, P. Cell walls of normal and lysozyme-damaged blastoconidia of Candida albicans: Localization of surface factor 4 antigen and vicinal-glycol staining. *Infect. Immun.* **1991**, *59*, 1312–1318. [CrossRef]

29.   Lee-Huang, S.; Huang, P.L.; Sun, Y.; Kung, H.F.; Blithe, D.L.; Chen, H.C. Lysozyme and RNases as anti-HIV components in beta-core preparations of human chorionic gonadotropin. *Proc. Natl. Acad. Sci. USA* **1999**, *96*, 2678–2681. [CrossRef]

30.   Tagashira, A.; Nishi, K.; Matsumoto, S.; Sugahara, T. Anti-inflammatory effect of lysozyme from hen egg white on mouse peritoneal macrophages. *Cytotechnology* **2018**, *70*, 929. [CrossRef]

31.   Flo, T.H.; Smith, K.D.; Sato, S.; Rodriguez, D.J.; Holmes, M.A.; Strong, R.K.; Akira, S.; Aderem, A. Lipocalin 2 mediates an innate immune response to bacterial infection by sequestrating iron. *Nature* **2004**, *432*, 917–921. [CrossRef]

32.   Yang, J.; Goetz, D.; Li, J.Y.; Wang, W.; Mori, K.; Setlik, D.; Du, T.; Erdjument-Bromage, H.; Tempst, P.; Strong, R.; et al. An iron delivery pathway mediated by a lipocalin. *Mol. Cell* **2002**, *10*, 1045–1056. [CrossRef]

33.   Du, Z.P.; Wu, B.L.; Wu, X.; Lin, X.H.; Qiu, X.Y.; Zhan, X.F.; Wang, S.H.; Shen, J.H.; Zheng, C.P.; Wu, Z.Y.; et al. A systematic analysis of human lipocalin family and its expression in esophageal carcinoma. *Sci. Rep.* **2015**, *5*, 12010. [CrossRef] [PubMed]

34.   Bratt, T. Lipocalins and cancer. *Biochim. Biophys. Acta* **2000**, *1482*, 318–326. [CrossRef]

35.   Santiago-Sánchez, G.S.; Pita-Grisanti, V.; Quiñones-Díaz, B.; Gumpper, K.; Cruz-Monserrate, Z.; Vivas-Mejía, P.E. Biological Functions and Therapeutic Potential of Lipocalin 2 in Cancer. *Int. J. Mol. Sci.* **2020**, *21*, 4365. [CrossRef]

36.   Mesquita, S.D.; Ferreira, A.C.; Falcao, A.M.; Sousa, J.C.; Oliveira, T.G.; Correia-Neves, M.; Sousa, N.; Marques, F.; Palha, J.A. Lipocalin 2 modulates the cellular response to amyloid beta. *Cell Death Differ.* **2014**, *21*, 1588–1599. [CrossRef]

37. McKown, R.L.; Wang, N.; Raab, R.W.; Karnati, R.; Zhang, Y.; Williams, P.B.; Laurie, G.W. Lacritin and other new proteins of the lacrimal functional unit. *Exp. Eye Res.* **2009**, *88*, 848–858. [CrossRef]

38. Wang, J.; Wang, N.; Xie, J.; Walton, S.C.; McKown, R.L.; Raab, R.W.; Ma, P.; Beck, S.L.; Coffman, G.L.; Hussaini, I.M.; et al. Restricted epithelial proliferation by lacritin via PKCalpha-dependent NFAT and mTOR pathways. *J. Cell Biol.* **2006**, *174*, 689–700. [CrossRef]

39. Wang, W.; Despanie, J.; Shi, P.; Edman-Woolcott, M.C.; Lin, Y.-A.; Cui, H.; Heur, J.M.; Fini, M.E.; Hamm-Alvarez, S.F.; MacKay, J.A. Lacritin-mediated regeneration of the corneal epithelia by protein polymer nanoparticles. *J. Mater. Chem. B Mater. Biol. Med.* **2014**, *2*, 8131–8141. [CrossRef]

40. McKown, R.L.; Coleman Frazier, E.V.; Zadrozny, K.K.; Deleault, A.M.; Raab, R.W.; Ryan, D.S.; Sia, R.K.; Lee, J.K.; Laurie, G.W. A cleavage-potentiated fragment of tear lacritin is bactericidal. *J. Biol. Chem.* **2014**, *289*, 22172–22182. [CrossRef]

41. Vijmasi, T.; Chen, F.Y.T.; Balasubbu, S.; Gallup, M.; McKown, R.L.; Laurie, G.W.; McNamara, N.A. Topical Administration of Lacritin Is a Novel Therapy for Aqueous-Deficient Dry Eye Disease. *Investig. Ophthalmol. Vis. Sci.* **2014**, *55*, 5401–5409. [CrossRef] [PubMed]

42. Caputo, E.; Camarca, A.; Moharram, R.; Tornatore, P.; Thatcher, B.; Guardiola, J.; Martin, B.M. Structural study of GCDFP-15/gp17 in disease versus physiological conditions using a proteomic approach. *Biochemistry* **2003**, *42*, 6169–6178. [CrossRef] [PubMed]

43. Chiu, W.W.-C.; Chamley, L.W. Human seminal plasma prolactin-inducible protein is an immunoglobulin G-binding protein. *J. Reprod. Immunol.* **2003**, *60*, 97–111. [CrossRef]

44. Hassan, M.I.; Waheed, A.; Yadav, S.; Singh, T.P.; Ahmad, F. Prolactin inducible protein in cancer, fertility and immunoregulation: Structure, function and its clinical implications. *Cell. Mol. Life Sci.* **2009**, *66*, 447–459. [CrossRef] [PubMed]

45. Petrakis, N.L.; Lowenstein, J.M.; Wiencke, J.K.; Lee, M.M.; Wrensch, M.R.; King, E.B.; Hilton, J.F.; Miike, R. Gross cystic disease fluid protein in nipple aspirates of breast fluid of Asian and non-Asian women. *Cancer Epidemiol. Biomark. Prev.* **1993**, *2*, 573–579.

46. Edechi, C.A.; Ikeogu, N.M.; Akaluka, G.N.; Terceiro, L.E.L.; Machado, M.; Salako, E.S.; Barazandeh, A.F.; Kung, S.K.P.; Uzonna, J.E.; Myal, Y. The Prolactin Inducible Protein Modulates Antitumor Immune Responses and Metastasis in a Mouse Model of Triple Negative Breast Cancer. *Front. Oncol.* **2021**, *11*, 456. [CrossRef]

47. You, J.; Fitzgerald, A.; Cozzi, P.J.; Zhao, Z.; Graham, P.; Russell, P.J.; Walsh, B.J.; Willcox, M.; Zhong, L.; Wasinger, V.; et al. Post-translation modification of proteins in tears. *Electrophoresis* **2010**, *31*, 1853–1861. [CrossRef]

48. Rieg, S.; Garbe, C.; Sauer, B.; Kalbacher, H.; Schittek, B. Dermcidin is constitutively produced by eccrine sweat glands and is not induced in epidermal cells under inflammatory skin conditions. *Br. J. Dermatol.* **2004**, *151*, 534–539. [CrossRef]

49. Rieg, S.; Steffen, H.; Seeber, S.; Humeny, A.; Kalbacher, H.; Dietz, K.; Garbe, C.; Schittek, B. Deficiency of dermcidin-derived antimicrobial peptides in sweat of patients with atopic dermatitis correlates with an impaired innate defense of human skin in vivo. *J. Immunol.* **2005**, *174*, 8003–8010. [CrossRef]

50. Chang, W.C.; Huang, M.S.; Yang, C.J.; Wang, W.Y.; Lai, T.C.; Hsiao, M.; Chen, C.H. Dermcidin identification from exhaled air for lung cancer diagnosis. *Eur. Respir. J.* **2010**, *35*, 1182–1185. [CrossRef]

51. Stewart, G.D.; Skipworth, R.J.E.; Pennington, C.J.; Lowrie, A.G.; Deans, D.A.C.; Edwards, D.R.; Habib, F.K.; Riddick, A.C.P.; Fearon, K.C.H.; Ross, J.A. Variation in dermcidin expression in a range of primary human tumours and in hypoxic/oxidatively stressed human cell lines. *Br. J. Cancer* **2008**, *99*, 126–132. [CrossRef] [PubMed]

52. Zackular, J.P.; Chazin, W.J.; Skaar, E.P. Nutritional Immunity: S100 Proteins at the Host-Pathogen Interface. *J. Biol. Chem.* **2015**, *290*, 18991. [CrossRef] [PubMed]

53. Donato, R.; Cannon, B.R.; Sorci, G.; Riuzzi, F.; Hsu, K.; Weber, D.J.; Geczy, C.L. Functions of S100 Proteins. *Curr. Mol. Med.* **2013**, *13*, 24. [CrossRef] [PubMed]

54. Donato, R. Intracellular and extracellular roles of S100 proteins. *Microsc. Res. Tech.* **2003**, *60*, 540–551. [CrossRef]

55. Harder, J.; Schröder, J.M. RNase 7, a novel innate immune defense antimicrobial protein of healthy human skin. *J. Biol. Chem.* **2002**, *277*, 46779–46784. [CrossRef]

56. Rademacher, F.; Dreyer, S.; Kopfnagel, V.; Gläser, R.; Werfel, T.; Harder, J. The Antimicrobial and Immunomodulatory Function of RNase 7 in Skin. *Front. Immunol.* **2019**, *10*, 2553. [CrossRef]

57. Spencer, J.D.; Schwaderer, A.L.; Wang, H.; Bartz, J.; Kline, J.; Eichler, T.; Desouza, K.R.; Sims-Lucas, S.; Baker, P.; Hains, D.S. Ribonuclease 7, an antimicrobial peptide upregulated during infection, contributes to microbial defense of the human urinary tract. *Kidney Int.* **2013**, *83*, 615–625. [CrossRef]

58. Canny, G.; Levy, O. Bactericidal/permeability-increasing protein (BPI) and BPI homologs at mucosal sites. *Trends Immunol.* **2008**, *29*, 541–547. [CrossRef]

59. Wang, L.; Liu, Z.; Dong, Z.; Pan, J.; Ma, X. Azurocidin-induced inhibition of oxygen metabolism in mitochondria is antagonized by heparin. *Exp. Ther. Med.* **2014**, *8*, 1473. [CrossRef]

60. Khan, M.; Carmona, S.; Sukhumalchandra, P.; Roszik, J.; Philips, A.; Perakis, A.A.; Kerros, C.; Zhang, M.; Qiao, N.; St. John, L.S.; et al. Cathepsin G is expressed by acute lymphoblastic leukemia and is a potential immunotherapeutic target. *Front. Immunol.* **2018**, *8*, 1975. [CrossRef]

61. Melino, S.; Santone, C.; Di Nardo, P.; Sarkar, B. Histatins: Salivary peptides with copper(II)- and zinc(II)-binding motifs. *FEBS J.* **2014**, *281*, 657–672. [CrossRef] [PubMed]

62. Mak, P. Hemocidins in a functional and structural context of human antimicrobial peptides. *Front. Biosci.* **2008**, *13*, 6859–6871. [CrossRef] [PubMed]

63. Posta, N.; Csősz, É.; Oros, M.; Pethő, D.; Potor, L.; Kalló, G.; Hendrik, Z.; Sikura, K.É.; Méhes, G.; Tóth, C.; et al. Hemoglobin oxidation generates globin-derived peptides in atherosclerotic lesions and intraventricular hemorrhage of the brain, provoking endothelial dysfunction. *Lab. Investig.* **2020**, *100*, 986–1002. [CrossRef] [PubMed]

64. Mak, P.; Wójcik, K.; Silberring, J.; Dubin, A. Antimicrobial peptides derived from heme-containing proteins: Hemocidins. *Antonie Van Leeuwenhoek* **2000**, *77*, 197–207. [CrossRef]

65. Naveed, M.; Nadeem, F.; Mehmood, T.; Bilal, M.; Anwar, Z.; Amjad, F. Protease—A Versatile and Ecofriendly Biocatalyst with Multi-Industrial Applications: An Updated Review. *Catal. Lett.* **2021**, *151*, 307–323. [CrossRef]

66. Feng, Y.; Li, Q.; Chen, J.; Yi, P.; Xu, X.; Fan, Y.; Cui, B.; Yu, Y.; Li, X.; Du, Y.; et al. Salivary protease spectrum biomarkers of oral cancer. *Int. J. Oral Sci.* **2019**, *11*, 7. [CrossRef]

67. Fu, R.; Klinngam, W.; Heur, M.; Edman, M.C.; Hamm-Alvarez, S.F. Tear proteases and protease inhibitors: Potential biomarkers and disease drivers in ocular surface disease. *Eye Contact Lens* **2020**, *46*, S70. [CrossRef]

68. Magalhães, B.; Trindade, F.; Barros, A.S.; Klein, J.; Amado, F.; Ferreira, R.; Vitorino, R. Reviewing Mechanistic Peptidomics in Body Fluids Focusing on Proteases. *Proteomics* **2018**, *18*, 1800187. [CrossRef]

69. Vandooren, J.; Itoh, Y. Alpha-2-Macroglobulin in Inflammation, Immunity and Infections. *Front. Immunol.* **2021**, *12*, 5411. [CrossRef]

70. Ranasinghe, S.L.; McManus, D.P. Protease Inhibitors of Parasitic Flukes: Emerging Roles in Parasite Survival and Immune Defence. *Trends Parasitol.* **2017**, *33*, 400–413. [CrossRef]

71. Janciauskiene, S.; Wrenger, S.; Immenschuh, S.; Olejnicka, B.; Greulich, T.; Welte, T.; Chorostowska-Wynimko, J. The multifaceted effects of Alpha1-Antitrypsin on neutrophil functions. *Front. Pharmacol.* **2018**, *9*, 341. [CrossRef] [PubMed]

72. Roemisch, J.; Gray, E.; Hoffmann, J.N.; Wiedermann, C.J.; Kalina, U. Antithrombin: A new look at the actions of a serine protease inhibitor. *Blood Coagul. Fibrinolysis* **2002**, *13*, 657–670. [CrossRef] [PubMed]

73. Agbowuro, A.A.; Huston, W.M.; Gamble, A.B.; Tyndall, J.D.A. Proteases and protease inhibitors in infectious diseases. *Med. Res. Rev.* **2018**, *38*, 1295–1331. [CrossRef] [PubMed]

74. Xia, J.; Gao, J.; Tang, W. Nosocomial infection and its molecular mechanisms of antibiotic resistance. *Biosci. Trends* **2016**, *10*, 14–21. [CrossRef] [PubMed]

75. Sousa, S.A.; Feliciano, J.R.; Pita, T.; Soeiro, C.F.; Mendes, B.L.; Alves, L.G.; Leitão, J.H. Bacterial Nosocomial Infections: Multidrug Resistance as a Trigger for the Development of Novel Antimicrobials. *Antibiotics* **2021**, *10*, 942. [CrossRef]

76. Wieler, L.H.; Ewers, C.; Guenther, S.; Walther, B.; Lübke-Becker, A. Methicillin-resistant staphylococci (MRS) and extended-spectrum beta-lactamases (ESBL)-producing Enterobacteriaceae in companion animals: Nosocomial infections as one reason for the rising prevalence of these potential zoonotic pathogens in clinical samples. *Int. J. Med. Microbiol.* **2011**, *301*, 635–641. [CrossRef]

77. Nguyen, M.; Joshi, S.G. Carbapenem resistance in Acinetobacter baumannii, and their importance in hospital-acquired infections: A scientific review. *J. Appl. Microbiol.* **2021**, *131*, 2715–2738. [CrossRef]

78. Mak, S.; Xu, Y.; Nodwell, J.R. The expression of antibiotic resistance genes in antibiotic-producing bacteria. *Mol. Microbiol.* **2014**, *93*, 391–402. [CrossRef]

79. Avershina, E.; Shapovalova, V.; Shipulin, G. Fighting Antibiotic Resistance in Hospital-Acquired Infections: Current State and Emerging Technologies in Disease Prevention, Diagnostics and Therapy. *Front. Microbiol.* **2021**, *12*, 2044. [CrossRef]

80. Magana, M.; Pushpanathan, M.; Santos, A.L.; Leanse, L.; Fernandez, M.; Ioannidis, A.; Giulianotti, M.A.; Apidianakis, Y.; Bradfute, S.; Ferguson, A.L.; et al. The value of antimicrobial peptides in the age of resistance. *Lancet. Infect. Dis.* **2020**, *20*, e216–e230. [CrossRef]

81. Kosikowska, P.; Lesner, A. Antimicrobial peptides (AMPs) as drug candidates: A patent review (2003–2015). *Expert Opin. Ther. Pat.* **2016**, *26*, 689–702. [CrossRef] [PubMed]

82. Wach, A.; Dembowsky, K.; Dale, G.E. Pharmacokinetics and Safety of Intravenous Murepavadin Infusion in Healthy Adult Subjects Administered Single and Multiple Ascending Doses. *Antimicrob. Agents Chemother.* **2018**, *62*, e02355-17. [CrossRef] [PubMed]

83. Otvos, L.; Ostorhazi, E.; Szabo, D.; Zumbrun, S.D.; Miller, L.L.; Halasohoris, S.A.; Desai, P.D.; Veldt, S.M.I.; Kraus, C.N. Synergy Between Proline-Rich Antimicrobial Peptides and Small Molecule Antibiotics Against Selected Gram-Negative Pathogens in vitro and in vivo. *Front. Chem.* **2018**, *6*, 309. [CrossRef] [PubMed]

84. Mukhopadhyay, S.; Bharath Prasad, A.S.; Mehta, C.H.; Nayak, U.Y. Antimicrobial peptide polymers: No escape to ESKAPE pathogens—A review. *World J. Microbiol. Biotechnol.* **2020**, *36*, 131. [CrossRef] [PubMed]

85. Van der Plas, M.J.A.; Cai, J.; Petrlova, J.; Saleh, K.; Kjellström, S.; Schmidtchen, A. Method development and characterisation of the low-molecular-weight peptidome of human wound fluids. *Elife* **2021**, *10*, e66876. [CrossRef]

86. Hartman, E.; Wallblom, K.; van der Plas, M.J.A.; Petrlova, J.; Cai, J.; Saleh, K.; Kjellström, S.; Schmidtchen, A. Bioinformatic Analysis of the Wound Peptidome Reveals Potential Biomarkers and Antimicrobial Peptides. *Front. Immunol.* **2021**, *11*, 620707. [CrossRef]

87. Dahlman, A.; Puthia, M.; Petrlova, J.; Schmidtchen, A.; Petruk, G. Thrombin-Derived C-Terminal Peptide Reduces *Candida*-Induced Inflammation and Infection In Vitro and In Vivo. *Antimicrob. Agents Chemother.* **2021**, *65*, e01032-21. [CrossRef]

88. Hu, S.; Loo, J.A.; Wong, D.T. Human body fluid proteome analysis. *Proteomics* **2006**, *6*, 6326. [CrossRef]

89. Leeman, M.; Choi, J.; Hansson, S.; Storm, M.U.; Nilsson, L. Proteins and antibodies in serum, plasma, and whole blood—size characterization using asymmetrical flow field-flow fractionation (AF4). *Anal. Bioanal. Chem.* **2018**, *410*, 4867. [CrossRef]

90. Huang, L.; Shao, D.; Wang, Y.; Cui, X.; Li, Y.; Chen, Q.; Cui, J. Human body-fluid proteome: Quantitative profiling and computational prediction. *Brief. Bioinform.* **2021**, *22*, 315–333. [CrossRef]

91. Bellei, E.; Bergamini, S.; Monari, E.; Fantoni, L.I.; Cuoghi, A.; Ozben, T.; Tomasi, A. High-abundance proteins depletion for serum proteomic analysis: Concomitant removal of non-targeted proteins. *Amino Acids* **2011**, *40*, 145–156. [CrossRef] [PubMed]

92. Chien, S.C.; Chen, C.Y.; Lin, C.F.; Yeh, H.I. Critical appraisal of the role of serum albumin in cardiovascular disease. *Biomark. Res.* **2017**, *5*, 31. [CrossRef] [PubMed]

93. Jain, S.; Gautam, V.; Naseem, S. Acute-phase proteins: As diagnostic tool. *J. Pharm. Bioallied Sci.* **2011**, *3*, 118. [CrossRef] [PubMed]

94. Gum, E.T.; Swanson, R.A.; Alano, C.; Liu, J.; Hong, S.; Weinstein, P.R.; Panter, S.S. Human serum albumin and its N-terminal tetrapeptide (DAHK) block oxidant-induced neuronal death. *Stroke* **2004**, *35*, 590–595. [CrossRef] [PubMed]

95. Gomme, P.T.; McCann, K.B. Transferrin: Structure, function and potential therapeutic actions. *Drug Discov. Today* **2005**, *10*, 267–273. [CrossRef]

96. Von Bonsdorff, L.; Sahlstedt, L.; Ebeling, F.; Ruutu, T.; Parkkinen, J. Apotransferrin administration prevents growth of Staphylococcus epidermidis in serum of stem cell transplant patients by binding of free iron. *FEMS Immunol. Med. Microbiol.* **2003**, *37*, 45–51. [CrossRef]

97. Ardehali, R.; Shi, L.; Janatova, J.; Mohammad, S.F.; Burns, G.L. The inhibitory activity of serum to prevent bacterial adhesion is mainly due to apo-transferrin. *J. Biomed. Mater. Res. A* **2003**, *66*, 21–28. [CrossRef]

98. Alayash, A.I.; Andersen, C.B.F.; Moestrup, S.K.; Bülow, L. Haptoglobin: The hemoglobin detoxifier in plasma. *Trends Biotechnol.* **2013**, *31*, 2–3. [CrossRef]

99. Kristiansen, M.; Graversen, J.H.; Jacobsen, C.; Sonne, O.; Hoffman, H.J.; Law, S.K.A.; Moestrup, S.K. Identification of the haemoglobin scavenger receptor. *Nature* **2001**, *409*, 198–201. [CrossRef]

100. MacKellar, M.; Vigerust, D.J. Role of Haptoglobin in Health and Disease: A Focus on Diabetes. *Clin. Diabetes* **2016**, *34*, 148. [CrossRef]

101. Dominiczak, M.H.; Caslake, M.J. Apolipoproteins: Metabolic role and clinical biochemistry applications. *Ann. Clin. Biochem.* **2011**, *48*, 498–515. [CrossRef] [PubMed]

102. Tada, N.; Sakamoto, T.; Kagami, A.; Mochizuki, K.; Kurosaka, K. Antimicrobial activity of lipoprotein particles containing apolipoprotein Al. *Mol. Cell. Biochem.* **1993**, *119*, 171–178. [CrossRef] [PubMed]

103. Dell'Olmo, E.; Gaglione, R.; Sabbah, M.; Schibeci, M.; Cesaro, A.; Di Girolamo, R.; Porta, R.; Arciello, A. Host defense peptides identified in human apolipoprotein B as novel food biopreservatives and active coating components. *Food Microbiol.* **2021**, *99*, 103804. [CrossRef] [PubMed]

104. Nokhoijav, E.; Guba, A.; Kumar, A.; Kunkli, B.; Kalló, G.; Káplár, M.; Somodi, S.; Garai, I.; Csutak, A.; Tóth, N.; et al. Metabolomic Analysis of Serum and Tear Samples from Patients with Obesity and Type 2 Diabetes Mellitus. *Int. J. Mol. Sci.* **2022**, *23*, 4534. [CrossRef]

105. Prabha, J.L. Tear Secretion-A Short Review. *J. Pharm. Sci. Res.* **2014**, *6*, 155–157.

106. Fullard, R.J.; Snyder, C. Protein levels in nonstimulated and stimulated tears of normal human subjects. *Investig. Ophthalmol. Vis. Sci.* **1990**, *31*, 1119–1126.

107. Tiffany, J.M. Tears in health and disease. *Eye* **2003**, *17*, 923–926. [CrossRef]

108. Janeway, J.C.A.; Travers, P.; Walport, M.; Shlomchik, M.J. *The Front Line of Host Defense*, 5th ed.; Garland Science: New York, NY, USA, 2001.

109. Shao, D.; Huang, L.; Wang, Y.; Cui, X.; Li, Y.; Wang, Y.; Ma, Q.; Du, W.; Cui, J. HBFP: A new repository for human body fluid proteome. *Database* **2021**, *2021*, baab065. [CrossRef]

110. Janssen, P.T.; van Bijsterveld, O.P. Origin and biosynthesis of human tear fluid proteins. *Investig. Ophthalmol. Vis. Sci.* **1983**, *24*, 623–630.

111. Humphrey, S.P.; Williamson, R.T. A review of saliva: Normal composition, flow, and function. *J. Prosthet. Dent.* **2001**, *85*, 162–169. [CrossRef]

112. Shaila, M.; Pai, G.P.; Shetty, P. Salivary protein concentration, flow rate, buffer capacity and pH estimation: A comparative study among young and elderly subjects, both normal and with gingivitis and periodontitis. *J. Indian Soc. Periodontol.* **2013**, *17*, 42–46. [CrossRef] [PubMed]

113. Henskens, Y.M.; van der Velden, U.; Veerman, E.C.; Nieuw Amerongen, A.V. Protein, albumin and cystatin concentrations in saliva of healthy subjects and of patients with gingivitis or periodontitis. *J. Periodontal Res.* **1993**, *28*, 43–48. [CrossRef] [PubMed]

114. Schulz, B.L.; Cooper-White, J.; Punyadeera, C.K. Saliva proteome research: Current status and future outlook. *Crit. Rev. Biotechnol.* **2013**, *33*, 246–259. [CrossRef] [PubMed]

115. Csősz, É.; Kalló, G.; Márkus, B.; Deák, E.; Csutak, A.; Tőzsér, J. Quantitative body fluid proteomics in medicine—A focus on minimal invasiveness. *J. Proteom.* **2016**, *153*, 30364–30365. [CrossRef]

116. Lahiri, D.; Nag, M.; Banerjee, R.; Mukherjee, D.; Garai, S.; Sarkar, T.; Dey, A.; Sheikh, H.I.; Pathak, S.K.; Edinur, H.A.; et al. Amylases: Biofilm Inducer or Biofilm Inhibitor? *Front. Cell. Infect. Microbiol.* **2021**, *11*, 355. [CrossRef]

117. Bechler, J.; Bermudez, L.E. Investigating the Role of Mucin as Frontline Defense of Mucosal Surfaces against Mycobacterium avium Subsp. hominissuis. *J. Pathog.* **2020**, *2020*, 9451591. [CrossRef]

118. Linden, S.K.; Sutton, P.; Karlsson, N.G.; Korolik, V.; McGuckin, M.A. Mucins in the mucosal barrier to infection. *Mucosal Immunol.* **2008**, *1*, 183–197. [CrossRef]

119. Eaves-Pyles, T.; Patel, J.; Arigi, E.; Cong, Y.; Cao, A.; Garg, N.; Dhiman, M.; Pyles, R.B.; Arulanandam, B.; Miller, A.L.; et al. Immunomodulatory and antibacterial effects of cystatin 9 against Francisella tularensis. *Mol. Med.* **2013**, *19*, 263–275. [CrossRef]

120. Hajishengallis, G.; Russell, M.W. Innate Humoral Defense Factors. *Mucosal Immunol. Fourth Ed.* **2015**, *1–2*, 251–270. [CrossRef]

121. Levine, M. Susceptibility to Dental Caries and the Salivary Proline-Rich Proteins. *Int. J. Dent.* **2011**, *2011*, 953412. [CrossRef]

122. Metz-Boutigue, M.-H.; Shooshtarizadeh, P.; Prevost, G.; Haikel, Y.; Chich, J.-F. Antimicrobial peptides present in mammalian skin and gut are multifunctional defence molecules. *Curr. Pharm. Des.* **2010**, *16*, 1024–1039. [CrossRef] [PubMed]

123. ROBINSON, S.; ROBINSON, A.H. Chemical composition of sweat. *Physiol. Rev.* **1954**, *34*, 202–220. [CrossRef] [PubMed]

124. Harder, J.; Schröder, J.-M.; Gläser, R. The skin surface as antimicrobial barrier: Present concepts and future outlooks. *Exp. Dermatol.* **2013**, *22*, 1–5. [CrossRef]

125. Park, J.H.; Park, G.T.; Cho, I.H.; Sim, S.M.; Yang, J.M.; Lee, D.Y. An antimicrobial protein, lactoferrin exists in the sweat: Proteomic analysis of sweat. *Exp. Dermatol.* **2011**, *20*, 369–371. [CrossRef] [PubMed]

126. Jeong, S.; Ledee, D.R.; Gordon, G.M.; Itakura, T.; Patel, N.; Martin, A.; Fini, M.E. Interaction of clusterin and matrix metalloproteinase-9 and its implication for epithelial homeostasis and inflammation. *Am. J. Pathol.* **2012**, *180*, 2028–2039. [CrossRef]

127. Rizzi, F.; Bettuzzi, S. The clusterin paradigm in prostate and breast carcinogenesis. *Endocr. Relat. Cancer* **2010**, *17*, R1–R17. [CrossRef]

128. Sueno, K.; Nakaima, N.; Shingaki, K.; Ura, M.; Noda, Y.; Kosugi, T.; Eichner, H. Total protein concentration in selectively collected secretions from the middle and inferior meatus of the nose. *Auris. Nasus. Larynx* **1986**, *13* (Suppl. 1), S85–S88. [CrossRef]

129. Ruocco, L.; Fattori, B.; Romanelli, A.; Martelloni, M.; Casani, A.; Samolewska, M.; Rezzonico, R. A new collection method for the evaluation of nasal mucus proteins. *Clin. Exp. Allergy* **1998**, *28*, 881–888. [CrossRef]

130. Cole, A.M.; Dewan, P.; Ganz, T. Innate antimicrobial activity of nasal secretions. *Infect. Immun.* **1999**, *67*, 3267–3275. [CrossRef]

131. Meredith, S.D.; Raphael, G.D.; Baraniuk, J.N.; Banks, S.M.; Kaliner, M.A. The pathophysiology of rhinitis. III. The control of IgG secretion. *J. Allergy Clin. Immunol.* **1989**, *84*, 920–930. [CrossRef]

132. Saieg, A.; Brown, K.J.; Pena, M.T.; Rose, M.C.; Preciado, D. Proteomic analysis of pediatric sinonasal secretions shows increased MUC5B mucin in CRS. *Pediatr. Res.* **2015**, *77*, 356–362. [CrossRef]

133. Casado, B.; Pannell, L.K.; Viglio, S.; Iadarola, P.; Baraniuk, J.N. Analysis of the sinusitis nasal lavage fluid proteome using capillary liquid chromatography interfaced to electrospray ionization-quadrupole time of flight- tandem mass spectrometry. *Electrophoresis* **2004**, *25*, 1386–1393. [CrossRef] [PubMed]

134. Wang, H.; Chavali, S.; Mobini, R.; Muraro, A.; Barbon, F.; Boldrin, D.; Åberg, N.; Benson, M. A pathway-based approach to find novel markers of local glucocorticoid treatment in intermittent allergic rhinitis. *Allergy Eur. J. Allergy Clin. Immunol.* **2011**, *66*, 132–140. [CrossRef] [PubMed]

135. Tomazic, P.V.; Darnhofer, B.; Birner-Gruenberger, R. Nasal mucus proteome and its involvement in allergic rhinitis. *Expert Rev. Proteom.* **2020**, *17*, 191–199. [CrossRef]

136. Brunzel, N.A. *Fundamentals of Urine and Body Fluid Analysis*, 3rd ed.; Elsevier/Saunders: St. Louis, MO, USA, 2013.

137. Zhao, M.; Li, M.; Yang, Y.; Guo, Z.; Sun, Y.; Shao, C.; Li, M.; Sun, W.; Gao, Y. A comprehensive analysis and annotation of human normal urinary proteome. *Sci. Rep.* **2017**, *7*, 3024. [CrossRef]

138. Devuyst, O.; Olinger, E.; Rampoldi, L. Uromodulin: From physiology to rare and complex kidney disorders. *Nat. Rev. Nephrol.* **2017**, *13*, 525–544. [CrossRef]

139. Weiss, G.L.; Stanisich, J.J.; Sauer, M.M.; Lin, C.W.; Eras, J.; Zyla, D.S.; Trück, J.; Devuyst, O.; Aebi, M.; Pilhofer, M.; et al. Architecture and function of human uromodulin filaments in urinary tract infections. *Science* **2020**, *369*, 1005–1010. [CrossRef]

140. Kukulski, W. A glycoprotein in urine binds bacteria and blocks infections. *Science* **2020**, *369*, 917–918. [CrossRef]

141. Wu, T.H.; Li, K.J.; Yu, C.L.; Tsai, C.Y. Tamm–Horsfall Protein is a Potent Immunomodulatory Molecule and a Disease Biomarker in the Urinary System. *Molecules* **2018**, *23*, 200. [CrossRef]

142. Bergwik, J.; Kristiansson, A.; Allhorn, M.; Gram, M.; Åkerström, B. Structure, Functions, and Physiological Roles of the Lipocalin α1-Microglobulin (A1M). *Front. Physiol.* **2021**, *12*, 251. [CrossRef]

143. Åkerström, B.; Gram, M. A1M, an extravascular tissue cleaning and housekeeping protein. *Free Radic. Biol. Med.* **2014**, *74*, 274–282. [CrossRef] [PubMed]

144. Gunnarsson, R.; Åkerström, B.; Hansson, S.R.; Gram, M. Recombinant alpha-1-microglobulin: A potential treatment for preeclampsia. *Drug Discov. Today* **2017**, *22*, 736–743. [CrossRef] [PubMed]

145. Kristiansson, A.; Gram, M.; Flygare, J.; Hansson, S.R.; Åkerström, B.; Storry, J.R. The Role of α 1-Microglobulin (A1M) in Erythropoiesis and Erythrocyte Homeostasis-Therapeutic Opportunities in Hemolytic Conditions. *Int. J. Mol. Sci.* **2020**, *21*, 7234. [CrossRef] [PubMed]

146. Gonzalez-Calero, L.; Martin-Lorenzo, M.; Ramos-Barron, A.; Ruiz-Criado, J.; Maroto, A.S.; Ortiz, A.; Gomez-Alamillo, C.; Arias, M.; Vivanco, F.; Alvarez-Llamas, G. Urinary Kininogen-1 and Retinol binding protein-4 respond to Acute Kidney Injury: Predictors of patient prognosis? *Sci. Rep.* **2016**, *6*, 19667. [CrossRef]

147. Wong, M.K.-S. *Kininogen*; Academic Press: Cambridge, MA, USA, 2021.

148. Weinberg, M.S.; Azar, P.; Trebbin, W.M.; Solomon, R.J. The role of urinary kininogen in the regulation of kinin generation. *Kidney Int.* **1985**, *28*, 975–981. [CrossRef]

149. Ben Nasr, A.; Herwald, H.; Muller-Esterl, W.; Bjorck, L. Human kininogens interact with M protein, a bacterial surface protein and virulence determinant. *Biochem. J.* **1995**, *305*, 173–180. [CrossRef]

150. Rapala-Kozik, M.; Karkowska, J.; Jacher, A.; Golda, A.; Barbasz, A.; Guevara-Lora, I.; Kozik, A. Kininogen adsorption to the cell surface of *Candida* spp. *Int. Immunopharmacol.* **2008**, *8*, 237–241. [CrossRef]

151. Sonesson, A.; Nordahl, E.A.; Malmsten, M.; Schmidtchen, A. Antifungal activities of peptides derived from domain 5 of high-molecular-weight kininogen. *Int. J. Pept.* **2011**, *2011*, 1–12. [CrossRef]

152. Hitti, J.; Lapidus, J.A.; Lu, X.; Reddy, A.P.; Jacob, T.; Dasari, S.; Eschenbach, D.A.; Gravett, M.G.; Nagalla, S.R. Noninvasive diagnosis of intraamniotic infection: Proteomic biomarkers in vaginal fluid. *Am. J. Obstet. Gynecol.* **2010**, *203*, 32.e1–32.e8. [CrossRef]

153. Pizzorno, J.E.; Murray, M.T.; Joiner-Bey, H. *Vaginitis*, 3rd ed.; Churchill Livingstone: London, UK, 2016; ISBN 978-0-7020-5514-0.

154. Zegels, G.; Van Raemdonck, G.A.A.; Tjalma, W.A.A.; Van Ostade, X.W.M. Use of cervicovaginal fluid for the identification of biomarkers for pathologies of the female genital tract. *Proteome Sci.* **2010**, *8*, 63. [CrossRef]

155. Starodubtseva, N.L.; Brzhozovskiy, A.G.; Bugrova, A.E.; Kononikhin, A.S.; Indeykina, M.I.; Gusakov, K.I.; Chagovets, V.V.; Nazarova, N.M.; Frankevich, V.E.; Sukhikh, G.T.; et al. Label-free cervicovaginal fluid proteome profiling reflects the cervix neoplastic transformation. *J. Mass Spectrom.* **2019**, *54*, 693–703. [CrossRef] [PubMed]

156. Muytjens, C.M.J.; Yu, Y.; Diamandis, E.P. Discovery of Antimicrobial Peptides in Cervical-Vaginal Fluid from Healthy Nonpregnant Women via an Integrated Proteome and Peptidome Analysis. *Proteomics* **2017**, *17*, 1600461. [CrossRef] [PubMed]

157. Schjenken, J.E.; Robertson, S.A. The Female Response to Seminal Fluid. *Physiol. Rev.* **2020**, *100*, 1077–1117. [CrossRef] [PubMed]

158. Schjenken, J.E.; Robertson, S.A. Seminal fluid and immune adaptation for pregnancy–comparative biology in mammalian species. *Reprod. Domest. Anim.* **2014**, *49* (Suppl. 3), 27–36. [CrossRef]

159. Pilch, B.; Mann, M. Large-scale and high-confidence proteomic analysis of human seminal plasma. *Genome Biol.* **2006**, *7*, 1–10. [CrossRef]

160. Klavert, J.; van der Eerden, B.C.J. Fibronectin in Fracture Healing: Biological Mechanisms and Regenerative Avenues. *Front. Bioeng. Biotechnol.* **2021**, *9*, 274. [CrossRef]

161. Speziale, P.; Arciola, C.R.; Pietrocola, G. Fibronectin and Its Role in Human Infective Diseases. *Cells* **2019**, *8*, 1516. [CrossRef]

162. Hohenester, E. Structural biology of laminins. *Essays Biochem.* **2019**, *63*, 285–295. [CrossRef]

163. Bosler, J.S.; Davies, K.P.; Neal-Perry, G.S. Peptides in seminal fluid and their role in infertility: A potential role for opiorphin inhibition of neutral endopeptidase activity as a clinically relevant modulator of sperm motility: A review. *Reprod. Sci.* **2014**, *21*, 1334–1340. [CrossRef]

164. Lundwall, Å.; Bjartell, A.; Olsson, A.Y.; Malm, J. Semenogelin I and II, the predominant human seminal plasma proteins, are also expressed in non-genital tissues. *Mol. Hum. Reprod.* **2002**, *8*, 805–810. [CrossRef]

165. De Lamirande, E. Semenogelin, the main protein of the human semen coagulum, regulates sperm function. *Semin. Thromb. Hemost.* **2007**, *33*, 60–68. [CrossRef] [PubMed]

166. Zhao, H.; Lee, W.H.; Shen, J.H.; Li, H.; Zhang, Y. Identification of novel semenogelin I-derived antimicrobial peptide from liquefied human seminal plasma. *Peptides* **2008**, *29*, 505–511. [CrossRef] [PubMed]

167. Bourgeon, F.; Evrard, B.; Brillard-Bourdet, M.; Colleu, D.; Jégou, B.; Pineau, C. Involvement of semenogelin-derived peptides in the antibacterial activity of human seminal plasma. *Biol. Reprod.* **2004**, *70*, 768–774. [CrossRef] [PubMed]

168. Sakka, L.; Coll, G.; Chazal, J. Anatomy and physiology of cerebrospinal fluid. *Eur. Ann. Otorhinolaryngol. Head Neck Dis.* **2011**, *128*, 309–316. [CrossRef]

169. McPherson RA, P.M. *Henry's Clinical Diagnosis and Management by Laboratory Methods*, 23rd ed.; W.B. Saunders: Philadelphia, PA, USA, 2006.

170. McComb, J.G. Recent research into the nature of cerebrospinal fluid formation and absorption. *J. Neurosurg.* **1983**, *59*, 369–383. [CrossRef] [PubMed]

171. Begcevic, I.; Brinc, D.; Drabovich, A.P.; Batruch, I.; Diamandis, E.P. Identification of brain-enriched proteins in the cerebrospinal fluid proteome by LC-MS/MS profiling and mining of the Human Protein Atlas. *Clin. Proteom.* **2016**, *13*, 11. [CrossRef]

172. Jankovska, E.; Svitek, M.; Holada, K.; Petrak, J. Affinity depletion versus relative protein enrichment: A side-by-side comparison of two major strategies for increasing human cerebrospinal fluid proteome coverage. *Clin. Proteom.* **2019**, *16*, 9. [CrossRef]

173. Redzic, Z.B.; Preston, J.E.; Duncan, J.A.; Chodobski, A.; Szmydynger-Chodobska, J. The choroid plexus-cerebrospinal fluid system: From development to aging. *Curr. Top. Dev. Biol.* **2005**, *71*, 1–52. [CrossRef]

174. Reiber, H. Dynamics of brain-derived proteins in cerebrospinal fluid. *Clin. Chim. Acta* **2001**, *310*, 173–186. [CrossRef]

175. Liz, M.A.; Coelho, T.; Bellotti, V.; Fernandez-Arias, M.I.; Mallaina, P.; Obici, L. A Narrative Review of the Role of Transthyretin in Health and Disease. *Neurol. Ther.* **2020**, *9*, 395–402. [CrossRef]

176. Sharma, M.; Khan, S.; Rahman, S.; Singh, L.R. The extracellular protein, transthyretin is an oxidative stress biomarker. *Front. Physiol.* **2019**, *10*, 5. [CrossRef] [PubMed]

177. Jain, N.; Ådén, J.; Nagamatsu, K.; Evans, M.L.; Li, X.; McMichael, B.; Ivanova, M.I.; Almqvist, F.; Buxbaum, J.N.; Chapman, M.R. Inhibition of curli assembly and Escherichia coli biofilm formation by the human systemic amyloid precursor transthyretin. *Proc. Natl. Acad. Sci. USA* **2017**, *114*, 12184–12189. [CrossRef] [PubMed]

178. Hunt, J.M.; Tuder, R. Alpha 1 anti-trypsin: One protein, many functions. *Curr. Mol. Med.* **2012**, *12*, 827–835. [CrossRef] [PubMed]
179. Janciauskiene, S.M.; Bals, R.; Koczulla, R.; Vogelmeier, C.; Köhnlein, T.; Welte, T. The discovery of α1-antitrypsin and its role in health and disease. *Respir. Med.* **2011**, *105*, 1129–1139. [CrossRef] [PubMed]
180. Urade, Y. Biochemical and Structural Characteristics, Gene Regulation, Physiological, Pathological and Clinical Features of Lipocalin-Type Prostaglandin D2 Synthase as a Multifunctional Lipocalin. *Front. Physiol.* **2021**, *12*, 1627. [CrossRef]
181. Ricciotti, E.; Fitzgerald, G.A. Prostaglandins and inflammation. *Arterioscler. Thromb. Vasc. Biol.* **2011**, *31*, 986–1000. [CrossRef]
182. Kanda, N.; Ishikawa, T.; Watanabe, S. Prostaglandin D2 induces the production of human beta-defensin-3 in human keratinocytes. *Biochem. Pharmacol.* **2010**, *79*, 982–989. [CrossRef]
183. Guo, T.; Noble, W.; Hanger, D.P. Roles of tau protein in health and disease. *Acta Neuropathol.* **2017**, *133*, 665. [CrossRef]
184. Michalicova, A.; Majerova, P.; Kovac, A. Tau Protein and Its Role in Blood–Brain Barrier Dysfunction. *Front. Mol. Neurosci.* **2020**, *13*, 178. [CrossRef]
185. Kobayashi, N.; Masuda, J.; Kudoh, J.; Shimizu, N.; Yoshida, T. Binding sites on tau proteins as components for antimicrobial peptides. *Biocontrol Sci.* **2008**, *13*, 49–56. [CrossRef]

 *biomedicines*

Review

Cathelicidin LL-37 in Health and Diseases of the Oral Cavity

Joanna Tokajuk [1,2], Piotr Deptuła [1], Ewelina Piktel [3], Tamara Daniluk [1], Sylwia Chmielewska [1], Tomasz Wollny [4], Przemysław Wolak [5], Krzysztof Fiedoruk [1] and Robert Bucki [1,*]

1. Department of Medical Microbiology and Nanobiomedical Engineering, Medical University of Białystok, Mickiewicza 2C, 15-222 Białystok, Poland; asiatokajuk@gmail.com (J.T.); piotr.deptula@umb.edu.pl (P.D.); tamara.daniluk@umb.edu.pl (T.D.); sylwia.chmielewska@umb.edu.pl (S.C.); krzysztof.fiedoruk@umb.edu.pl (K.F.)
2. Dentistry and Medicine Tokajuk, Zelazna 9/7, 15-297 Bialystok, Poland
3. Independent Laboratory of Nanomedicine, Medical University of Białystok, Mickiewicza 2B, 15-222 Białystok, Poland; ewelina.piktel@wp.pl
4. Holy Cross Oncology Center of Kielce, Artwińskiego 3, 25-734 Kielce, Poland; tomwollny@gmail.com
5. Institute of Medical Science, Collegium Medicum, Jan Kochanowski University of Kielce, IX Wieków Kielc 19A, 25-317 Kielce, Poland; przemyslaw.wolak@ujk.edu.pl
* Correspondence: buckirobert@gmail.com

Abstract: The mechanisms for maintaining oral cavity homeostasis are subject to the constant influence of many environmental factors, including various chemicals and microorganisms. Most of them act directly on the oral mucosa, which is the mechanical and immune barrier of the oral cavity, and such interaction might lead to the development of various oral pathologies and systemic diseases. Two important players in maintaining oral health or developing oral pathology are the oral microbiota and various immune molecules that are involved in controlling its quantitative and qualitative composition. The LL-37 peptide is an important molecule that upon release from human cathelicidin (hCAP-18) can directly perform antimicrobial action after insertion into surface structures of microorganisms and immunomodulatory function as an agonist of different cell membrane receptors. Oral LL-37 expression is an important factor in oral homeostasis that maintains the physiological microbiota but is also involved in the development of oral dysbiosis, infectious diseases (including viral, bacterial, and fungal infections), autoimmune diseases, and oral carcinomas. This peptide has also been proposed as a marker of inflammation severity and treatment outcome.

Keywords: oral cavity; human cathelicidin; antimicrobial peptides; immunomodulation

Citation: Tokajuk, J.; Deptuła, P.; Piktel, E.; Daniluk, T.; Chmielewska, S.; Wollny, T.; Wolak, P.; Fiedoruk, K.; Bucki, R. Cathelicidin LL-37 in Health and Diseases of the Oral Cavity. *Biomedicines* **2022**, *10*, 1086. https://doi.org/10.3390/biomedicines10051086

Academic Editor: Jitka Petrlova

Received: 16 April 2022
Accepted: 2 May 2022
Published: 7 May 2022

Publisher's Note: MDPI stays neutral with regard to jurisdictional claims in published maps and institutional affiliations.

1. Introduction

The regular flow of nutrients in the temperature-stable and hydrated environment of the oral cavity, along with its spatial diversity, creates a perfect ecological habitat for various microorganisms, which shortly after the birth of the host become their integral part as the oral microbiota, or more broadly, microbiome (i.e., bacterio-, archae-, myco-, viro-, and protozoome) [1]. In normal healthy conditions, these highly biodiversified microbial populations, represented by ~1000 species, live in metabolically cooperative and self-regulating, e.g., via *quorum sensing*, biofilm communities coexisting in a symbiotic/commensal relationship with the human host (in a eubiosis state) and due to their colonization resistance role, are considered as a part of a nonspecific defense system. Therefore, the term "oralome" has been proposed to encompass all inter-microbial as well as host–microbiome interactions underlying the oral cavity ecosystem [1]. However, any disturbance of this homeostasis, resulting in uncontrolled overgrowth of certain bacterial species (dysbiosis or "unbalanced microbiome") and, in turn, their transition from a commensal to parasitic status, is responsible for the development of pathological conditions such as infectious and autoimmune diseases, along with the possible longitudinal consequences, such as oral carcinomas [2]. The most common oral cavity diseases associated with microorganism-induced inflammation include: dental caries, pulpitis, refractory apical periodontitis, periodontal diseases, implant-associated infection, candidiasis, and diseases of the oral mucosa. Yet, the pathogenesis of these remarkably polymicrobial and successively progressing diseases is known in a general outline only. Although the major implicated bacterial species are identified and classified into six color-coded complexes according to the role (the Socransky complexes) [3,4], underlying host factors, in particular immunity, are poorly understood. Briefly, the members of yellow, green, and purple complexes (e.g., *Streptococcus, Veilonella,* and *Actinomyces*) are considered as the primary or early colonizers of the oral cavity, which allow the succession of the secondary, the orange complex, e.g., *Fusobacterium nucleatum*, which acts a "bridge" for the late, red complex, colonizers. It is believed that the increased proportion of the latter two, in particular the red one, involving three obligate anerobic Gram-negative species, namely *Porphyromonas gingivalis, Tannerella forsythia,* and *Treponema denticola*, considered as the major periodontal pathogens, leads to pathological, associated with excessive immune response, changes resulting in gingival connective tissue and alveolar bone damage manifested clinically as periodontal diseases [3]. Hence, the distinction between these "good" (or beneficial) and "bad" (or useless/pathogenic) commensals may allow the development of new therapies for certain oral diseases. Specifically, their imbalanced proportion must be reconciled with nonspecific, i.e., the relying on microbial-associated molecular patterns (MAMPs) and pattern recognition receptors (PRRs), nature of the innate response and implicates a fine-tuned strategy of its coordination, e.g., via positive/negative feedback loops with the adaptive immune system, to appropriately respond to qualitative and quantitative changes in the composition of oral microbiota. Endogenous antimicrobial peptides (AMPs), such as cathelicidins or defensins, appear to be vital elements of this strategy [5–7]. In fact, the functional versatility and structural diversity—represented by over 100 various human AMPs—suggest that the protection of a specific body's microenvironments, rather than the systemic action, as well as maintaining homeostasis with microbiota, are the major function of these small, <100 amino acids, peptides [8]. For instance, over 45 AMPs are found in the human saliva and the gingival crevicular fluid (GCF), which are classified into six major functional classes, namely (i) cationic peptides (CAMPs), (ii) mediating bacterial agglutination and adhesion, (iii) metal ion chelators, (iv) peroxidases, (v) protease inhibitors, and (vi) peptides, with activity against bacterial cell walls [5,6]. Indeed, in the recent literature review by Silva et al. focused on AMPs in controlling the oral pathogens [9], the cathelicidin LL-37 and β-defensin-2 are two AMPs most commonly linked with the periodontal and cariogenic pathogens, respectively. Similarly, the elimination of CAMPs from human airway fluid, via a cation-exchange chromatography, was correlated with a substantial reduction in its antibacterial activity [10].

Hence, the term "host defense peptides" (HDPs) better reflects their role [11]. However, over a decade after the legitimation of their importance for oral health stability by the Seventh European Workshop on Periodontology [12], we are still far from understanding the exact role of these immensely pleiotropic and diverse peptides in this process. For instance, it is believed that the microenvironment-specific action is mediated via different, synergistically acting, combinations (or "cocktails") of AMPs, allowing efficient elimination of the target pathogens and/or their toxic products at concentrations not affecting the host cells and/or commensal microbes [13]. Indeed, the expression of AMPs is unevenly regulated by various bacterial species, including periodontal and oral mucosa microorganism(s). Thus, the composition of AMP cocktails may reflect the physiological response to specific pathogen(s). It is consistent with the observation that PRRs, such as Toll-like (TLRs) and NOD-like (NLRs) receptors, of oral epithelial cells may induce antibacterial actions, e.g., β-defensins production without concomitant inflammatory response, or the secretion of pro-inflammatory cytokines, such as interleukin-8 (IL-8) [14,15]. Accordingly, a mechanism where AMP-resistant, "good" commensal bacteria induce their expression to control "bad" or potentially pathogenic species has been proposed as a form of adaptive mutualistic co-evolution between the oral microbiota and the host [16].

Conversely, AMPs may act as natural buffers neutralizing the pro-inflammatory stimuli, e.g., LPS or (lipo)teichoic acids, produced by commensal bacteria in order to maintain the oral microbiota–tissues equilibrium. This hypothesis is supported by the observation that the concentrations of most AMPs found in body fluids, e.g., in saliva or GCF, do not reach minimum inhibitory concentration (MIC) values recorded by in vitro tests. Hence, biological activities other than bactericidal are fulfilled by AMPs [14,15]. In fact, many AMPs are classified as damage-associated molecular patterns (DAMPs), i.e., molecules released from damaged or dying cells, and initiate a diverse range of physiological and pathophysiological functions, acting as "alarmins" in the immune surveillance system [14,15]. Moreover, to make it more complicated, the recent metagenomic studies indicate the existence of populational variations in the oral microbiota composition, the so-called stomatotypes, characterized, for example, by the predominance of Proteobacteria (stomatotype 1) and Bacteroides (stomatotype 2) in the study by Willis et al. [2].

A compelling amount of evidence demonstrates that the LL-37 peptide is one of the AMPs that appear to be vital for maintaining the eubiosis in the oral cavity and, hence, it may serve as a potential agent for pharmaceutical prophylaxis or the treatment of its various dysbiotic conditions, e.g., through direct application, the application of its synthetic derivatives, or the modulation of its expression [17,18]. In fact, the therapeutic potential of LL-37 in other diseases has been already recognized by biopharmaceutical companies, e.g., in the treatment of venous leg ulcers (Promore Pharma AB, Sweden) [19] or chronic suppurative otitis media (OctoPlus BV) [20]. Currently, 69 clinical trials are recorded in the ClinicalTrials database (www.clinicaltrials.gov; search term: LL-37 or cathelicidin; accessed on 25 March 2022) evaluating its role and therapeutic/diagnostic potential in >30 different diseases/disorders, ranging from (i) bacterial infections, sepsis, periodontitis, tuberculosis, and HIV to (ii) various skin (e.g., psoriasis and atopic dermatitis), intestinal tract (e.g., Crohn disease), or respiratory tract conditions (e.g., asthma and cystic fibrosis) and (iii) melanoma, chronic kidney disease, and vitamin D deficiency. In particular, the role of vitamin D, as a potent cathelicidin inducer, is frequently studied in this context. However, due to a wide distribution and abundance of LL-37 across the human body (Figure 1), as well as its association with the pathogenesis of several infectious and non-infectious diseases [21,22], it seems to be relevant to consider potential medical and safety implications of LL-37-based therapies.

In the current review, we discuss possible mechanisms behind the protective role of LL-37 in the oral cavity, supplementing the literature with data from the constantly growing "-omics" and "-ome," databases e.g., genomic, proteomic, transcriptomic, interactome, or reactome databases, as a source of comprehensive information about protein expression, distribution, and interaction networks in the human body.

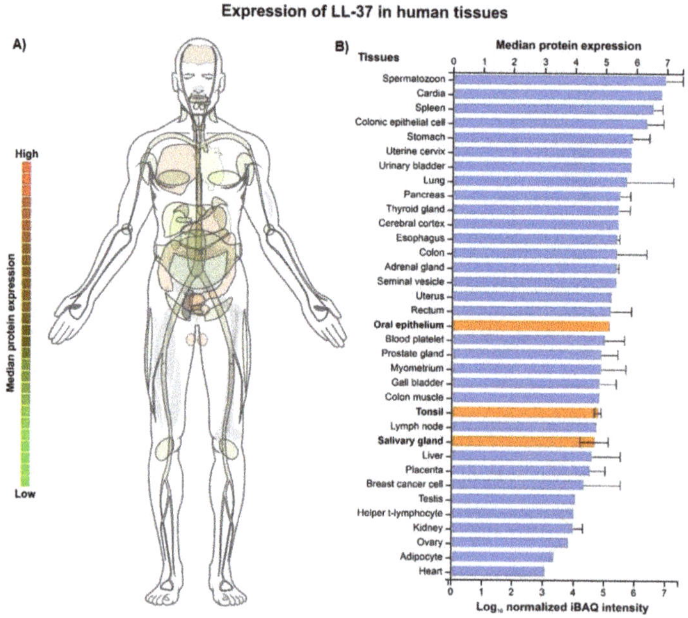

Figure 1. Expression of LL-37 in 35 human tissues, including oral epithelium, tonsils, and salivary glands (highlighted in orange). The data were obtained from ProteomicsDB (https://www.ProteomicsDB.org) last accessed 25 March 2022 [23] and represent mass spectrometry (MS1-level) proteome quantification; the intensity-based absolute quantification (iBAQ) value is a measure of protein abundance and corresponds to the sum of all the peptide intensities divided by the number of observable peptides of a protein (panel **A**). Relative abundance of CAMPs, including LL-37 (highlighted in red), present in the oral cavity (whole organism—integrated; LL-37 is highlighted in red). The data were obtained from PAXdb (Protein Abundance Database; https://pax-db.org/) last accessed 25 March 2022) [24] and are based on a mass-spectrometry-based study of the human proteome by Wilhelm et al. (2014) [25]; parts-per-million (ppm) values represent the abundance of each protein with reference to the entire expressed proteome, i.e., each protein entity is enumerated relative to all other protein molecules in the sample (panel **B**).

2. Mechanism of LL-37 Expression in the Oral Cavity and Its Action at the Molecular Level

Perceiving the human cathelicidin, encoded by the CAMP gene (cathelicidin antimicrobial peptide, i.e., containing the cathelin-like domain), as a single protein is oversimplification. In fact, due to its biological functional diversity, directed by multiple size variants and concentration- or microenvironment-dependent action, LL-37 is synonymous with "versatile," "pleiotropic," "multifunctional," "multifaceted," "factotum," or even "moonlighting" protein [18,21,26,27]. Although, the latter term is a misnomer, since cathelicidin does not meet the condition of no post-translational modifications, it shares the dark side of protein moonlighting, namely implication in various diseases, including cancers [28].

According to the Gene Ontology (GO) database, the CAMP gene is implicated in 24 biological processes (BPs), of which 16 are associated with immune responses against microorganisms, e.g., cell lysis and cellular response to LPS or peptidoglycan, whereas the remaining 9 represent various immune modulatory functions (both pro- and anti-inflammatory), e.g., neutrophil activation, cellular response to interleukins, chronic inflammatory response, or angiogenesis regulation (Figure 2A). These immunomodulatory actions are exerted by LL-37 interaction with various cellular receptors, e.g., chemotaxis (angiogenesis and wound healing) via the formyl receptor-like 1 (FPRL1) or alternatively by the CXC chemokine receptor 2 (CXCR2), cytokine release via the purinergic receptor R2X7, and IL-8 release in the lung epithelial cells through the epidermal growth factor receptor (EGFR), as reviewed by Verjans et al. [29].

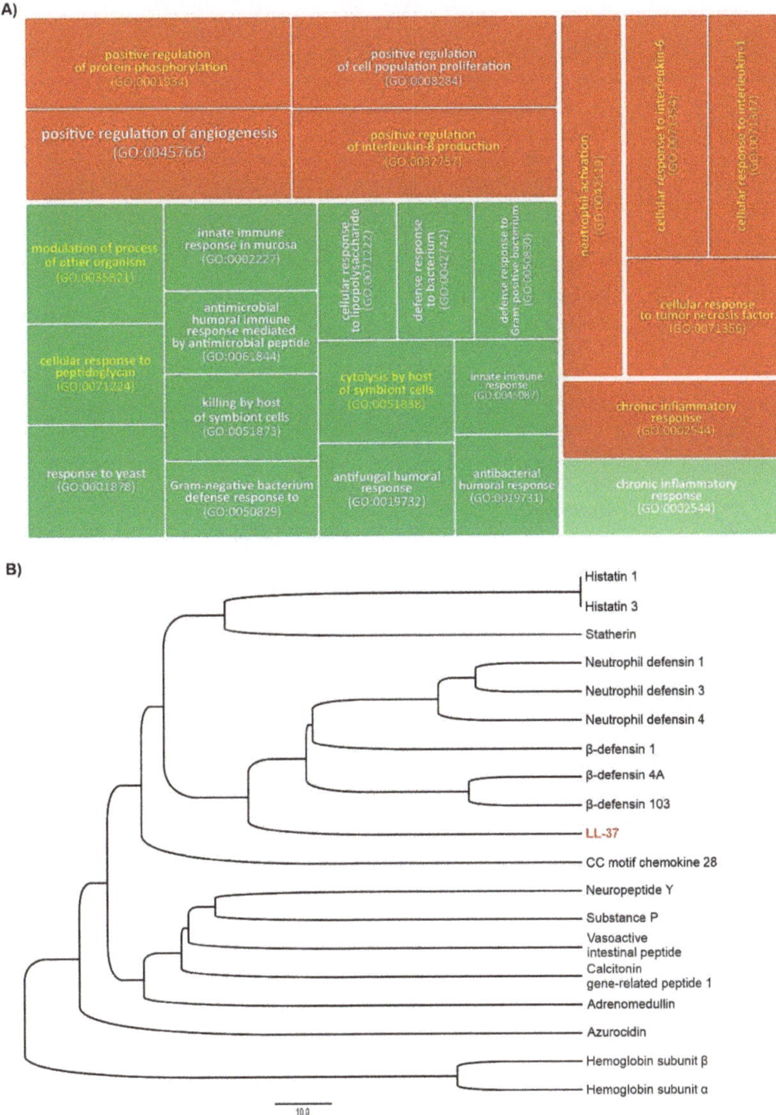

Figure 2. Tree map of the Biological Process (BP) Gene Ontology (GO) category for human cathelicidin LL-37 (panel **A**); green and red boxes show processes associated with immune responses against microorganisms and immunomodulatory action, respectively. Functional similarity of 20 human cationic antimicrobial peptides (CAMPs) present in the oral cavity inferred from comparing their Biological Process (BP) Gene Ontology (GO) category (panel **B**); the BPs unique for human cathelicidin LL-37 are indicated by yellow color in panel A. The dendrogram was constructed using Dice coefficient and UPGMA clustering with NTSYSpc software ver. 2.1 (Exeter Software) based on the GOs collected from the QuickGO server (www.ebi.ac.uk/QuickGO/).

Remarkably, in the "immunomodulatory" group, 7 categories are unique for LL-37, in comparison to 19 other AMPs from the oral cavity, revealing its extraordinary potential to modulate the immune response and, in general, its functional dissimilarity (Figure 2B). Furthermore, according to the data from the Protein Abundance Database (PAXdb), among these AMPs, LL-37 is the fourth- and third-most-abundant AMP in the whole body and saliva, respectively (Figure 3). This relative abundance of LL-37 in the oral cavity is supported also by other proteomic and transcriptomic data (Figure 4). It is noteworthy that in contrast to the microbicidal activity, immunomodulatory functions of LL-37, such as anti-endotoxic, wound-healing, or angiogenic ones, are unaffected by the physiological salt concentrations and, in general, are exerted at low concentrations (<1 μM) [27].

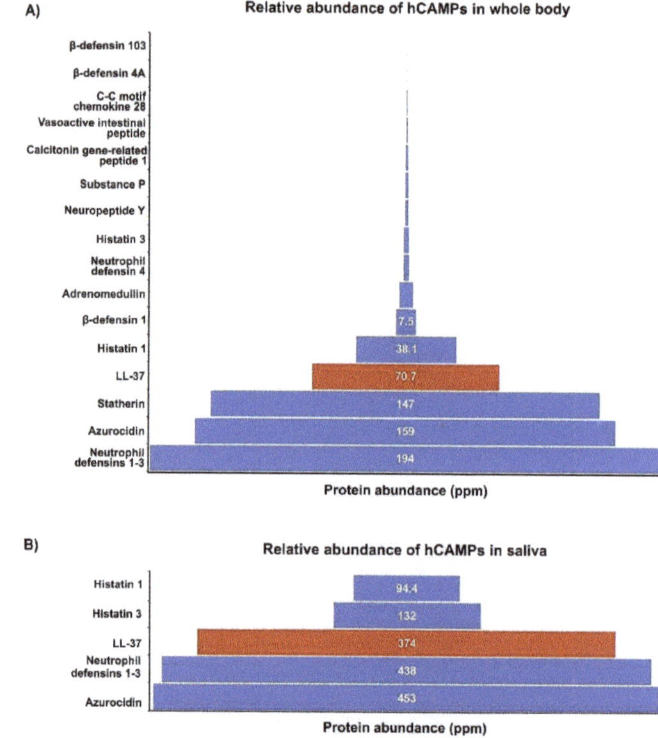

Figure 3. Relative abundance of CAMPs, including LL-37 (highlighted in red), in whole organism (panel **A**) and saliva (panel **B**). The data were obtained from PAXdb (Protein Abundance Database; https://pax-db.org/, accessed on 25 March 2022) [24] and are based on a mass-spectrometry-based study of the human proteome by Wilhelm et al. (2014) [25]; parts-per-million (ppm) values represent the abundance of each protein with reference to the entire expressed proteome, i.e., each protein entity is enumerated relative to all other protein molecules in the sample.

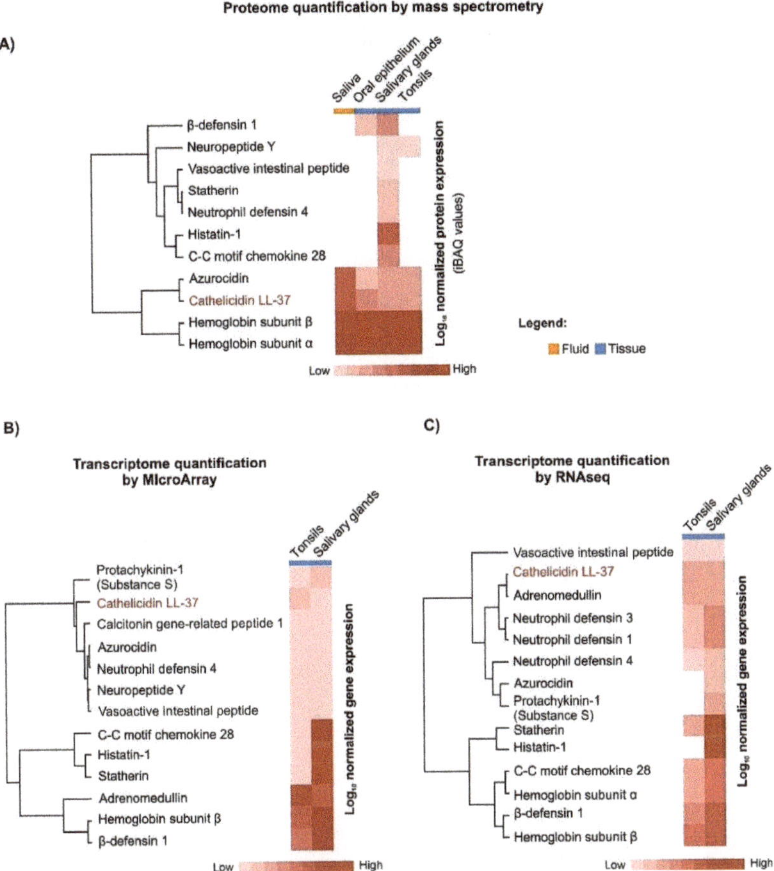

Figure 4. Comparison of CAMPs expression of in saliva, oral epithelium, tonsils and salivary glands. The data were obtained from ProteomicsDB (https://www.proteomicsDB.org, accessed on 25 March 2022) [23], and represent mass spectrometry (MS1-level) human proteome quantification; the iBAQ (intensity-based absolute quantification) value is a measure of protein abundance, and corresponds to the sum of all the peptides intensities divided by the number of observable peptides of a protein (panel **A**). Correspondingly, the gene expression results inferred from MicroArray and RNAseq transcriptomic studies are shown in panel **B** and **C**, respectively.

Furthermore, both activities are mediated by distinct regions of the LL-37 peptide; for example, amino acid residues 13–32 and 17–29 are a part of the central and core antimicrobial region, respectively. The first twelve N-terminal residues are crucial for chemotaxis and LPS neutralization, although in the latter cases, both N- and C-terminal residues are relevant (Figure 5). The possibility to separate both activities, e.g., toward microbicidal or immunomodulatory, via amino acid sequence manipulation, illustrates the therapeutic potential of LL-37 (see below). Correspondingly, although the LL-37 peptide—the product of proteolytic cleavage of 18 kDa cathelicidin precursor protein (hCAP-18) by proteinase-3 in neutrophils—is, in general, synonymous with the human cathelicidin, depending on the hCAP-18 processing proteases, other variants can be distinguished. For instance, in an acidic vaginal environment, gastricsin cleaves hCAP-18 into ALL-38 peptide, and TLN-58 variant, found in the lesion vesicle of palmoplantar pustulosis, is a product of skin proteases (kallikreins) [30]. In addition, the latter proteases can split the mature LL-37 into several peptides, such as RK-31, KR-30, KS-20, KS-22, KS-27, or LL-29, characterized by the same or even superior (RK-31 and KS-30) antimicrobial, but attenuated immunomodulatory, action than the native LL-37 (Figure 5) [31].

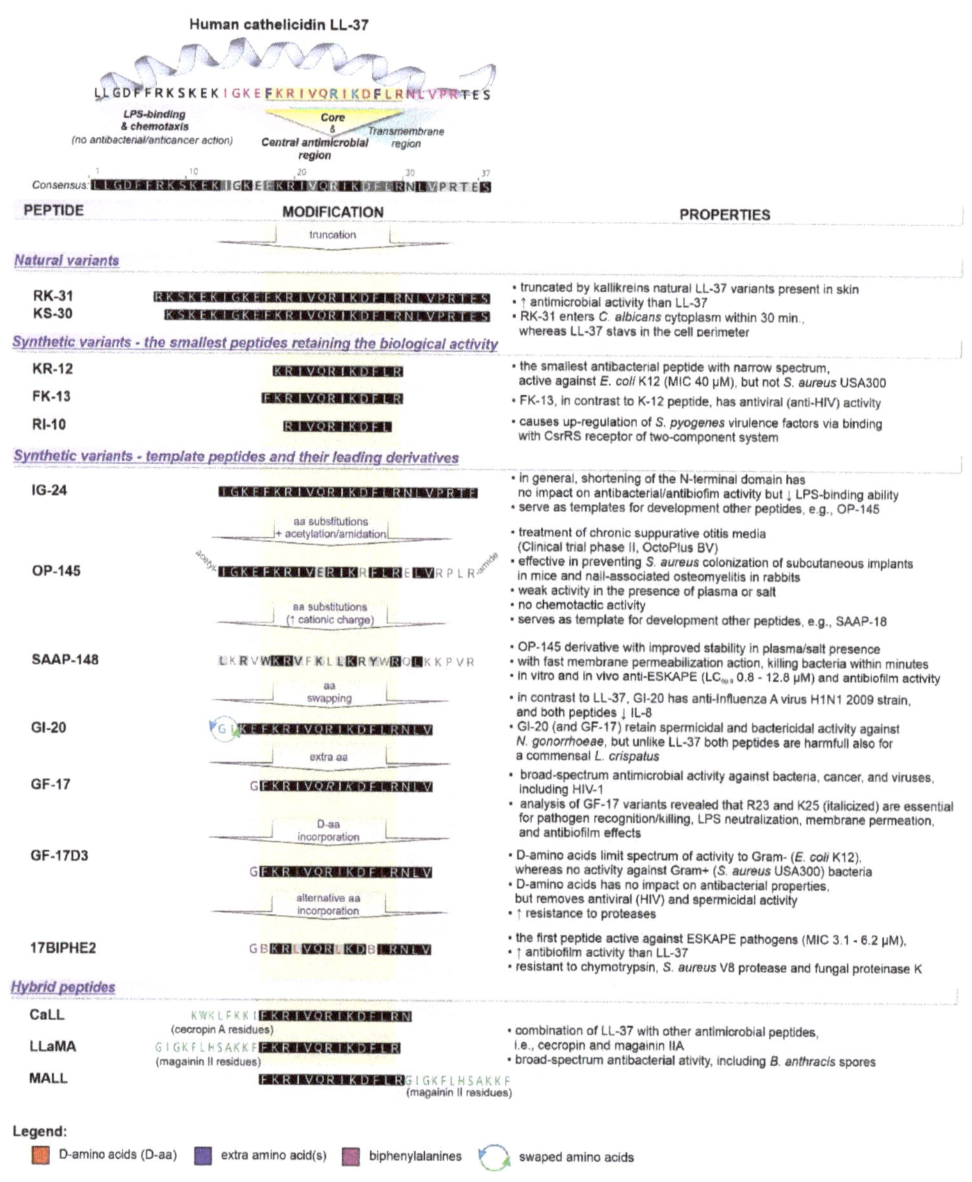

Figure 5. LL-37 and its selected natural and synthetic variants. Amino acids identical to the residues in native LL-37 are illustrated on a black background (with white letters); the others represent substitutions and/or modifications. The core antimicrobial region LL-37 residues essential for its interaction with cell membranes, i.e., phenylalanines F17/F27, and pathogen recognition/killing; LPS neutralization; membrane permeation; and antibiofilm effects, i.e., arginine—R23 and lysine—K25, are in bold and highlighted in violet and green, respectively.

Structurally, LL-37 is an amphipathic peptide with two helical regions separated by a loop and a short unstructured C-terminal tail [32] (Figure 5). It was reported that the net charge +6 directs LL-37 activity toward negatively charged bacterial surfaces, rather than human cells, leading to higher values of cytotoxic concentrations for eukaryotic cells (1–10 μM vs. 13–25 μM) [33]. In the physiological salt concentrations, ~35% content of hydrophobic residues allows LL-37 to adopt an amphipathic helical structure, which is further stabilized by interaction with cell membranes. However, the exact mechanisms of membrane permeation by LL-37 and its selectivity toward microbial membranes are still debated [33,34]. Remarkably, the recent studies in this area have revealed impressive conformational plasticity of LL-37 and its truncated variants, reflected by the formation of various oligomeric and supramolecular fiber-like structures (Figure 6). Importantly, such LL-37 fibers appear to affect cell membranes differentially than classical "barrel-stave" or "carpet/toroidal"-type mechanisms [35,36] (Figure 6). For example, Sancho-Vaello et al. proposed a model of pore formation via LPS extraction from the outer membrane by LL-37, with its subsequent diffusion into the periplasmic space, followed by interaction with the inner membrane involving oligomerization and fiber formation [37]. Furthermore, the authors identified a single arginine residue (Arg$_{23}$) as crucial in "eukaryotic" lipid, i.e., saturated lipid, recognition and suggested that the LL-37-fibrils mediated the recruitment of the host lipids as a novel, "armor-like," mechanism of antibacterial defense [37,38]. Similarly, depending on the lipid composition, Shahmiri et al. described two distinct LL-37 membrane interaction pathways, characterized by (i) pore formation in bilayers of unsaturated phospholipids and (ii) membrane modulation with saturated phospholipids, followed by the formation of fibrous peptide-lipid superstructures in the latter pathway [33]. Finally, LL-37 fibrils, via binding with DNA and forming spatially periodic DNA nanocrystalline immunocomplexes, have been described as potent simulators of the TLR-9-mediated response [39]. In general, binding extracellular DNA/RNA is a basic function of LL-37 that stabilizes nucleic acids and prevents degradation by the host and bacterial DNases/RNases, as well as enhances their uptake by various cells, such as macrophages, dendritic cells, and B cells, leading to TLR9 and TLR7 activation [40]. The same function of LL-37 is observed in neutrophil extracellular traps (NETs) [41]. However, at an antimicrobial and inflammatory focus, citrullination of its arginine residues, catalyzed by peptidyl-arginine deiminases, compromises the bactericidal activity of LL-37 and abrogates its TLR signaling and, hence, immunomodulatory functions [42]. However, citrullination of LL-37, and in turn impairing pro-inflammatory responses in macrophages, may be a potential treatment method of sepsis [13].

Assuming antimicrobial activity, it should be noted that LL-37 has a unique, among helical AMPs, ability to produce a helical structure in the absence of cell membranes. Therefore, LL-37 is prone to oligomerization or binding with other hydrophobic molecules, which in turn decreases its membrane permeation potential and explains a lower antimicrobial activity in the presence of body fluids, e.g., saliva, plasma, or rich media [27,43]. The antimicrobial activity of LL-37 is also reduced (2- to 8-fold) in the presence of 100 mM Na$^+$, as well as at low pH values; at the same time, saliva provides protection against LL-37 degradation by proteases secreted by periodontal pathogens [44]. Additionally, concentrations of LL-37 in saliva (0.14–3 μg/mL) [5] and GCF (0.01–10.8 μg/mL) [45] are lower than the MIC for periodontal pathogens, suggesting existing in vivo higher local concentrations of LL-37 or that other biological functions are important in the oral cavity [5]. However, laboratory susceptibility tests may not reflect in vivo situations. For instance, bacteria in the presence of carbonate, e.g., in saliva, are more susceptible to LL-37, and likewise in hypoxic conditions [5,6]. Therefore, the direct antimicrobial killing action of LL-37, alone or in combination with other AMPs, may be an important mechanism of defense in oxygen-deficient environments, such as inflamed gingiva, where O$_2$-dependent bactericidal activity of phagocytes is reduced [5,6]. Accordingly, Wuersching et al. have recently reported that O$_2$ availability alters the antimicrobial activity of LL-37 [46] (see below). Conversely, the major direct antimicrobial action of LL-37 may be the prevention of biofilms formation, which occurs at subinhibitory concentrations as low as 0.5 µg/mL, and represents 1/128 of the MIC value [47].

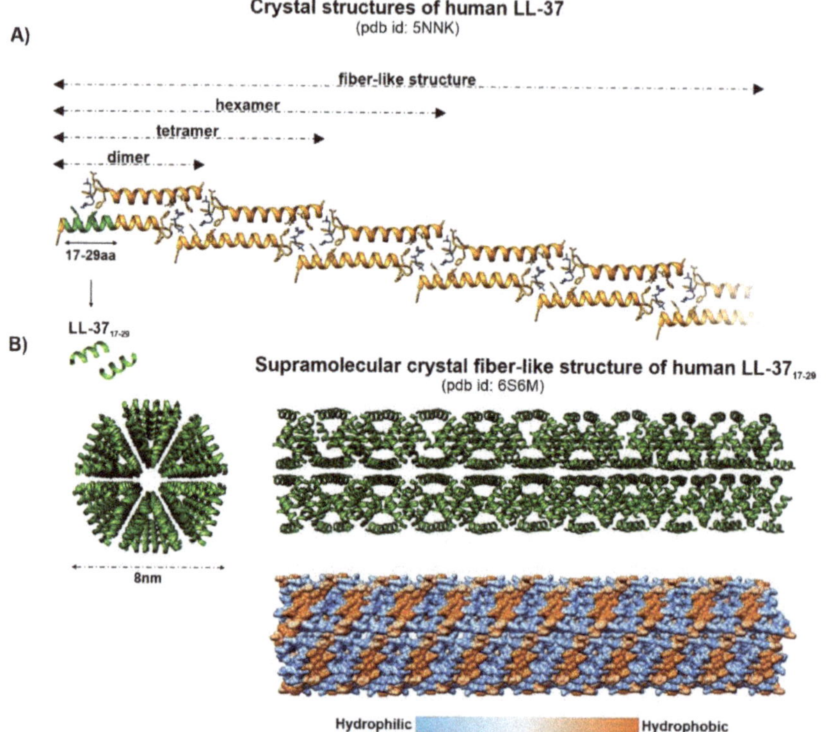

Figure 6. Human LL-37 oligomer structures interacting with lipid membranes (panel **A**) and its antimicrobial core fragment (residues 17–29), LL-37$_{17-29}$, forming supramolecular fiber-like structures (panel **B**), described by Sancho-Vaello et al. [38] and Engelberg and Landau [36], respectively.

Nevertheless, controlling of the intensity of the immune response induced by LPS/LTA, flagellin, and other molecules released by the oral microbiota (MAMPs) occurs in the physiological range of LL-37 concentrations (1–5 µg/mL) and appears to be a fundamental protective function of LL-37 [48]. Several studies have reported dismissing by LL-37 of various LPS-induced inflammatory responses, including protection against periodontal bone resorption [49], via blocking the binding of LPS to CD14 and lipopolysaccharide-binding protein (LBP) and/or indirect effects on immune cells [50]. In general, modulation of inflammation by LL-37 in response to the bacterial challenge is mediated through balancing TLR activation [40,51]. For instance, bacteria killed by LL-37 or the addition of LL-37 to non-viable bacteria strongly inhibits TLR activation. This "silent" killing reduces the inflammatory response and, in turn, tissue damage by unnecessary inflammation when the bacteria are no longer harmful [40]. It is also crucial from the perspective of concentration-dependent activity of LL-37.

Since the expression of LL-37 is not directly regulated by LPS and other MAMPs but is TLR dependent, it can be affected by other factors that directly coordinate the CAMP gene expression, thereby affecting its ultimate antimicrobial action. Vitamin D metabolite 1,25-dihydroxyvitamin D3 (1,25(OH)D3) has been recognized as the most important inducer of LL-37 expression, through binding with so-called vitamin D response elements (VDRE) for the vitamin D receptor (VDR) in the CAMP gene promotor. In fact, vitamin D links

LL-37 expression with MAMPs or other inflammatory stimuli because the expression of 1-α-hydroxylase (CYP27B1), which converts 25-hydroxyvitamin D3 to the active 1,25(OH)D3, can be also upregulated by the stimulation of TLRs [52]. Conversely, 1,25(OH)2D3 has no impact on LL-37 production in colonic epithelial cells, where butyrate acts as the CAMP gene inducer. Likewise, TNF-α, IL-17A, TLR agonists, insulin-like growth factor, MUC2 mucin, simvastatin, injury and wounding, and endoplasmic reticulum stress upregulate LL-37 expression. Certain bacteria (*Shigella* and *Neisseria*), bacterial exotoxins, IFN-γ, IL-6, glucocorticoids, transmigration across activated endothelium, and calcipotriol have been identified as downregulating agents [22,27].

Accordingly, McMahon et al. have shown that 1,25(OH)2D3 can induce not only the expression of LL-37 but also the innate immune regulator TREM-1 (triggering receptor expressed on myeloid cells), which augments TLR-mediated antimicrobial responses and the production of proinflammatory chemokines and cytokines in response to bacterial and fungal infections [53].

3. Involvement of the LL-37 Peptide in Maintaining Homeostasis of Oral Microbiota

Resistance to killing and/or evasion of the host immune responses is an inherent characteristic of pathogenic microorganisms that discriminates them from commensal species. Accordingly, to recognize the role of immune response in periodontitis, Ji et al. [54–56] hypothesized the existence of different patterns of susceptibility among no periodontopathic, i.e., *Streptococcus, Actinomyces, and Veillonella*, and periodontopathic bacteria (Figure 7) to the major antibacterial defense mechanisms in the gingival sulcus, namely neutrophil-mediated phagocytosis and bactericidal action of two AMPs—LL-37 and β-defensin-3. In general, the results revealed a rather high variability across the species and an increased resistance to phagocytosis in the periodontopathic group but no significant differences in susceptibility to AMPs. However, more evident patterns were observed after splitting bacteria into the Socransky complexes, i.e., early colonizers (the yellow/green/purple complexes) versus the orange and red complexes as well as *A. actinomycetemcomitans*. Specifically, the members of the red complex (*P. gingivalis, T. forsythia*, and *T. denticola*) were recognized as the most resistant to both phagocytosis and LL-37. In Figure 7, we adapted the LL-37 susceptibility patterns from the study by Ji et al. [55] and supplemented them with MIC values from four other studies [57–60]. Indeed, the resistance to LL-37 in *P. gingivalis* has been recently connected with the major surface glycoproteins (Pgm6 and Pgm7), also called outer-membrane-protein-A-like proteins (OmpALPs) [61].

It should be noted that in the original paper linking periodontitis in patients with Kostman's disease with *A. actinomycetemcomitans* [62], the authors reasoned its susceptibility to LL-37 based on a 3–4 log reduction in colony counts after 90 min treatment of 10^5 bacteria with 20 µg/mL of LL-37 in 10 mmol/L phosphate buffer at pH = 7.2 with 1% yeast extract (TSBY). The methodology was derived from Tanaka et al. [63], who also reported the susceptibility of three tested *A. actinomycetemcomitans* strains, based on a 99% effective dose (ED$_{99}$), i.e., the theoretical concentration of LL-37 at which there is a two $\log_{10}$ decrease in survivors after 1 h incubation with LL-37 at 37 °C, ranging from 8.2 to 11.6 µg/mL. Accordingly, substantial discrepancies between the MIC values for the same *A. actinomycetemcomitans* strains are reported by various studies, e.g., 62.5 vs. >200 (strain Y4) [55,58,59], 100 vs. >200 µg/mL (strain Y4) [55], and 50 vs. >200 µg/mL (strain ATCC 29523) [58,59]. Furthermore, susceptibility of three *T. denticola* strains to LL-37 was assessed by Rosenfeld et al. based on 85% growth inhibition following 1 h of exposure of 10^8 bacteria to LL-37 at 50 µg/mL [64], which correlates with 62.5 µg/mL MIC value for this species by Ji et al. [55]. Remarkably, these values are in contrast with high LL-37 MICs (449.4 µg/mL) for two other spirochaetes—*Borrelia* and *T. pallidum* [65]. Overall, these discrepancies impede drawing definite conclusions regarding the role of direct bactericidal action of LL-37 in the eradication of the leading periodontopathogens.

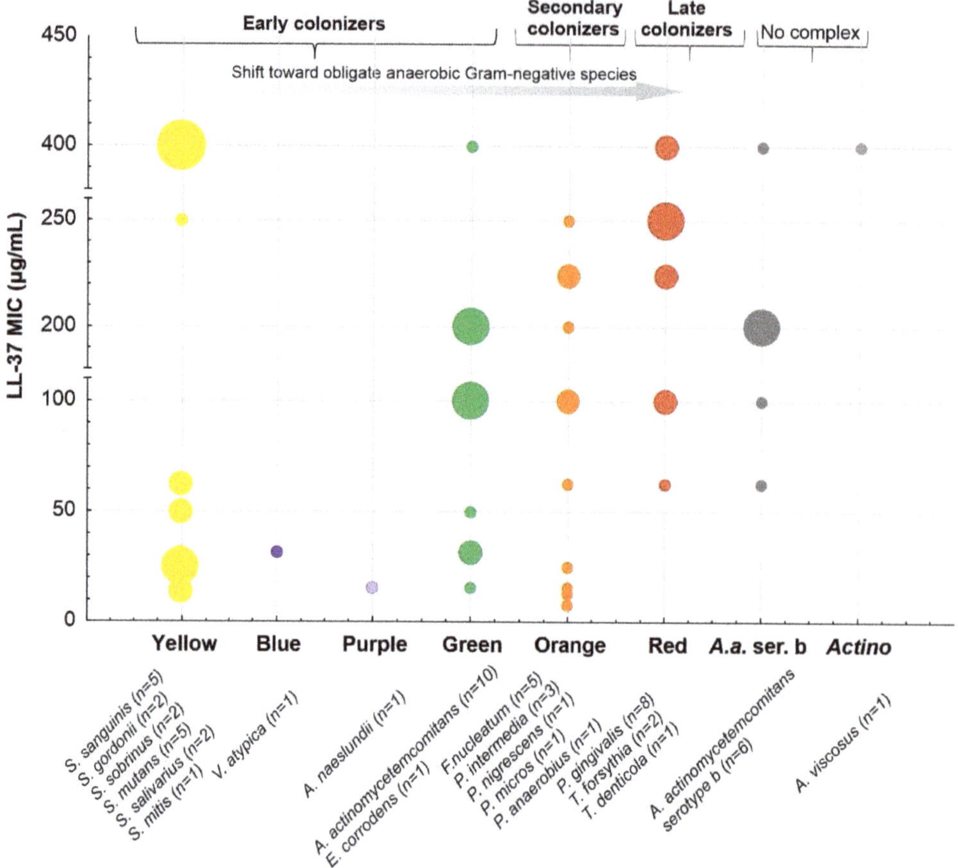

Figure 7. Susceptibility of selected bacteria representing the periodontal complexes (yellow, blue, purple, green, orange, and not-belonging-to-any-complex *A. actinomycetemcomitans* serotype b and *A. viscosus*) to LL-37 (MIC values, μg/mL) based on the references [55,57–60]; the repeated strains in various studies were analyzed individually, and for MICs expressed as ">" or not recorded "≤," for example, >100 or ≤100, the higher dilution value, i.e., 200, was used.

Conversely, the importance of LL-37 in *A. actinomycetemcomitans* eradication should be analyzed from a perspective of stimulation of LL-37 expression by vitamin D and synergistic action with responses mounted by TREM-1 as reported by McMahon et al. [17]. Briefly, the authors using a 3D air–liquid interface culture system observed that OKF6/TERT cells stimulated with 1,25(OH)2D3 for 24 h exhibit a significant increase in antibacterial activity against *A. actinomycetemcomitans* cells [17]. Accordingly, other immune responses promoted by LL-37, such as agglutination of *A. actinomycetemcomitans* and autophagy of *P. gingivalis*, might be responsible for their eradication [60,66]. Interestingly, the susceptibility of the oral bacteria to LL-37 has been recently connected with their oxygen requirements [46]. In detail, facultatively anaerobic bacteria such as *S. mutans*, *S. sanguinis*, or *Actinomyces naeslundii* appear to be less susceptible to LL-37 than the obligate anerobic species, e.g., *F. nucleatum*, *Veillonella parvula*, and *Parvimonas micra*. Hence, oxygen may represent a selective agent for LL-37-mediated killing. Furthermore, the authors reported concentration-dependent dynamics of planktonic growth inhibition characterized by two steps: (i) primary—requiring lower LL-37 concentration (25–100 μg/mL), followed by (ii) a second significant reduction in growth at 100 mg/mL (or 250 µg/mL for biofilms) of both groups of bacteria—denoted as "threshold concentration." Conversely, resistance to LL-37 may reflect an extraordinary "commensal" role of certain species, e.g., *F. nucleatum*, that due to the binding ability with other species, i.e., multigeneric coaggregation, is traditionally considered a "bridge" between the early and late colonizers [67]. In line with this, recent studies have shown that the *F. nucleatum* inducing expression of IL-8 and multiple AMPs may protect the host from "bad" commensals or non-beneficial/pathogenic bacteria [16]. For instance, *F. nucleatum subsp. nucleatum* (strain ATCC 25586 and 23726) and *F. nucleatum* subsp. *vincentii* (ATCC 49256) as naturally resistant to β-defensin 2 and 3, via inducing the expression of β-defensin 2 in oral epithelial cells through TLR-1/2 and TLR-2/6, along with LL-37 and CCL20, may act as homeostatic agents, ensuring protection against the "red" pathogen, *P. gingivalis*, which is susceptible to this AMP (and cannot induce its expression) [16,68]. In contrast, a poor induction of β-defensin 2 was noted in oral epithelial cells sensitive to *F. nucleatum* subsp. *polymorphum* ATCC 10953. Additionally, Ji et al. described *F. nucleatum* (and *P. intermedia*) as "self-limiting" commensals, i.e., inducing the host response that eliminates them efficiently and prevents overgrowth [54], which is consistent with its role in keeping the epithelium in a "heightened state of readiness" without promoting notable inflammatory cytokine responses [16]. Notably, since this protective action is mediated via cell-wall-associated FAD-I protein (Fusobacterium-Associated Defensin Inducer), it has been proposed as a novel, relying on the stimulation of AMP expression, immunoregulatory therapeutic agent to treat dysbiotic conditions in the oral cavity [16,68]. Importantly, recently, the ability of *F. nucleatum* to promote the expression of LL-37 in the presence of vitamin D3 has been also identified [16]. Considered as a "good" commensal, the *S. gordonii* M5 strain upregulates the expression of LL-37 and β-defensin-3 and is highly resistant to both AMPs (MIC > 125µg/mL) [55]. Interestingly, *S. gordonii* is also an essential prerequisite for further *P. gingivalis* colonization [67]. However, *P. gingivalis* can diminish these beneficial activities by downregulating the expression of proinflammatory IL-8 in the host cells, causing the so-called chemokine paralysis [69], as well as the inhibition of IL-8 activation by *F. nucleatum* and other "good" commensal. However, the mechanism of this process remains to be elucidated [16]. It is reported that the induction of IL-8 by such "good" commensals is responsible for the constitutive gradient in normal oral mucosa that contributes to the entry of percolating PMNs into the oral mucosa. On the contrary, Bachrach et al. suggested an opposite relation between *F. nucleatum* and *P. gingivalis*, i.e., the protection of *F. nucleatum* against LL-37 killing via its degradation by *P. gingivalis* protease (gingipain), as an example of a "group protection" mechanism [44,70]. Moreover, proteases of periodontal pathogens due to the degradation of the host protease inhibitors (SLPI and elafin) and, in turn, the stimulation of the protease 3 release may be considered direct modulators of LL-37 processing [71]. It is also noteworthy that certain pathogenic bacteria can use LL-37 as a gene expression inducer to promptly adapt to new conditions. For example, LL-37 upregulates the virulence genes of *Streptococcus pyogenes* via binding with its CsrRS receptor [72]. Therefore, different scenarios of interactions between the oral bacteria and LL-37 and other AMPs have been proposed (Figure 8) [16]. Scheb-Wetzel et al. showed that *E. faecalis* is highly susceptible to the antimicrobial effect of cathelicidin LL-37, which has been suggested in other studies as well [73]. In addition to its antibacterial activity, LL-37 inhibits *C. albicans* adhesion and growth. LL-37 can interact with the cell surface of *C. albicans* through its binding to cell wall polysaccharides, especially mannans, as well as exoglucanase Xog1. Notably, LL-37 causes cell aggregation, cell wall remodeling, and β-glucan exposure in *C. albicans* cells. Consequently, LL-37 strongly reduces *C. albicans* adhesion to plastic surfaces and oral epidermoid OECM-1 cells. LL-37 induces cell wall stress and the UPR, and Sfp1 contributes to the regulation of these stress responses [74].

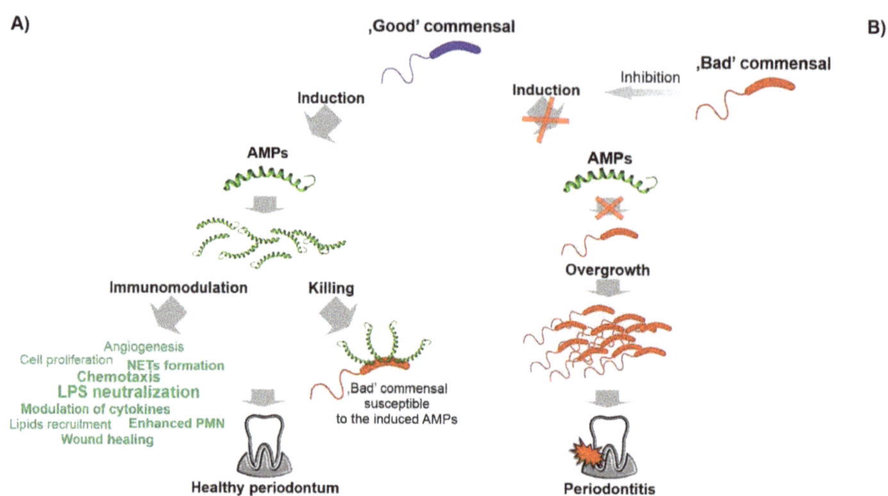

Figure 8. Scenarios of interactions between the oral bacteria and LL-37 and other AMPs. (i) "Good" commensals induce AMPs (and are resistant to the induced AMPs), while "bad" commensals are susceptible to the induced AMPs; (ii) "good" commensals, by inducing AMPs, enable the host to protect themselves from potential attack by pathogenic bacteria and to control eubiosis (panel **A**); (iii) "bad" commensals inhibit the pro-AMPs action of "good" commensals (panel **B**), adapted from [16].

It is worth also underlining that oral epithelial cells represent the first line of defense against viruses that are spread via saliva, including KSHV, herpesviruses (such as herpes simplex virus type 1), human cytomegalovirus, Epstein–Barr virus, and HPV. LL-37 might perform antiviral activities against those microorganisms due to its cationic nature and amphipathic structure, leading to binding to microorganism surfaces and subsequent virion perturbation [75]. LL-37, after virus envelope/capsid disruption, gains the ability to irreversibly bind to the virus DNA or RNA. LL-37 specifically has also been shown to enhance other methods of innate antiviral immunity, namely interferon expression. The analysis of docking studies and the display of positive interfacial hydrophobicity of LL-37 resulting in the disruption of COVID-19 viral membrane elucidate the fact that LL-37 could be effective against all variants of SARS-CoV-2. [76]. Overall, the induction of endogenously expressed LL-37 in oral epithelial cells may help to restrict oral infection.

4. LL-37 Peptide in Dental Caries, Pulpitis, Refractory Apical Periodontitis, and Periodontal Diseases

The oral cavity has a unique environment—hard tissues are not completely covered with epithelium but pierce through it. Such anatomy creates a specific environment that is constantly exposed to bacteria and associated infections that might take place within the tooth and the epithelium/junctional epithelium [77]. Periodontitis, pulpitis, periapical periodontitis, and caries are the most common diseases of the oral cavity. All of them have the features of opportunistic infections, which depend on biofilm formation and external factors, predisposing and modifying this process by changes in the host's defense system. Periodontitis concerns the tissues supporting the tooth, i.e., the periodontium. These include the gums, bones, periodontal ligaments, and root cementum. The disease begins with the gums and is called gingivitis—gum disease disappears after the triggers are removed but, if left untreated, gingivitis turns into periodontitis, in which we observe irreversible changes in the periodontium—loss of connective tissue and bone. The progression of the disease leads to the mobility and subsequent loss of the tooth [46,78]. The main bacteria associated with periodontitis development are *P. gingivalis*, *T. denticola*, *T. forsythia*, and *A. actinomycetemcomitans*. These microorganisms have different virulence factors, such as bacterial collagenases, sulfides, and endotoxins (lipopolysaccharides). Moreover, *A. actinomycetemcomitans* releases leukotoxin, while *P. gingivalis*, *T. denticola*, and *T. forsythia*, release trypsin-like peptidases. Regardless of the biofilm, the occurrence and course of periodontitis may be influenced by modifying and predisposing factors such as dry mouth, biofilm retention factors, smoking, metabolic factors, poor nutrition, medications, various systemic diseases (e.g., hematological disorders and endocrinological disorders (diabetes)), and stress. Periodontitis, a chronic inflammatory disease, occurs when the balance between the patient's immune system and the pathogenicity of the bacteria is upset. Its development depends on whether the immune system (neutrophils) complements and antibodies neutralize the metabolites of the bacterial plaque and its metabolites in the periodontal connective tissue. Neutrophils prevent bacteria and their metabolic products from getting into the connective tissue, but if they get there, pro-inflammatory cytokines IL-1 and PGE2 are released, which increase inflammation and thus the destruction of connective tissue (including bones). Periodontal pockets are formed, which provides even easier conditions for the development of periopathogens [79]. In 2017, a new periodontal classification was created. The European Federation and the American Academy of Periodontology have established based on pathophysiology that there are three types of periodontal disease, periapical periodontitis, peri-implantitis, and periodontitis, characterized by inflammation and the destruction of alveolar bone and connective tissues.

Caries represent a dysbiotic state of the oral microbiota related to the frequent consumption of sugars, poor hygiene, and bad patient compliance. *S. mutans* is a human cariogenic pathogen that can ferment sugar into lactic acid, acidifying the environment and promoting the development of acid-resistant bacteria and the subsequent disease. This disease refers to hard teeth tissue and begins with the demineralization of the enamel. The progression of the disease causes the breakdown of dentin (the tissue under the enamel), followed by the involvement of the dental pulp filling the tooth chamber and its death. Caries treatment consists of preventing bacterial plaque deposition at an early stage; in the case of enamel involvement conservative treatment by filling the cavity with composite material; and when reaching the dentin or pulp, using materials that stimulate the cells of these tissues to regenerate the dentin pulp complex, i.e., for the production of repair dentin known as the dentine bridge [80–82]. According to research by Davidopoulou, LL-37 is present in saliva in early childhood and its concentration increases with age to achieve equilibrium in maturity. Among others, the concentration of cathelicidin in saliva depends on milk consumption, age, number of teeth, and various mechanical stimuli [83–85]. The expression of LL-37 in unstimulated saliva in children with a high caries index is low compared to similar age patients with average caries activity. In general, the LL-37 concentration is significantly higher in patients without and with low caries activity [58,83]. In addition, the protective role of LL-37 in dental caries was suggested in patients with edentulism [86]. However, the opposite results were presented by Colombo et al. [87], who observed that only combinations of LL-37 with β-defensins and histatin-5, but not LL-37 alone, were positively associated with the caries levels. For example, Phattarataratip et al. observed that *S. mutans* strains isolated from caries-free patients were more susceptible to LL-37 than those isolated from caries-active subjects, though without statistical significance [88]. It is worth underlining that LL-37 achieves an antibiofilm effect by reducing the adherence of bacteria to the tooth surface, as well as by interfering with the thickness of the biofilm. Moreover, conventional antibiotic therapy usually is not effective in eliminating biofilm bacteria, mainly due to the high resistance of bacteria to antibiotics and the incompetence of antibiotics to penetrate the biofilm, while LL-37 handles those difficulties. The direct bactericidal effect is based on the lysis of bacteria and the inhibition of the pro-inflammatory activity of bacteria wall components, especially LPS and teichoic acid. As a result, the production of cytokines (IL-1, IL-6, and TNF-alfa) and other inflammatory responses is reduced. Importantly, LL-37 provides the maximum inhibitory effect on primary colonizers, i.e., the yellow–orange complex, whereas the pathogenic red complex is resistant to the LL-37, as mentioned above. Moreover, some bacteria can produce enzymes, mostly proteases, that inactivate LL-37. *P. gingivalis* can produce arginine-specific gingipains, cysteine proteases that inactivate the LL-37 peptide (arg-gingipains and lys-gingipains) [44]. Similar proteases are secreted by the red complex periodontal pathogens [89]. *T. forsythia* releases various MMP-carinase, which use LL-37 as a substrate. Mirolase, a subtilisin-like serine protease, degrades cathelicidin [90]. LL-37 might also behave similar to an opsonin because it facilitates the removal of antigens by the immune system. Apart from these actions, LL-37 directly affects fibroblasts (cells with the greatest number in periodontal tissues)—it enables their migration to the site of inflammation [91] and stimulates the release of IL-8, IL-6, and TIMP-1, as well as beta-FGF, HGF, and KGF [91], which may stimulate the regeneration of periodontal tissues and thus the remission of the disease [92]. Overall, the impairing protective role of LL-37 in maintaining the oral health stability by direct (e.g., enzymatic degradation by bacterial proteases) and indirect factors related to dysbiotic conditions is responsible for the development and progression of caries.

Chronic apical periodontitis is a lesion formed by the periradicular host defense as a response to microorganisms present in the root canal system. Although no specific microorganism has been identified as the principal etiologic agent of pulpal and periapical pathosis, some species have been more frequently reported in the root canal space. Previous studies have shown that such species as *E. faecalis*, *Eubacterium*, *Fusobacterium*, *Peptostreptococcus*, *Porphyromonas*, *Prevotella*, and *C. albicans* are commonly encountered in endodontic infection [93,94]. Interestingly, Jonsson et al. observed that the treatment of periodontal ligament cells with LL-37 in the range 0.1-1 μM completely inhibited the expression of LPS-induced monocyte chemoattractant protein-1 (MCP-1) and reduced the production of interleukin-6 by 50–70% [95]. On the contrary, pro-apoptotic activity, reflected by elevated levels of caspase 3 (an apoptosis mediator), required 1 μM doses of LL-37, whereas anti-proliferative effects were exerted at an 8 μM concentration of LL-37. Remarkably, the obtained values are well correlated with levels of LL-37 in GCF patients with chronic and aggressive periodontitis (Figure 9). Another study illustrates the dual, concentration-dependent, nature of LL-37, that on crossing a certain threshold, instead of promoting tissue and bone regeneration or wound healing, contributes to their destruction [96].

In addition to the above-reported discoveries, LL-37 is characterized by anti-osteoclast properties. As evidenced, cathelicidin LL-37 limits bone resorption by inhibiting TLR ligand and reduces LPS-promoted osteoclast production, as well as stimulates angiogenesis and bone regeneration. Moreover, LL-37 inhibits LPS-induced inflammation, promotes BMSC proliferation and migration, and stimulates the osteogenic differentiation of BMSCs in both normal and inflammatory microenvironments via the P2X7 receptor and MAPK signaling pathway [97]. Interestingly, it was recently reported that osteo-inductive biomaterials regulate gene expression through the MAPK pathway. They are used in periodontal microsurgery to regenerate periodontal tissue [97]. Accordingly, osteogenic differentiation and increase in calcium mineral deposition mediated by LL-37 were demonstrated by Liu et al. [98]. The LL-37-linked differentiation processes are also crucial for the continuity of permanent teeth development when immature teeth suffer from periodontitis or progressive carious lesions and thus pulpal necrosis occurs. In such cases, LL-37 was reported to stimulate migration and odonto/osteogenic differentiation of stem cells from the apical papilla (SCAPs), being the main source of cells for the primary dentin formation in the roots and seed cells for pulp regeneration, through the Akt/Wnt/b-catenin signaling pathway [99]. Recently, there has been a lot of research on the regenerative treatment of the dental pulp, which would reduce the number of root canal treatments and increase the prognosis for long-term maintenance of the teeth in the oral cavity. LL-37 can stimulate odontoblastic cells to form reparative dentin, as well as fibroblast-like stem cells that are multipotent and are likely to polarize, among others, to odontoblasts or osteoblasts; nevertheless, the whole mechanism requires further study. It is believed that cathelicidin LL-37 will benefit as a pulp capping agent as the forerunner for cell differentiation and the creation of a dentin bridge [100–102]. The LL-37-mediated increase in dentine sialophosphoprotein (DSPP) production and *DSPP* gene expression contributes possibly to the differentiation of dental pulp stem cells into odontoblast-like cells [103]. Cathelicidin-related antimicrobial peptide (CRAMP, encoded by the Camp gene) was found to be expressed in rat odontoblasts at the early dentinogenesis stage, and formyl peptide receptor 2 (FPR2), i.e., one of the transmembrane G-protein-coupled receptors to which CRAMP/LL-37 binds to cause various physiological effects, is constantly expressed in the sub-odontoblastic layer. Additionally, both CRAMP and FPR2 were recorded to be expressed in response to cavity formation, indicating that both of these agents appear during physiological and reparative dentin formation [104]. Moreover, in the report by Kajiya et al., LL-37 was demonstrated to enhance the migration of human pulp cells and thus increase pulp–dentin complex regeneration by the activation of epidermal growth factor receptor (RGFR) and c-Jun N-terminal kinase by the induction of heparin-binding cell migration [100].

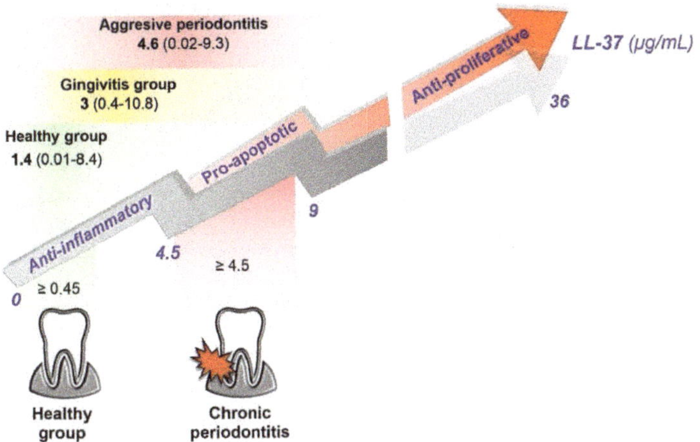

Figure 9. Illustration of concentration-dependent anti-inflammatory (exerted at a concentration ≥ 0.45 µg/mL or 0.1 µM), pro-apoptotic (exerted at a concentration ≥ 4.5 µg/mL or 1 µM), and anti-proliferative (exerted at a concentration ≥ 36 µg/mL or 8 µM) activity of LL-37 in gingival crevicular fluid (GCF) postulated by Jonsson at al. [95]. The authors correlated the anti-inflammatory and pro-apoptotic values with concentrations of LL-37 in GCF patients with chronic periodontitis and the healthy control group from the reference [105]—the lower part of the image (the size of areas highlighted in green and red represent the upper and lower quartiles of LL-37 concentrations in the healthy and chronic periodontitis group, respectively). The upper part of the image shows LL-37 concentrations in GCF patients with gingivitis and aggressive periodontitis in comparison to the healthy control group from the reference [45] (the size areas of highlighted in green, yellow, and red and the higher color intensity represent the range of LL-37 concentrations (the values in parentheses) and the means, respectively).

As widely recognized, the inflammatory environment and vitamin D intensify the production of the LL-37 peptide [106]. Furthermore, McMahon et al. have observed that due to the endogenous expression of 1-α-hydroxylase in the gingival epithelial cells, systemic vitamin D treatment may be sufficient to activate the innate immune response in these cells [17]. Interestingly, a strong association of the VDR gen polymorphism with chronic periodontitis has been revealed by a recent meta-analysis [107]. Recently, triggering TREM-1 has also been proposed as a potential target for the treatment or prevention of oral cavity diseases. Indeed, several other studies have underlined the importance of TREM-1 in periodontal diseases [108–110]. For instance, increased TREM-1 expression in gingival tissue was observed in both chronic and aggressive periodontitis and correlated with the levels of the red complex species, whereas a significantly elevated load of *A. actinomycetemcomitans* was noted only in the aggressive form [111].

It was evidenced that patients susceptible to caries and children with high carious activity have lower levels of LL-3 and even a lack of peptide might be observed in patients with some neutrophil deficiency or Kostmann syndrome, Papillon–Lefèvre (PLS) syndrome, or Haim–Munk syndrome (Figure 10) [62,113–115]. A lack or impaired proteolytic processing of hCAP-18 is observed in patients with Papillon–Lefèvre and Haim–Munk syndromes, caused by the mutational inactivation of the cathepsin C gene. In turn, the reduced activation of neutrophil serine proteases, i.e., proteinase-3 results in the development of severe periodontal disease [115,116]. Interestingly, the periodontal pathogens may exert the opposite action, i.e., increase the level of the proteinase 3 [71]. To confirm, significantly elevated levels of the proteinase 3 in GCF samples from patients with periodontitis and gingivitis, along with a positive correlation with clinical periodontal parameters of sampling sites, have been reported [117]. To sum up, LL-37 is responsible for limiting the periodontal and caries pathogens and inhibiting the progression of periodontal disease. LL-37 might also act as a diagnostic tool (early marker of inflamed tissues), a prognostic tool in periodontitis, and also a potential agent for the prevention and treatment of periodontal diseases.

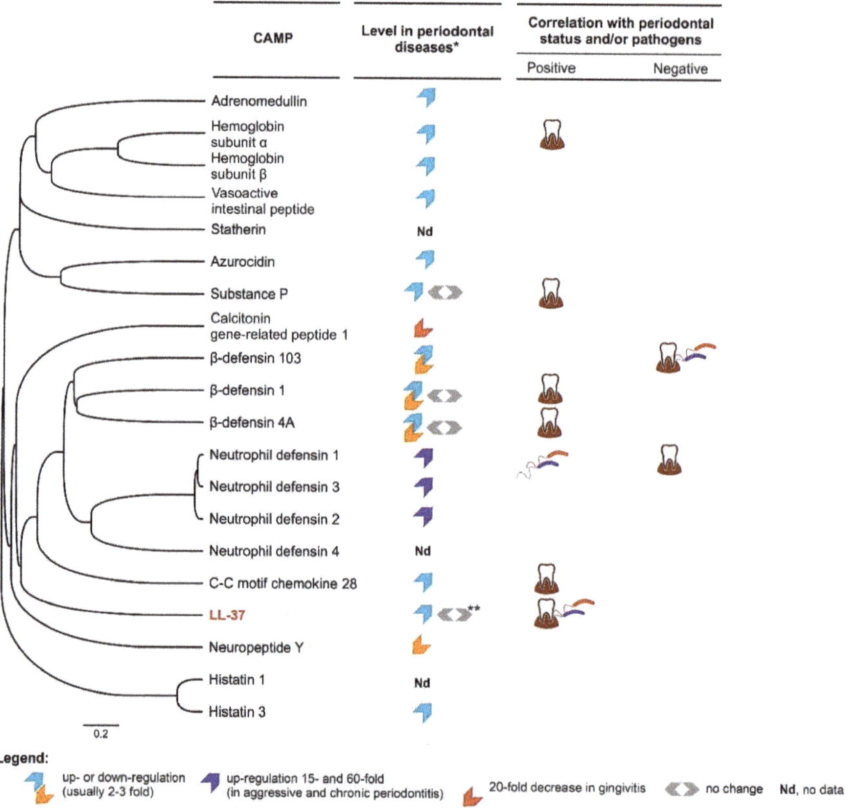

Figure 10. Comparison of 20 human cationic antimicrobial peptides (CAMPs) present in oral cavity, including LL-37 (highlighted in red), based on the amino acid sequence similarity as well as changes in their expression in periodontal conditions and correlation with the periodontal status and/or bacterial pathogens. The alignment of hCAMP sequences (only aa representing active peptides were

aligned, i.e., without signal sequences and/or proteolytically cleaved fragments) and their phylogenetic tree was performed using MAFFT version 7 server (based on the L-INS-I alignment algorithm and average linkage—UPGMA, respectively) [112]. The data regarding the hCAMP expression levels (, i.e., up- or downregulation) and their correlation with the periodontal status (, represented by probing depth, PD; bleeding on probing, BOP; clinical attachment loss, CAL; plaque index, PI; gingival index, GI; papillary bleeding index, PBI) and microbiological parameters (, i.e., changes in occurrence of *P. gingivalis*, *T. forsythia*, and *T. denticola*) were obtained from the references [5,6,113].

5. LL-37 Peptide in Diseases of Oral Mucosa and Its Implication in Oral Cancer Development

Some of the pathogens responsible for periodontal and pulp diseases, as well as mucosa diseases, may also be the etiological factors of pre-cancerous conditions and oral cavity cancers, e.g., *P. gingivalis* or *C. albicans* were isolated from the keratinocytes of patients with squamous cell carcinoma of the oral cavity [118]. The participation of yeasts in the process of oncogenesis has been known for a long time, and, in part, it was related to the *Candida* ability to produce nitrosamine [119]. Interestingly the mutual interaction between *P. gingivalis* and *C. albicans* might be of importance since *P. gingivalis* causes the formation of elongated hyphae in *C. albicans*, while the fungus increases the expression of genes responsible for the virulence of this periopathogen [119,120]. In this setting, LL-37 may prevent the formation of neoplastic chambers by promoting *P. gingivalis* autophagy in keratinocytes and influencing TRIM22 and LAMP3 [66]. As for *C. albicans*, LL-37 influences cell wall remodeling, reducing adhesion of *C. albicans* to the surface cathelicidin, which can bind to mannans and Xog1 exoglucanase, stimulating the Mkc1 MAP kinase pathway to maintain wall integrity [74]. *Candida*'s epithelial invasion can cause hyperplastic conditions probably only when the infection is chronic, deep, and associated with risk factors such as tobacco and alcohol, or others [119]. Moreover, *F. nucleatum* has been implicated in oral cancers since enriched fusobacterial organisms have been found in leukoplakic lesions; i.e., precursors of OSCC [121]. Oral ulcers, one of the clinical conditions representing a sign of Behcet's disease (BD) associated with cancer development, are linked to the adhesion of *S. sanguis* to buccal epithelial cells. A study by Mumcu et al. has shown a positive correlation between salivary levels of LL-37 and the number of monthly oral ulcers in patients with Behcet's disease [122].

Importantly, altered expression of hCAP18/LL-37 was observed in oral inflammatory lesions with and without microbial infection or oral cancer. Noteworthy, inflammatory conditions in the oral mucosa might raise the concentration of LL-37 in saliva, which might directly promote carcinogenesis, considering that inflammation has now been confirmed as the seventh hallmark of tumors. One observation that links LL-37 ability to cancer development is the LL-37 ability to stimulate cell proliferation by decreasing cell apoptosis [123]. LL-37 was also reported to be engaged in oral submucous fibrosis (OSF) development, which is a chronic disease significantly contributing to mortality since a high malignant transformation rate was observed in subjects that were diagnosed with this condition (1.5–15%) [124]. Pathological characteristics of OSF include chronic inflammation, excessive collagen deposition in the connective tissues below the oral mucosal epithelium, local inflammation in the lamina propria or deep connective tissues, and degenerative changes. Causative factors of OSF include autoimmunity; vitamins B, C, and iron deficiencies; chewing betel nut; consumption of spicy foods; human papillomavirus (HPV) infection; and genetic mutations [124]. Oral lichen planus (OLP) is another chronic inflammatory disease of the oral mucosa that might lead to changes in LL-37 expression, but its etiology is unknown. It is characterized by the lysis of the basal keratinocytes after being damaged by cell-mediated immune reactions [125]. In patients with OLP, the salivary concentration of LL-37 was significantly higher than in healthy subjects. The oral lichen planus (OLP) expresses more hCAP18/LL-37 peptide, as detected by intense immunohistochemical staining, compared to the healthy epithelium, and this increased expression is not related to microbial infections [126]. Since it is estimated that 1 in 6 cancers in the oral cavity is attributable to either viral or bacterial infections, the positive loop between infection, inflammation that manifests by host increase of hCAP-18/LL-37 expression, and oral tumor development should be considered as the molecular background of many different malignancies. The molecular nature of different microorganism might be of importance as well, since some bacteria are considered more likely to be involved in oncogenesis, as mentioned above.

To date, LL-37 has been established to display variable, tumor-biology-dependent effects on cancerous cells, both pro- and antitumorigenic [127]. As demonstrated in a broad spectrum of studies, LL-37 suppresses colon and gastric cancer progression, as well as exerts anti-cancer effects against hematological malignancies via the molecular mechanisms that involve triggering of caspase-independent apoptosis, a decrease in proteasome activity, and inhibition of angiogenesis processes by FPR1 activation [128–130]. At the same time, by recruiting MSCs and increasing their invasiveness, activating EGFR and MEK/ERK1/2 signaling pathways, as well as stimulating ERB-family receptors, human cathelicidin acts as a tumor promoter supporting the invasive phenotype of cancer cells [131–133]. For this reason, the role of LL-37 in cancer expansion is not clearly defined and unified for all cancers, since its effect is tissue specific. In the aspect of oral cavity tissues, the amount of data on LL-37's impact is unsatisfactory to date. CAMP/LL-37 expression is significantly downregulated in OSCC, and its low expression correlates with histological differentiation and lymph node metastasis. More importantly, a cell-specific methylation pattern in the promoter region of human CAMP was detected, which suggests that LL-37 plays a suppressive role in the progression of this cancer [134]. Indeed, LL-37 was reported to trigger caspase-dependent apoptosis by the P53-Bcl-2/BAX signaling pathway in the human OSCC HSC-3 cells [135]. In another study, the anti-tumor effects of the synthetic fragment of the LL-37 peptide, KI-21-3, against oral squamous cell carcinoma were demonstrated in in vivo settings and concluded as resulting from antiproliferative and proapoptotic activities of this compound [136]. Similarly, the C-terminal domain of human CAP18(109–135) was reported to induce mitochondrial depolarization and apoptosis in tongue squamous cell carcinoma SAS-H1 cells [137]. Importantly, the cytotoxic effect was not reproduced in healthy human gingival fibroblasts (HGF) and human keratinocyte line (HaCaT) [137], which could be attributed to the different structure and physicochemical features of cancerous cells when compared to the healthy ones (such as the presence of cholesterol-rich lipid rafts or different surface charge) [138]. Nevertheless, it was not characterized in detail. In contrast to these promising reports, a comparative analysis of the biological effects of LL-37 against three oral tongue squamous cell carcinomas revealed its fluctuating effect and variable impact on malignant phenotype. Although exogenous LL-37 decreased mostly proliferation of cancer cells, it stimulated their migration and invasion at the same time, as well as promoted MMP-2 and MMP-9 expression, suggesting a pro-tumorigenic effect. Interestingly, it enhanced the total amount of EGFR, but the EGFR pathways were recorded to be mostly decreased. Finally, the expression of hCAP18/LL-37 decreased with the severity of oral dysplasia and was lower in cancer tissue samples, but it did not correlate with the clinical outcomes of the patients [139]. These data suggest that determining the exact role of LL-37 in the development of oral cancers may be hampered due to the difficulty in correlating the expression of this peptide with the aggressiveness of a given cancer. Certainly, the LL-37 peptide shows many beneficial effects in protecting oral tissues from malignant transformation—one of the most relevant reports on this aspect is the study by Brice et al., indicating the restriction of Kaposi's sarcoma-associated herpesvirus (KSHV) infectivity, which results from direct disruption by LL-37 of the viral envelope, which inhibits its entry into oral epithelial cells, limiting ultimately KSHV-associated disease occurrence [140]. The crucial role of anti-inflammatory properties of LL-37 should also be recognized considering the ever-increasing number of reports linking the severity of inflammation in tissues to the development of tumors within them [141]. Nevertheless, given the multifactorial nature of cancer transformation and the spectrum of factors that influence the invasiveness of cancer tumors, some caution is needed in determining the true role of this peptide in this aspect of oral health. However, considering the pleiotropic biological activities of LL-37 resulting from the ability to interact with a spectrum of different membrane receptors, the exploration of the potential effect of this peptide on the development of cancers in the oral cavity as well is justified.

6. LL-37 Peptide as a Guardian of Oral Mucosa Mechanical Properties (LL-37 in Saliva and Saliva–Mucosal Surface Interference)

In recent years, new evidence has indicated that the dysfunction of physiological processes during infection, development of inflammation, and carcinogenesis generate structural changes in cells, cellular organelles, and tissues. These alterations translate into changes in the rheological properties of these biological structures. [142–148]. New techniques that allow us to measure the rheological properties of biological samples (cells and tissues) have made it possible to determine the rheological properties of different biological structures. [143,145,149]. Recent work has also shown the modifications in the mechanical properties of tissues within the oral cavity due to pathological changes [150]. Some lesions, such as condylomas, and focal epithelial hyperplasia are at least in part associated with HPV infections. If left undiagnosed and untreated, they might lead to cancers [151]. Changes in tissue structures can be visible using microscopic observations, but some of these changes can only manifest themselves as changes in tissue mechanical properties, which nowadays are often described as mechanomarkers. Interestingly, some mechanical changes can be observed earlier than changes in standard histopathological examinations [146] (Figure 11).

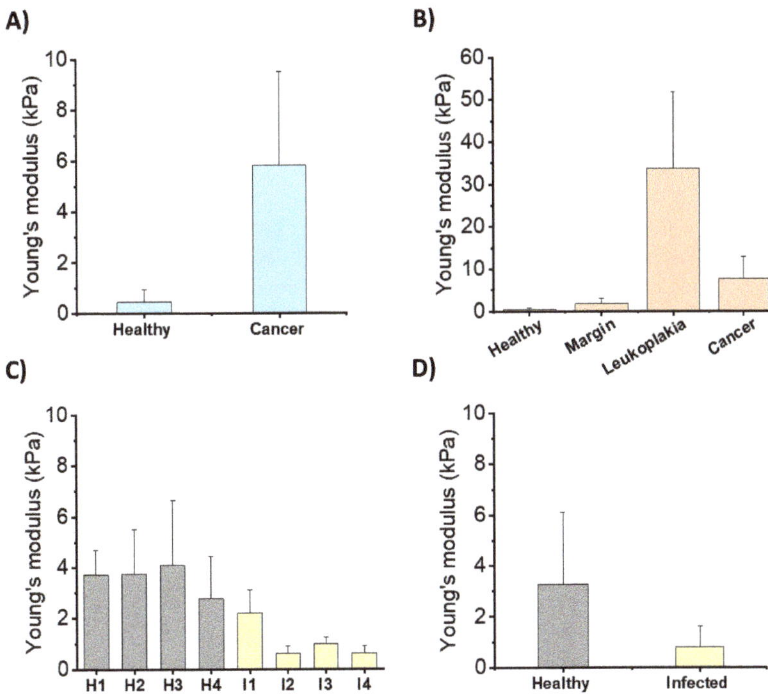

Figure 11. Young's modulus values obtained for healthy and diseased tissues using the AFM indentation technique: (**A**) Average Young's modulus values of healthy and colon cancer tissues. A

significantly higher Young's modulus of neoplastic tissues as compared to healthy tissues was shown. The greater stiffness of neoplastic tissues is associated with the presence of an increased amount of extracellular matrix protein. Collagen overexpression, matrix fibrosis, cross-linking, and vascularization occur during tumor progression [146]. (**B**) Young's modulus value distribution for healthy mouth mucosa and diseased tissues (leukoplakia and cancer). Differences between healthy tissue and tissue outside of the leukoplakia area are noticed. The stiffness of the leukoplakia samples was higher compared to the surrounding mucosa. Inhomogeneity of stiffness within leukoplakia samples was observed, which might act as a mechano-agonist that promotes oncogenesis. Stiffness of cancer samples was significantly lower than that within the precancerous ones [152]. (**C**) Mechanical properties of healthy stomach tissues (H1-H4) and those infected with *Helicobacter pylori* (I1-I4). The mean values of tissues' Young's modulus ± standard deviation for each healthy and infected tissue [147]. (**D**) Rheological difference between healthy stomach tissue and tissue during inflammation caused by *H. pylori* infection. The mean values of tissues' Young's modulus ± standard deviation [147].

Since the first in vitro observation indicating the ability of the LL-37 peptide to increase the cell stiffness of epithelial cells, the new mechanism of LL-37 action might also be based on its ability to modulate viscoelastic properties of the cells and tissues. It has been shown that changes in cell and tissue mechanics under the influence of chronic microbial colonization can, in the long term, disrupt transcriptional processes in the cell nuclei of epithelial cells and thus cause cancer risk [147,153]. Studies of gastric tissues from children with confirmed *H. pylori*-induced inflammation indicated that the infected tissues had lower stiffness compared to healthy tissues. Since *H. pylori* can colonize the oral cavity, a similar effect on oral tissue as the response of tissue mechanics during the inflammatory process might be expected. Interestingly, some changes in tissue's mechanical properties might be mediated by *H. pylori's* virulence factors [147]. It is tempting to assume that within the oral cavity, we will also observe similar changes. Panel 1B shows the results of extracellular matrix stiffness within leukoplakia of the human oral mucosa, which as previously mentioned here is known for the increased expression of hCAP-18/LL-37. In our study, we observed heterogeneity in stiffness within leukoplakia samples, reflecting an increase in collagen regeneration and accumulation (increasing density) in the extracellular matrix (ECM) and an increase in the stiffness of leukoplakia samples compared to the surrounding mucus. These results indicate that the changes occurring also in the oral cavity under the influence of pathological processes, inflammatory or neoplastic conditions, change the rheological properties of tissues. The influence of saliva components, including LL-37, on the mucosa plays an important role in these processes. As already mentioned here, saliva rich in antimicrobial peptides, including LL-37, plays a major role in the pathophysiology of mucosa, representing a natural barrier of the oral cavity [83,154,155]. In addition to LL-37, saliva contains a number of components with different actions, including histatin, a polypeptide with antifungal and antimicrobial activities; mucins, a group of glycoproteins that contribute to the viscoelastic nature of mucosal secretions; proline-rich proteins (PRPs); and other salivary proteins, such as cystatins, amylase, and kallikrein. In addition, saliva contains proteins such as lysozyme and secretory immunoglobulin A (sIgA), which are also part of the immune system [156]. These saliva components interact to a greater or lesser extent with oral tissues as well as biomaterials that may be present in the oral cavity [157].

The LL-37 peptide, through its broad protective properties, may act as a guardian of the normal mechanics of oral tissues, influencing the pathological factors causing its mechanical changes (Figure 2). This is particularly important due to the fact that mechanical properties are essential for the proper functioning of the human body [147]. In previous work [157], the anti-inflammatory effect of LL-37 at the site of infection was associated with LL-37-mediated increase in HUVECs epithelial cell stiffness, which prevents increased pericellular permeability. This effect is associated with inflammation that can lead to local accumulation of tissue fluid. This homeostatic performance of the LL-37 peptide accounts for the stabilization of tissue mechanics at values that characterize healthy and normal tissue. Changes in cell stiffness in the presence of the LL-37 peptide [158] indicate that after adding LL-37, sphingosine-1-phosphate, and LPS or infection with *P. aeruginosa*, an increase in stiffness and F-actin content was observed in the cortical area of A549 cells and primary human lung epithelial cells. The LL-37-induced increase in cell stiffness was accompanied by a decrease in the permeability and uptake of *P. aeruginosa*. It can be concluded that the antibacterial activity of LL-37, in addition to its direct action on bacterial cells, involves the interaction between LL-37 and the multicomponent lipid membrane [30,36,158]. It consists in increasing the rigidity of the cells on which bacterial cells are deposited, which reduces the ability of bacteria to translocate into the epithelium [127]. The reduced ability of bacteria to invade protects tissues from changes in their mechanics, which in the long term may protect against more severe changes in tissue structure. It has been shown that changes in cell and tissue mechanics under the influence of chronic microbial colonization can, in the long term, disrupt transcriptional processes in the cell nuclei of epithelial cells and thus cause cancer risk [148,159].

7. Conclusions

The sum of the above-cited papers indicates that human cathelicidin LL-37 plays a crucial role in oral cavity homeostasis by the modulation of the inflammatory cytokine content and, thus, its cellular effects on dental tissues, providing a suitable microenvironment for vascularization, promoting the mesenchymal stem cells' differentiation and migration, and limiting the effects of bacteria-derived inflammatory factors due to its bactericidal activities. In might be predicted that in the future, immobilization of the LL-37 peptide or its peptidomimetics in a resin composite filling or a polymerizable root canal sealer would achieve the purpose of using cationic peptides in combating some oral cavity pathologies, such as caries and pulpal infections (Figure 12).

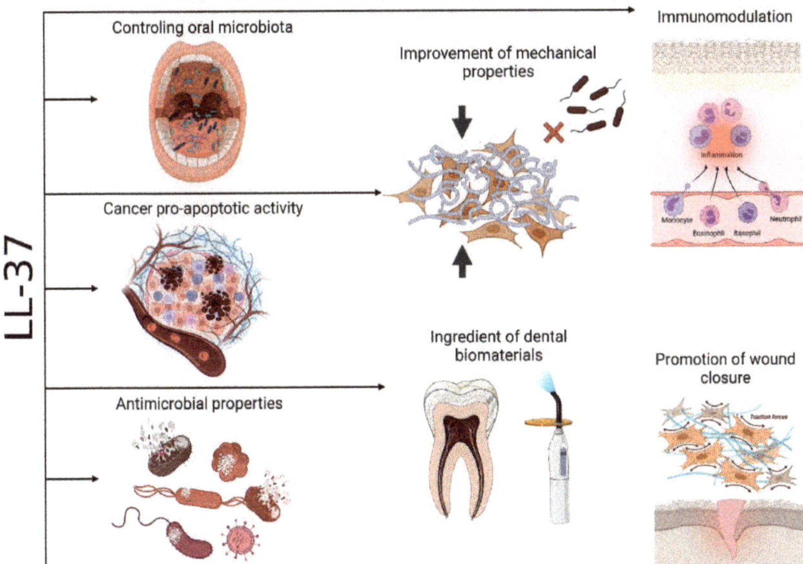

Figure 12. Cathelicidin LL-37 as a pivotal factor in maintaining homeostasis of the oral cavity.

Author Contributions: Conceptualization, R.B.; writing—original draft preparation, J.T., P.D., E.P., T.D., S.C., and K.F.; writing—review and editing, T.W., P.W., and R.B.; visualization, K.F. and P.D.; supervision, R.B. All authors have read and agreed to the published version of the manuscript.

Funding: This work was supported by the National Science Center, Poland, under Grant UMO-2018/31/B/NZ6/02476 (to RB) and the Medical University of Bialystok (SUB/1/DN/21/003/1122 to R.B.).

Data Availability Statement: All sources of data reported are provided as references in figure descriptions.

Conflicts of Interest: The funders had no role in the design of the study; in the collection, analyses, or interpretation of data; in the writing of the manuscript; or in the decision to publish the results.

References

1. Radaic, A.; Kapila, Y.L. The oralome and its dysbiosis: New insights into oral microbiome-host interactions. *Comput. Struct. Biotechnol. J.* **2021**, *19*, 1335–1360. [CrossRef] [PubMed]
2. Willis, J.R.; Gabaldón, T. The Human Oral Microbiome in Health and Disease: From Sequences to Ecosystems. *Microorganisms* **2020**, *8*, 308. [CrossRef] [PubMed]
3. Socransky, S.S.; Haffajee, A.D. Periodontal microbial ecology. *Periodontology 2000* **2005**, *38*, 135–187. [CrossRef] [PubMed]
4. Darveau, R.P. Periodontitis: A polymicrobial disruption of host homeostasis. *Nat. Rev. Microbiol.* **2010**, *8*, 481–490. [CrossRef]
5. Gorr, S.U. Antimicrobial peptides in periodontal innate defense. *Front. Oral Biol.* **2012**, *15*, 84–98. [CrossRef]
6. Gorr, S.U.; Abdolhosseini, M. Antimicrobial peptides and periodontal disease. *J. Clin. Periodontol.* **2011**, *38* (Suppl. 11), 126–141. [CrossRef]
7. Lin, B.; Li, R.; Handley, T.N.G.; Wade, J.D.; Li, W.; O'Brien-Simpson, N.M. Cationic Antimicrobial Peptides Are Leading the Way to Combat Oropathogenic Infections. *ACS Infect. Dis.* **2021**, *7*, 2959–2970. [CrossRef]
8. Prasad, S.V.; Fiedoruk, K.; Daniluk, T.; Piktel, E.; Bucki, R. Expression and Function of Host Defense Peptides at Inflammation Sites. *Int. J. Mol. Sci.* **2019**, *21*, 104. [CrossRef]
9. da Silva, B.R.; de Freitas, V.A.; Nascimento-Neto, L.G.; Carneiro, V.A.; Arruda, F.V.; de Aguiar, A.S.; Cavada, B.S.; Teixeira, E.H. Antimicrobial peptide control of pathogenic microorganisms of the oral cavity: A review of the literature. *Peptides* **2012**, *36*, 315–321. [CrossRef]
10. Cole, A.M.; Liao, H.I.; Stuchlik, O.; Tilan, J.; Pohl, J.; Ganz, T. Cationic polypeptides are required for antibacterial activity of human airway fluid. *J. Immunol.* **2002**, *169*, 6985–6991. [CrossRef]
11. Brown, K.L.; Hancock, R.E. Cationic host defense (antimicrobial) peptides. *Curr. Opin. Immunol.* **2006**, *18*, 24–30. [CrossRef] [PubMed]
12. Tonetti, M.S.; Chapple, I.L.; Working Group 3 of Seventh European Workshop on Periodontology. Biological approaches to the development of novel periodontal therapies–consensus of the Seventh European Workshop on Periodontology. *J. Clin. Periodontol.* **2011**, *38* (Suppl. 11), 114–118. [CrossRef] [PubMed]
13. Mookherjee, N.; Anderson, M.A.; Haagsman, H.P.; Davidson, D.J. Antimicrobial host defence peptides: Functions and clinical potential. *Nat. Rev. Drug Discov.* **2020**, *19*, 311–332. [CrossRef] [PubMed]
14. Uehara, A.; Fujimoto, Y.; Fukase, K.; Takada, H. Various human epithelial cells express functional Toll-like receptors, NOD1 and NOD2 to produce anti-microbial peptides, but not proinflammatory cytokines. *Mol. Immunol.* **2007**, *44*, 3100–3111. [CrossRef] [PubMed]
15. Liang, W.; Diana, J. The Dual Role of Antimicrobial Peptides in Autoimmunity. *Front. Immunol.* **2020**, *11*, 2077. [CrossRef]
16. Ghosh, S.K.; Feng, Z.; Fujioka, H.; Lux, R.; McCormick, T.S.; Weinberg, A. Conceptual Perspectives: Bacterial Antimicrobial Peptide Induction as a Novel Strategy for Symbiosis with the Human Host. *Front. Microbiol.* **2018**, *9*, 302. [CrossRef]
17. McMahon, L.; Schwartz, K.; Yilmaz, O.; Brown, E.; Ryan, L.K.; Diamond, G. Vitamin D-mediated induction of innate immunity in gingival epithelial cells. *Infect. Immun.* **2011**, *79*, 2250–2256. [CrossRef]
18. Wang, G.; Narayana, J.L.; Mishra, B.; Zhang, Y.; Wang, F.; Wang, C.; Zarena, D.; Lushnikova, T.; Wang, X. Design of Antimicrobial Peptides: Progress Made with Human Cathelicidin LL-37. *Adv. Exp. Med. Biol.* **2019**, *1117*, 215–240. [CrossRef]
19. Mahlapuu, M.; Sidorowicz, A.; Mikosinski, J.; Krzyżanowski, M.; Orleanski, J.; Twardowska-Saucha, K.; Nykaza, A.; Dyaczynski, M.; Belz-Lagoda, B.; Dziwiszek, G.; et al. Evaluation of LL-37 in healing of hard-to-heal venous leg ulcers: A multicentric prospective randomized placebo-controlled clinical trial. *Wound Repair Regen.* **2021**, *29*, 938–950. [CrossRef]
20. Peek, N.; Nell, M.J.; Brand, R.; Jansen-Werkhoven, T.; van Hoogdalem, E.J.; Verrijk, R.; Vonk, M.J.; Wafelman, A.R.; Valentijn, A.; Frijns, J.H.M.; et al. Ototopical drops containing a novel antibacterial synthetic peptide: Safety and efficacy in adults with chronic suppurative otitis media. *PLoS ONE* **2020**, *15*, e0231573. [CrossRef]
21. Vandamme, D.; Landuyt, B.; Luyten, W.; Schoofs, L. A comprehensive summary of LL-37, the factotum human cathelicidin peptide. *Cell Immunol.* **2012**, *280*, 22–35. [CrossRef] [PubMed]
22. Yang, B.; Good, D.; Mosaiab, T.; Liu, W.; Ni, G.; Kaur, J.; Liu, X.; Jessop, C.; Yang, L.; Fadhil, R.; et al. Significance of LL-37 on Immunomodulation and Disease Outcome. *Biomed. Res. Int.* **2020**, *2020*, 8349712. [CrossRef] [PubMed]

23. Samaras, P.; Schmidt, T.; Frejno, M.; Gessulat, S.; Reinecke, M.; Jarzab, A.; Zecha, J.; Mergner, J.; Giansanti, P.; Ehrlich, H.C.; et al. ProteomicsDB: A multi-omics and multi-organism resource for life science research. *Nucleic Acids Res.* **2020**, *48*, D1153–D1163. [CrossRef] [PubMed]

24. Wang, M.; Herrmann, C.J.; Simonovic, M.; Szklarczyk, D.; von Mering, C. Version 4.0 of PaxDb: Protein abundance data, integrated across model organisms, tissues, and cell-lines. *Proteomics* **2015**, *15*, 3163–3168. [CrossRef] [PubMed]

25. Wilhelm, M.; Schlegl, J.; Hahne, H.; Gholami, A.M.; Lieberenz, M.; Savitski, M.M.; Ziegler, E.; Butzmann, L.; Gessulat, S.; Marx, H.; et al. Mass-spectrometry-based draft of the human proteome. *Nature* **2014**, *509*, 582–587. [CrossRef] [PubMed]

26. Bucki, R.; Leszczyńska, K.; Namiot, A.; Sokołowski, W. Cathelicidin LL-37: A multitask antimicrobial peptide. *Arch. Immunol. Exp.* **2010**, *58*, 15–25. [CrossRef]

27. Xhindoli, D.; Pacor, S.; Benincasa, M.; Scocchi, M.; Gennaro, R.; Tossi, A. The human cathelicidin LL-37–A pore-forming antibacterial peptide and host-cell modulator. *Biochim. Biophys. Acta* **2016**, *1858*, 546–566. [CrossRef]

28. Henderson, B.; Martin, A.C. Protein moonlighting: A new factor in biology and medicine. *Biochem. Soc. Trans.* **2014**, *42*, 1671–1678. [CrossRef]

29. Verjans, E.T.; Zels, S.; Luyten, W.; Landuyt, B.; Schoofs, L. Molecular mechanisms of LL-37-induced receptor activation: An overview. *Peptides* **2016**, *85*, 16–26. [CrossRef]

30. Dang, X.; Wang, G. Spotlight on the Selected New Antimicrobial Innate Immune Peptides Discovered During 2015-2019. *Curr. Top. Med. Chem.* **2020**, *20*, 2984–2998. [CrossRef]

31. Murakami, M.; Kameda, K.; Tsumoto, H.; Tsuda, T.; Masuda, K.; Utsunomiya, R.; Mori, H.; Miura, Y.; Sayama, K. TLN-58, an Additional hCAP18 Processing Form, Found in the Lesion Vesicle of Palmoplantar Pustulosis in the Skin. *J. Investig. Derm.* **2017**, *137*, 322–331. [CrossRef] [PubMed]

32. Li, X.; Li, Y.; Han, H.; Miller, D.W.; Wang, G. Solution Structures of Human LL-37 Fragments and NMR-Based Identification of a Minimal Membrane-Targeting Antimicrobial and Anticancer Region. *J. Am. Chem. Soc.* **2006**, *128*, 5776–5785. [CrossRef] [PubMed]

33. Shahmiri, M.; Enciso, M.; Adda, C.G.; Smith, B.J.; Perugini, M.A.; Mechler, A. Membrane Core-Specific Antimicrobial Action of Cathelicidin LL-37 Peptide Switches Between Pore and Nanofibre Formation. *Sci. Rep.* **2016**, *6*, 38184. [CrossRef] [PubMed]

34. Zeth, K.; Sancho-Vaello, E. Structural Plasticity of LL-37 Indicates Elaborate Functional Adaptation Mechanisms to Bacterial Target Structures. *Int. J. Mol. Sci.* **2021**, *22*. [CrossRef] [PubMed]

35. Majewska, M.; Zamlynny, V.; Pieta, I.S.; Nowakowski, R.; Pieta, P. Interaction of LL-37 human cathelicidin peptide with a model microbial-like lipid membrane. *Bioelectrochemistry* **2021**, *141*, 107842. [CrossRef] [PubMed]

36. Engelberg, Y.; Landau, M. The Human LL-37(17-29) antimicrobial peptide reveals a functional supramolecular structure. *Nat. Commun.* **2020**, *11*, 3894. [CrossRef] [PubMed]

37. Sancho-Vaello, E.; Gil-Carton, D.; François, P.; Bonetti, E.-J.; Kreir, M.; Pothula, K.R.; Kleinekathöfer, U.; Zeth, K. The structure of the antimicrobial human cathelicidin LL-37 shows oligomerization and channel formation in the presence of membrane mimics. *Sci. Rep.* **2020**, *10*, 17356. [CrossRef] [PubMed]

38. Sancho-Vaello, E.; François, P.; Bonetti, E.J.; Lilie, H.; Finger, S.; Gil-Ortiz, F.; Gil-Carton, D.; Zeth, K. Structural remodeling and oligomerization of human cathelicidin on membranes suggest fibril-like structures as active species. *Sci. Rep.* **2017**, *7*, 15371. [CrossRef]

39. Lee, E.Y.; Zhang, C.; Di Domizio, J.; Jin, F.; Connell, W.; Hung, M.; Malkoff, N.; Veksler, V.; Gilliet, M.; Ren, P.; et al. Helical antimicrobial peptides assemble into protofibril scaffolds that present ordered dsDNA to TLR9. *Nat. Commun.* **2019**, *10*, 1012. [CrossRef]

40. Scheenstra, M.R.; van Harten, R.M.; Veldhuizen, E.J.A.; Haagsman, H.P.; Coorens, M. Cathelicidins Modulate TLR-Activation and Inflammation. *Front. Immunol.* **2020**, *11*, 1137. [CrossRef]

41. Neumann, A.; Völlger, L.; Berends, E.T.M.; Molhoek, E.M.; Stapels, D.A.C.; Midon, M.; Friães, A.; Pingoud, A.; Rooijakkers, S.H.M.; Gallo, R.L.; et al. Novel Role of the Antimicrobial Peptide LL-37 in the Protection of Neutrophil Extracellular Traps against Degradation by Bacterial Nucleases. *J. Innate Immun.* **2014**, *6*, 860–868. [CrossRef] [PubMed]

42. Bryzek, D.; Golda, A.; Budziaszek, J.; Kowalczyk, D.; Wong, A.; Bielecka, E.; Shakamuri, P.; Svoboda, P.; Pohl, J.; Potempa, J.; et al. Citrullination-Resistant LL-37 Is a Potent Antimicrobial Agent in the Inflammatory Environment High in Arginine Deiminase Activity. *Int. J. Mol. Sci.* **2020**, *21*, 9126. [CrossRef] [PubMed]

43. Bucki, R.; Namiot, D.B.; Namiot, Z.; Savage, P.B.; Janmey, P.A. Salivary mucins inhibit antibacterial activity of the cathelicidin-derived LL-37 peptide but not the cationic steroid CSA-13. *J. Antimicrob. Chemother.* **2008**, *62*, 329–335. [CrossRef] [PubMed]

44. Gutner, M.; Chaushu, S.; Balter, D.; Bachrach, G. Saliva enables the antimicrobial activity of LL-37 in the presence of proteases of Porphyromonas gingivalis. *Infect. Immun.* **2009**, *77*, 5558–5563. [CrossRef] [PubMed]

45. Turkoglu, O.; Emingil, G.; Eren, G.; Atmaca, H.; Kutukculer, N.; Atilla, G. Gingival crevicular fluid and serum hCAP18/LL-37 levels in generalized aggressive periodontitis. *Clin. Oral Investig.* **2017**, *21*, 763–769. [CrossRef]

46. Wuersching, S.N.; Huth, K.C.; Hickel, R.; Kollmuss, M. Inhibitory effect of LL-37 and human lactoferricin on growth and biofilm formation of anaerobes associated with oral diseases. *Anaerobe* **2021**, *67*, 102301. [CrossRef]

47. Overhage, J.; Campisano, A.; Bains, M.; Torfs, E.C.; Rehm, B.H.; Hancock, R.E. Human host defense peptide LL-37 prevents bacterial biofilm formation. *Infect. Immun.* **2008**, *76*, 4176–4182. [CrossRef]

48. Dürr, U.H.; Sudheendra, U.S.; Ramamoorthy, A. LL-37, the only human member of the cathelicidin family of antimicrobial peptides. *Biochim. Biophys. Acta* **2006**, *1758*, 1408–1425. [CrossRef]

49. Nakamichi, Y.; Horibe, K.; Takahashi, N.; Udagawa, N. Roles of cathelicidins in inflammation and bone loss. *Odontology* **2014**, *102*, 137–146. [CrossRef]

50. Scott, A.; Weldon, S.; Buchanan, P.J.; Schock, B.; Ernst, R.K.; McAuley, D.F.; Tunney, M.M.; Irwin, C.R.; Elborn, J.S.; Taggart, C.C. Evaluation of the Ability of LL-37 to Neutralise LPS In Vitro and Ex Vivo. *PLoS ONE* **2011**, *6*, e26525. [CrossRef]

51. Mookherjee, N.; Brown, K.L.; Bowdish, D.M.; Doria, S.; Falsafi, R.; Hokamp, K.; Roche, F.M.; Mu, R.; Doho, G.H.; Pistolic, J.; et al. Modulation of the TLR-mediated inflammatory response by the endogenous human host defense peptide LL-37. *J. Immunol.* **2006**, *176*, 2455–2464. [CrossRef] [PubMed]

52. Jagelavičienė, E.; Vaitkevičienė, I.; Šilingaitė, D.; Šinkūnaitė, E.; Daugėlaitė, G. The Relationship between Vitamin D and Periodontal Pathology. *Medicina* **2018**, *54*, 45. [CrossRef] [PubMed]

53. Colonna, M.; Facchetti, F. TREM-1 (triggering receptor expressed on myeloid cells): A new player in acute inflammatory responses. *J. Infect. Dis.* **2003**, *187* (Suppl. 2), S397–S401. [CrossRef]

54. Ji, S.; Choi, Y. Innate immune response to oral bacteria and the immune evasive characteristics of periodontal pathogens. *J. Periodontal Implant Sci.* **2013**, *43*, 3–11. [CrossRef] [PubMed]

55. Ji, S.; Hyun, J.; Park, E.; Lee, B.L.; Kim, K.K.; Choi, Y. Susceptibility of various oral bacteria to antimicrobial peptides and to phagocytosis by neutrophils. *J. Periodontal Res.* **2007**, *42*, 410–419. [CrossRef] [PubMed]

56. Ji, S.; Kim, Y.; Min, B.M.; Han, S.H.; Choi, Y. Innate immune responses of gingival epithelial cells to nonperiodontopathic and periodontopathic bacteria. *J. Periodontal Res.* **2007**, *42*, 503–510. [CrossRef]

57. Leszczynska, K.; Namiot, D.; Byfield, F.J.; Cruz, K.; Zendzian-Piotrowska, M.; Fein, D.E.; Savage, P.B.; Diamond, S.; McCulloch, C.A.; Janmey, P.A.; et al. Antibacterial activity of the human host defence peptide LL-37 and selected synthetic cationic lipids against bacteria associated with oral and upper respiratory tract infections. *J. Antimicrob. Chemother.* **2013**, *68*, 610–618. [CrossRef]

58. Ouhara, K.; Komatsuzawa, H.; Yamada, S.; Shiba, H.; Fujiwara, T.; Ohara, M.; Sayama, K.; Hashimoto, K.; Kurihara, H.; Sugai, M. Susceptibilities of periodontopathogenic and cariogenic bacteria to antibacterial peptides, β-defensins and LL37, produced by human epithelial cells. *J. Antimicrob. Chemother.* **2005**, *55*, 888–896. [CrossRef]

59. Altman, H.; Steinberg, D.; Porat, Y.; Mor, A.; Fridman, D.; Friedman, M.; Bachrach, G. In vitro assessment of antimicrobial peptides as potential agents against several oral bacteria. *J. Antimicrob. Chemother.* **2006**, *58*, 198–201. [CrossRef]

60. Sol, A.; Ginesin, O.; Chaushu, S.; Karra, L.; Coppenhagen-Glazer, S.; Ginsburg, I.; Bachrach, G. LL-37 opsonizes and inhibits biofilm formation of Aggregatibacter actinomycetemcomitans at subbactericidal concentrations. *Infect. Immun.* **2013**, *81*, 3577–3585. [CrossRef]

61. Horie, T.; Inomata, M.; Into, T. OmpA-Like Proteins of *Porphyromonas gingivalis* Mediate Resistance to the Antimicrobial Peptide LL-37. *J. Pathog.* **2018**, *2018*, 2068435. [CrossRef] [PubMed]

62. Pütsep, K.; Carlsson, G.; Boman, H.G.; Andersson, M. Deficiency of antibacterial peptides in patients with morbus Kostmann: An observation study. *Lancet* **2002**, *360*, 1144–1149. [CrossRef]

63. Tanaka, D.; Miyasaki, K.T.; Lehrer, R.I. Sensitivity of Actinobacillus actinomycetemcomitans and Capnocytophaga spp. to the bactericidal action of LL-37: A cathelicidin found in human leukocytes and epithelium. *Oral Microbiol. Immunol.* **2000**, *15*, 226–231. [CrossRef]

64. Rosenfeld, Y.; Papo, N.; Shai, Y. Endotoxin (lipopolysaccharide) neutralization by innate immunity host-defense peptides. Peptide properties and plausible modes of action. *J. Biol. Chem.* **2006**, *281*, 1636–1643. [CrossRef]

65. Sambri, V.; Marangoni, A.; Giacani, L.; Gennaro, R.; Murgia, R.; Cevenini, R.; Cinco, M. Comparative in vitro activity of five cathelicidin-derived synthetic peptides against Leptospira, Borrelia and Treponema pallidum. *J. Antimicrob. Chemother.* **2002**, *50*, 895–902. [CrossRef] [PubMed]

66. Yang, X.; Niu, L.; Pan, Y.; Feng, X.; Liu, J.; Guo, Y.; Pan, C.; Geng, F.; Tang, X. LL-37-Induced Autophagy Contributed to the Elimination of Live Porphyromonas gingivalis Internalized in Keratinocytes. *Front. Cell. Infect. Microbiol.* **2020**, *10*, 561761. [CrossRef]

67. Sbordone, L.; Bortolaia, C. Oral microbial biofilms and plaque-related diseases: Microbial communities and their role in the shift from oral health to disease. *Clin. Oral Investig.* **2003**, *7*, 181–188. [CrossRef]

68. Ahmad, S.M.; Bhattacharyya, P.; Jeffries, N.; Gisselbrecht, S.S.; Michelson, A.M. Two Forkhead transcription factors regulate cardiac progenitor specification by controlling the expression of receptors of the fibroblast growth factor and Wnt signaling pathways. *Development* **2016**, *143*, 306–317. [CrossRef]

69. Darveau, R.P.; Belton, C.M.; Reife, R.A.; Lamont, R.J. Local chemokine paralysis, a novel pathogenic mechanism for Porphyromonas gingivalis. *Infect. Immun.* **1998**, *66*, 1660–1665. [CrossRef]

70. Bachrach, G.; Altman, H.; Kolenbrander, P.E.; Chalmers, N.I.; Gabai-Gutner, M.; Mor, A.; Friedman, M.; Steinberg, D. Resistance of Porphyromonas gingivalis ATCC 33277 to direct killing by antimicrobial peptides is protease independent. *Antimicrob. Agents Chemother.* **2008**, *52*, 638–642. [CrossRef]

71. Laugisch, O.; Schacht, M.; Guentsch, A.; Kantyka, T.; Sroka, A.; Stennicke, H.R.; Pfister, W.; Sculean, A.; Potempa, J.; Eick, S. Periodontal pathogens affect the level of protease inhibitors in gingival crevicular fluid. *Mol. Oral Microbiol.* **2012**, *27*, 45–56. [CrossRef] [PubMed]

72. Velarde, J.J.; Ashbaugh, M.; Wessels, M.R. The human antimicrobial peptide LL-37 binds directly to CsrS, a sensor histidine kinase of group A Streptococcus, to activate expression of virulence factors. *J. Biol. Chem.* **2014**, *289*, 36315–36324. [CrossRef] [PubMed]

73. Goldmann, O.; Scheb-Wetzel, M.; Rohde, M.; Bravo, A. New Insights into the Antimicrobial Effect. *Infect. Immun.* **2014**, *82*, 4496.

74. Hsu, C.-M.; Liao, Y.-L.; Chang, C.-K.; Lan, C.-Y. Candida albicans Sfp1 Is Involved in the Cell Wall and Endoplasmic Reticulum Stress Responses Induced by Human Antimicrobial Peptide LL-37. *Int. J. Mol. Sci.* **2021**, *22*, 10633. [CrossRef]

75. Lee, J.; Lee, D.G. Antimicrobial peptides (AMPs) with dual mechanisms: Membrane disruption and apoptosis. *J. Microbiol. Biotechnol.* **2015**, *25*, 759–764. [CrossRef]

76. Nireeksha; Varma, S.R.; Damdoum, M.; Alsaegh, M.A.; Hegde, M.N.; Kumari, S.N.; Ramamurthy, S.; Narayanan, J.; Imran, E.; Shabbir, J.; et al. Immunomodulatory Expression of Cathelicidins Peptides in Pulp Inflammation and Regeneration: An Update. *Curr. Issues Mol. Biol.* **2021**, *43*, 116–126. [CrossRef]

77. Dale, B.A.; Fredericks, L.P. Antimicrobial peptides in the oral environment: Expression and function in health and disease. *Curr. Issues Mol. Biol.* **2005**, *7*, 119–134.

78. Davidopoulou, S.; Diza, E.; Sakellari, D.; Menexes, G.; Kalfas, S. Salivary concentration of free LL-37 in edentulism, chronic periodontitis and healthy periodontium. *Arch. Oral Biol.* **2013**, *58*, 930–934. [CrossRef]

79. Offenbacher, S. Periodontal diseases: Pathogenesis. *Ann. Periodontol.* **1996**, *1*, 821–878. [CrossRef]

80. Li, Y.-H.; Huang, X.; Tian, X.-L. Recent advances in dental biofilm: Impacts of microbial interactions on the biofilm ecology and pathogenesis. *Aims Bioeng.* **2017**, *4*, 335–350. [CrossRef]

81. Marsh, P.D.; Moter, A.; Devine, D.A. Dental plaque biofilms: Communities, conflict and control. *Periodontology 2000* **2011**, *55*, 16–35. [CrossRef]

82. Guo, L.; He, X.; Shi, W. Intercellular communications in multispecies oral microbial communities. *Front. Microbiol.* **2014**, *5*, 328. [CrossRef] [PubMed]

83. Davidopoulou, S.; Diza, E.; Menexes, G.; Kalfas, S. Salivary concentration of the antimicrobial peptide LL-37 in children. *Arch. Oral Biol.* **2012**, *57*, 865–869. [CrossRef] [PubMed]

84. Thiruvanamalai Sivakumar, A.R.H.; Mednieks, M. Secretory proteins in the saliva of children. *J. Oral Sci.* **2009**, *51*, 573–580. [CrossRef] [PubMed]

85. Burke, J.C.; Evans, C.A.; Crosby, T.R.; Mednieks, M.I. Expression of secretory proteins in oral fluid after orthodontic tooth movement. *Am. J. Orthod. Dentofac. Orthop.* **2002**, *121*, 310–315. [CrossRef] [PubMed]

86. Malcolm, J.; Sherriff, A.; Lappin, D.F.; Ramage, G.; Conway, D.I.; Macpherson, L.M.; Culshaw, S. Salivary antimicrobial proteins associate with age-related changes in streptococcal composition in dental plaque. *Mol. Oral Microbiol.* **2014**, *29*, 284–293. [CrossRef]

87. Colombo, N.H.; Ribas, L.F.; Pereira, J.A.; Kreling, P.F.; Kressirer, C.A.; Tanner, A.C.; Duque, C. Antimicrobial peptides in saliva of children with severe early childhood caries. *Arch. Oral Biol.* **2016**, *69*, 40–46. [CrossRef]

88. Phattarataratip, E.; Olson, B.; Broffitt, B.; Qian, F.; Brogden, K.A.; Drake, D.R.; Levy, S.M.; Banas, J.A. Streptococcus mutans strains recovered from caries-active or caries-free individuals differ in sensitivity to host antimicrobial peptides. *Mol. Oral Microbiol.* **2011**, *26*, 187–199. [CrossRef]

89. Puklo, M.; Guentsch, A.; Hiemstra, P.S.; Eick, S.; Potempa, J. Analysis of neutrophil-derived antimicrobial peptides in gingival crevicular fluid suggests importance of cathelicidin LL-37 in the innate immune response against periodontogenic bacteria. *Oral Microbiol. Immunol.* **2008**, *23*, 328–335. [CrossRef]

90. Koziel, J.; Karim, A.Y.; Przybyszewska, K.; Ksiazek, M.; Rapala-Kozik, M.; Nguyen, K.-A.; Potempa, J. Proteolytic inactivation of LL-37 by karilysin, a novel virulence mechanism of Tannerella forsythia. *J. Innate Immun.* **2010**, *2*, 288–293. [CrossRef]

91. Oudhoff, M.J.; Blaauboer, M.E.; Nazmi, K.; Scheres, N.; Bolscher, J.G.; Veerman, E.C. The role of salivary histatin and the human cathelicidin LL-37 in wound healing and innate immunity. *Biol. Chem.* **2010**, *391*, 541–548. [CrossRef] [PubMed]

92. McCrudden, M.T.; O'Donnell, K.; Irwin, C.R.; Lundy, F.T. Effects of LL-37 on gingival fibroblasts: A role in periodontal tissue remodeling? *Vaccines* **2018**, *6*, 44. [CrossRef] [PubMed]

93. Stojanović, N.; Krunić, J.; Popović, B.; Stojičić, S.; Živković, S. Prevalence of Enterococcus faecalis and Porphyromonas gingivalis in infected root canals and their susceptibility to endodontic treatment procedures: A molecular study. *Srp. Arh. Za Celok. Lek.* **2014**, *142*, 535–541. [CrossRef] [PubMed]

94. Richards, D.; Davies, J.K.; Figdor, D. Starvation survival and recovery in serum of Candida albicans compared with Enterococcus faecalis. *Oral Surg. Oral Med. Oral Pathol. Oral Radiol. Endodontology* **2010**, *110*, 125–130. [CrossRef]

95. Jönsson, D.; Nilsson, B.O. The antimicrobial peptide LL-37 is anti-inflammatory and proapoptotic in human periodontal ligament cells. *J. Periodontal Res.* **2012**, *47*, 330–335. [CrossRef]

96. Harder, J.; Schröder, J.-M. *Antimicrobial Peptides: Role in Human Health and Disease*; Springer: Berlin/Heidelberg, Germany, 2016; p. 158.

97. Yu, X.; Quan, J.; Long, W.; Chen, H.; Wang, R.; Guo, J.; Lin, X.; Mai, S. LL-37 inhibits LPS-induced inflammation and stimulates the osteogenic differentiation of BMSCs via P2X7 receptor and MAPK signaling pathway. *Exp. Cell Res.* **2018**, *372*, 178–187. [CrossRef]

98. 10Liu, Z.; Yuan, X.; Liu, M.; Fernandes, G.; Zhang, Y.; Yang, S.; Ionita, C.N. Antimicrobial Peptide Combined with BMP2-Modified Mesenchymal Stem Cells Promotes Calvarial Repair in an Osteolytic Model. *Mol. Ther.* **2018**, *26*, 199–207. [CrossRef]

99. 10Cheng, Q.; Zeng, K.; Kang, Q.; Qian, W.; Zhang, W.; Gan, Q.; Xia, W. The antimicrobial peptide LL-37 promotes migration and odonto/osteogenic differentiation of stem cells from the apical papilla through the AKT/Wnt/β-catenin signaling pathway. *J. Endod.* **2020**, *46*, 964–972.

100. Kajiya, M.; Shiba, H.; Komatsuzawa, H.; Ouhara, K.; Fujita, T.; Takeda, K.; Uchida, Y.; Mizuno, N.; Kawaguchi, H.; Kurihara, H. The antimicrobial peptide LL37 induces the migration of human pulp cells: A possible adjunct for regenerative endodontics. *J. Endod.* **2010**, *36*, 1009–1013. [CrossRef]

101. Sarmiento, B.F.; Aminoshariae, A.; Bakkar, M.; Bonfield, T.; Ghosh, S.; Montagnese, T.A.; Mickel, A.K. The Expression of the Human Cathelicidin LL-37 in the Human Dental Pulp: An In Vivo St. *Int. J. Pharm.* **2016**, *1*, 5. [CrossRef]

102. Khung, R.; Shiba, H.; Kajiya, M.; Kittaka, M.; Ouhara, K.; Takeda, K.; Mizuno, N.; Fujita, T.; Komatsuzawa, H.; Kurihara, H. LL37 induces VEGF expression in dental pulp cells through ERK signalling. *Int. Endod. J.* **2015**, *48*, 673–679. [CrossRef] [PubMed]

103. Milhan, N.V.M.; de Barros, P.P.; de Lima Zutin, E.A.; de Oliveira, F.E.; Camargo, C.H.R.; Camargo, S.E.A. The Antimicrobial Peptide LL-37 as a Possible Adjunct for the Proliferation and Differentiation of Dental Pulp Stem Cells. *J. Endod.* **2017**, *43*, 2048–2053. [CrossRef] [PubMed]

104. Horibe, K.; Hosoya, A.; Hiraga, T.; NAkamura, H. Expression and localization of CRAMP in rat tooth germ and during reparative dentin formation. *Clin. Oral Investig.* **2018**, *22*, 2559–2566. [CrossRef] [PubMed]

105. Türkoğlu, O.; Emingil, G.; Kütükçüler, N.; Atilla, G. Gingival crevicular fluid levels of cathelicidin LL-37 and interleukin-18 in patients with chronic periodontitis. *J. Periodontol.* **2009**, *80*, 969–976. [CrossRef] [PubMed]

106. Andrukhov, A.B.; Behm, C. A Review of Antimicrobial Activity of Dental Mesenchymal Stromal Cells: Is There Any Potential? *Front. Oral Health* **2021**, *2*, 832976. [CrossRef]

107. Deng, H.; Liu, F.; Pan, Y.; Jin, X.; Wang, H.; Cao, J. BsmI, TaqI, ApaI, and FokI polymorphisms in the vitamin D receptor gene and periodontitis: A meta-analysis of 15 studies including 1338 cases and 1302 controls. *J. Clin. Periodontol.* **2011**, *38*, 199–207. [CrossRef]

108. Wu, D.; Weng, Y.; Feng, Y.; Liang, B.; Wang, H.; Li, L.; Wang, Z. Trem1 Induces Periodontal Inflammation via Regulating M1 Polarization. *J. Dent. Res.* **2021**, *101*, 437–447. [CrossRef]

109. Bostanci, N.; Abe, T.; Belibasakis, G.N.; Hajishengallis, G. TREM-1 Is Upregulated in Experimental Periodontitis, and Its Blockade Inhibits IL-17A and RANKL Expression and Suppresses Bone loss. *J. Clin. Med.* **2019**, *8*, 1579. [CrossRef]

110. Inanc, N.; Mumcu, G.; Can, M.; Yay, M.; Silbereisen, A.; Manoil, D.; Direskeneli, H.; Bostanci, N. Elevated serum TREM-1 is associated with periodontitis and disease activity in rheumatoid arthritis. *Sci. Rep.* **2021**, *11*, 2888. [CrossRef]

111. Willi, M.; Belibasakis, G.; Bostanci, N. Expression and regulation of TREM-1 in periodontal diseases. *Clin. Exp. Immunol.* **2014**, *178*, 190–200. [CrossRef]

112. Katoh, K.; Rozewicki, J.; Yamada, K.D. MAFFT online service: Multiple sequence alignment, interactive sequence choice and visualization. *Brief. Bioinform* **2019**, *20*, 1160–1166. [CrossRef] [PubMed]

113. Jourdain, M.L.; Velard, F.; Pierrard, L.; Sergheraert, J.; Gangloff, S.C.; Braux, J. Cationic antimicrobial peptides and periodontal physiopathology: A systematic review. *J. Periodontal Res.* **2019**, *54*, 589–600. [CrossRef] [PubMed]

114. Nilsson, B.-O. What can we learn about functional importance of human antimicrobial peptide LL-37 in the oral environment from severe congenital neutropenia (Kostmann disease)? *Peptides* **2020**, *128*, 170311. [CrossRef] [PubMed]

115. Eick, S.; Puklo, M.; Adamowicz, K.; Kantyka, T.; Hiemstra, P.; Stennicke, H.; Guentsch, A.; Schacher, B.; Eickholz, P.; Potempa, J. Lack of cathelicidin processing in Papillon-Lefèvre syndrome patients reveals essential role of LL-37 in periodontal homeostasis. *Orphanet J. Rare Dis.* **2014**, *9*, 148. [CrossRef] [PubMed]

116. Hart, T.C.; Hart, P.S.; Michalec, M.D.; Zhang, Y.; Firatli, E.; Van Dyke, T.E.; Stabholz, A.; Zlorogorski, A.; Shapira, L.; Soskolne, W.A. Haim-Munk syndrome and Papillon-Lefèvre syndrome are allelic mutations in cathepsin C. *J. Med. Genet.* **2000**, *37*, 88–94. [CrossRef] [PubMed]

117. Türkoğlu, O.; Azarsız, E.; Emingil, G.; Kütükçüler, N.; Atilla, G. Are Proteinase 3 and Cathepsin C Enzymes Related to Pathogenesis of Periodontitis? *Biomed. Res. Int.* **2014**, *2014*, 420830. [CrossRef]

118. Katz, J.; Onate, M.D.; Pauley, K.M.; Bhattacharyya, I.; Cha, S. Presence of Porphyromonas gingivalis in gingival squamous cell carcinoma. *Int. J. Oral Sci.* **2011**, *3*, 209–215. [CrossRef]

119. Di Cosola, M.; Cazzolla, A.P.; Charitos, I.A.; Ballini, A.; Inchingolo, F.; Santacroce, L. Candida albicans and oral carcinogenesis. A brief review. *J. Fungi* **2021**, *7*, 476. [CrossRef]

120. Sztukowska, M.N.; Dutton, L.C.; Delaney, C.; Ramsdale, M.; Ramage, G.; Jenkinson, H.F.; Nobbs, A.H.; Lamont, R.J. Community development between Porphyromonas gingivalis and Candida albicans mediated by InlJ and Als3. *MBio* **2018**, *9*, e00202–e00218. [CrossRef]

121. Amer, A.; Galvin, S.; Healy, C.; Moran, G.P. The microbiome of oral leukoplakia shows enrichment in Fusobacteria and Rothia species. *J. Oral Microbiol.* **2017**, *9*, 1325253. [CrossRef]

122. Mumcu, G.; Cimilli, H.; Karacayli, U.; Inanc, N.; Ture-Ozdemir, F.; Eksioglu-Demiralp, E.; Ergun, T.; Direskeneli, H. Salivary levels of antimicrobial peptides Hnp 1-3, Ll-37 and S100 in Behcet's disease. *Arch. Oral Biol.* **2012**, *57*, 642–646. [CrossRef] [PubMed]

123. Wang, Q.; Sztukowska, M.; Ojo, A.; Scott, D.A.; Wang, H.; Lamont, R.J. FOXO responses to P orphyromonas gingivalis in epithelial cells. *Cell. Microbiol.* **2015**, *17*, 1605–1617. [CrossRef] [PubMed]

124. Shih, Y.-H.; Wang, T.-H.; Shieh, T.-M.; Tseng, Y.-H. Oral submucous fibrosis: A review on etiopathogenesis, diagnosis, and therapy. *Int. J. Mol. Sci.* **2019**, *20*, 2940. [CrossRef] [PubMed]

125. Roopashree, M.; Gondhalekar, R.V.; Shashikanth, M.; George, J.; Thippeswamy, S.; Shukla, A. Pathogenesis of oral lichen planus–a review. *J. Oral Pathol. Med.* **2010**, *39*, 729–734. [CrossRef]

126. Okumura, K. Cathelicidins—therapeutic antimicrobial and antitumor host defense peptides for oral diseases. *Jpn. Dent. Sci. Rev.* **2011**, *47*, 67–81. [CrossRef]

127. Piktel, E.; Niemirowicz, K.; Wnorowska, U.; Wątek, M.; Wollny, T.; Głuszek, K.; Góźdź, S.; Leventhal, I.; Bucki, R. The Role of Cathelicidin LL-37 in Cancer Development. *Arch. Immunol. Exp.* **2016**, *64*, 33–46. [CrossRef]

128. Kuroda, K.; Fukuda, T.; Yoneyama, H.; Katayama, M.; Isogai, H.; Okumura, K.; Isogai, E. Anti-proliferative effect of an analogue of the LL-37 peptide in the colon cancer derived cell line HCT116 p53+/+ and p53-/-. *Oncol Rep.* **2012**, *28*, 829–834. [CrossRef]

129. Prevete, N.; Liotti, F.; Visciano, C.; Marone, G.; Melillo, R.M.; de Paulis, A. The formyl peptide receptor 1 exerts a tumor suppressor function in human gastric cancer by inhibiting angiogenesis. *Oncogene* **2014**. [CrossRef]

130. Wu, W.K.; Sung, J.J.; To, K.F.; Yu, L.; Li, H.T.; Li, Z.J.; Chu, K.M.; Yu, J.; Cho, C.H. The host defense peptide LL-37 activates the tumor-suppressing bone morphogenetic protein signaling via inhibition of proteasome in gastric cancer cells. *J. Cell Physiol.* **2010**, *223*, 178–186. [CrossRef]

131. Coffelt, S.B.; Marini, F.C.; Watson, K.; Zwezdaryk, K.J.; Dembinski, J.L.; LaMarca, H.L.; Tomchuck, S.L.; Honer zu Bentrup, K.; Danka, E.S.; Henkle, S.L.; et al. The pro-inflammatory peptide LL-37 promotes ovarian tumor progression through recruitment of multipotent mesenchymal stromal cells. *Proc. Natl. Acad. Sci. USA* **2009**, *106*, 3806–3811. [CrossRef]

132. von Haussen, J.; Koczulla, R.; Shaykhiev, R.; Herr, C.; Pinkenburg, O.; Reimer, D.; Wiewrodt, R.; Biesterfeld, S.; Aigner, A.; Czubayko, F.; et al. The host defence peptide LL-37/hCAP-18 is a growth factor for lung cancer cells. *Lung Cancer* **2008**, *59*, 12–23. [CrossRef] [PubMed]

133. Heilborn, J.D.; Nilsson, M.F.; Jimenez, C.I.; Sandstedt, B.; Borregaard, N.; Tham, E.; Sørensen, O.E.; Weber, G.; Ståhle, M. Antimicrobial protein hCAP18/LL-37 is highly expressed in breast cancer and is a putative growth factor for epithelial cells. *Int J. Cancer* **2005**, *114*, 713–719. [CrossRef] [PubMed]

134. Chen, X.; Qi, G.; Qin, M.; Zou, Y.; Zhong, K.; Tang, Y.; Guo, Y.; Jiang, X.; Liang, L.; Zou, X. DNA methylation directly downregulates human cathelicidin antimicrobial peptide gene (CAMP) promoter activity. *Oncotarget* **2017**, *8*, 27943–27952. [CrossRef] [PubMed]

135. Chen, X.; Ji, S.; Si, J.; Zhang, X.; Wang, X.; Guo, Y.; Zou, X. Human cathelicidin antimicrobial peptide suppresses proliferation, migration and invasion of oral carcinoma HSC-3 cells via a novel mechanism involving caspase-3 mediated apoptosis. *Mol. Med. Rep.* **2020**, *22*, 5243–5250. [CrossRef] [PubMed]

136. Açil, Y.; Torz, K.; Gülses, A.; Wieker, H.; Gerle, M.; Purcz, N.; Will, O.M.; Eduard Meyer, J.; Wiltfang, J. An experimental study on antitumoral effects of KI-21-3, a synthetic fragment of antimicrobial peptide LL-37, on oral squamous cell carcinoma. *J. Craniomaxillofac Surg* **2018**, *46*, 1586–1592. [CrossRef]

137. Okumura, K.; Itoh, A.; Isogai, E.; Hirose, K.; Hosokawa, Y.; Abiko, Y.; Shibata, T.; Hirata, M.; Isogai, H. C-terminal domain of human CAP18 antimicrobial peptide induces apoptosis in oral squamous cell carcinoma SAS-H1 cells. *Cancer Lett.* **2004**, *212*, 185–194. [CrossRef]

138. Wnorowska, U.; Fiedoruk, K.; Piktel, E.; Prasad, S.V.; Sulik, M.; Janion, M.; Daniluk, T.; Savage, P.B.; Bucki, R. Nanoantibiotics containing membrane-active human cathelicidin LL-37 or synthetic ceragenins attached to the surface of magnetic nanoparticles as novel and innovative therapeutic tools: Current status and potential future applications. *J. Nanobiotechnology* **2020**, *18*, 3. [CrossRef]

139. Vierthaler, M.; Rodrigues, P.C.; Sundquist, E.; Siponen, M.; Salo, T.; Risteli, M. Fluctuating role of antimicrobial peptide hCAP18/LL-37 in oral tongue dysplasia and carcinoma. *Oncol Rep.* **2020**, *44*, 325–338. [CrossRef]

140. Brice, D.C.; Toth, Z.; Diamond, G. LL-37 disrupts the Kaposi's sarcoma-associated herpesvirus envelope and inhibits infection in oral epithelial cells. *Antivir. Res.* **2018**, *158*, 25–33. [CrossRef]

141. Sgambato, A.; Cittadini, A. Inflammation and cancer: A multifaceted link. *Eur. Rev. Med. Pharm. Sci.* **2010**, *14*, 263–268.

142. Janmey, P.A.; Winer, J.P.; Murray, M.E.; Wen, Q. The hard life of soft cells. *Cell Motil Cytoskelet.* **2009**, *66*, 597–605. [CrossRef] [PubMed]

143. Pogoda, K.; Jaczewska, J.; Wiltowska-Zuber, J.; Klymenko, O.; Zuber, K.; Fornal, M.; Lekka, M. Depth-sensing analysis of cytoskeleton organization based on AFM data. *Eur. Biophys. J.* **2012**, *41*, 79–87. [CrossRef] [PubMed]

144. Stylianou, A.; Lekka, M.; Stylianopoulos, T. AFM assessing of nanomechanical fingerprints for cancer early diagnosis and classification: From single cell to tissue level. *Nanoscale* **2018**, *10*, 20930–20945. [CrossRef] [PubMed]

145. Hayashi, K.; Iwata, M. Stiffness of cancer cells measured with an AFM indentation method. *J. Mech. Behav. Biomed. Mater.* **2015**, *49*, 105–111. [CrossRef] [PubMed]

146. Deptuła, P.; Łysik, D.; Pogoda, K.; Cieśluk, M.; Namiot, A.; Mystkowska, J.; Król, G.; Głuszek, S.; Janmey, P.A.; Bucki, R. Tissue Rheology as a Possible Complementary Procedure to Advance Histological Diagnosis of Colon Cancer. *Acs Biomater. Sci. Eng.* **2020**, *6*, 5620–5631. [CrossRef] [PubMed]

147. Deptuła, P.; Suprewicz, Ł.; Daniluk, T.; Namiot, A.; Chmielewska, S.J.; Daniluk, U.; Lebensztejn, D.; Bucki, R. Nanomechanical Hallmarks of Helicobacter pylori Infection in Pediatric Patients. *Int. J. Mol. Sci.* **2021**, *22*, 5624. [CrossRef]

148. Song, D.; Shivers, J.L.; MacKintosh, F.C.; Patteson, A.E.; Janmey, P.A. Cell-induced confinement effects in soft tissue mechanics. *J. Appl. Phys.* **2021**, *129*, 140901. [CrossRef]

149. Stylianou, A.; Stylianopoulos, T. Atomic force microscopy probing of cancer cells and tumor microenvironment components. *BioNanoScience* **2016**, *6*, 33–46. [CrossRef]

150. Pogoda, K.; Cieśluk, M.; Deptuła, P.; Tokajuk, G.; Piktel, E.; Król, G.; Reszeć, J.; Bucki, R. Inhomogeneity of stiffness and density of the extracellular matrix within the leukoplakia of human oral mucosa as potential physicochemical factors leading to carcinogenesis. *Transl. Oncol.* **2021**, *14*, 101105. [CrossRef]
151. Ganesh, D.; Sreenivasan, P.; Öhman, J.; Wallström, M.; Braz-Silva, P.H.; Giglio, D.; Kjeller, G.; Hasseus, B. Potentially malignant oral disorders and cancer transformation. *Anticancer Res.* **2018**, *38*, 3223–3229. [CrossRef]
152. Cieśluk, M.; Pogoda, K.; Deptuła, P.; Werel, P.; Kułakowska, A.; Kochanowicz, J.; Mariak, Z.; Łysoń, T.; Reszeć, J.; Bucki, R. Nanomechanics and Histopathology as Diagnostic Tools to Characterize Freshly Removed Human Brain Tumors. *Int. J. Nanomed.* **2020**, *15*, 7509. [CrossRef] [PubMed]
153. Elosegui-Artola, A.; Oria, R.; Chen, Y.; Kosmalska, A.; Pérez-González, C.; Castro, N.; Zhu, C.; Trepat, X.; Roca-Cusachs, P. Mechanical regulation of a molecular clutch defines force transmission and transduction in response to matrix rigidity. *Nat. Cell Biol.* **2016**, *18*, 540–548. [CrossRef] [PubMed]
154. Ribeiro, A.E.R.A.; Lourenço, A.G.; Motta, A.C.F.; Komesu, M.C. Salivary Expression of Antimicrobial Peptide LL37 and Its Correlation with Pro-inflammatory Cytokines in Patients with Different Periodontal Treatment Needs. *Int. J. Pept. Res. Ther.* **2020**, *26*, 2547–2553. [CrossRef]
155. Grant, M.; Kilsgård, O.; Åkerman, S.; Klinge, B.; Demmer, R.T.; Malmström, J.; Jönsson, D. The human salivary antimicrobial peptide profile according to the oral microbiota in health, periodontitis and smoking. *J. Innate Immun.* **2019**, *11*, 432–444. [CrossRef]
156. Schenkels, L.C.; Veerman, E.C.; Nieuw Amerongen, A.V. Biochemical composition of human saliva in relation to other mucosal fluids. *Crit. Rev. Oral Biol. Med.* **1995**, *6*, 161–175. [CrossRef]
157. Łysik, D.; Mystkowska, J.; Markiewicz, G.; Deptuła, P.; Bucki, R. The Influence of Mucin-Based Artificial Saliva on Properties of Polycaprolactone and Polylactide. *Polymers* **2019**, *11*, 1880. [CrossRef]
158. Byfield, F.J.; Kowalski, M.; Cruz, K.; Leszczyńska, K.; Namiot, A.; Savage, P.B.; Bucki, R.; Janmey, P.A. Cathelicidin LL-37 increases lung epithelial cell stiffness, decreases transepithelial permeability, and prevents epithelial invasion by Pseudomonas aeruginosa. *J. Immunol.* **2011**, *187*, 6402–6409. [CrossRef]
159. Ridyard, K.E.; Overhage, J. The potential of human peptide LL-37 as an antimicrobial and anti-biofilm agent. *Antibiotics* **2021**, *10*, 650. [CrossRef]

MDPI
St. Alban-Anlage 66
4052 Basel
Switzerland
Tel. +41 61 683 77 34
Fax +41 61 302 89 18
www.mdpi.com

Biomedicines Editorial Office
E-mail: biomedicines@mdpi.com
www.mdpi.com/journal/biomedicines